Lecture Notes in Mathematics

Edited by A. Dold and B. Eckmann

Series: California Institute of Technology, Pasadena

Adviser: C. R. DePrima

593

Klaus Barbey

Heinz König

Abstract Analytic Function Theory and Hardy Algebras

Springer-Verlag

Berlin · Heidelberg · New York 1977

Authors

Klaus Barbey
Fachbereich Mathematik
Universität Regensburg
Universitätsstraße 31
8400 Regensburg/BRD

Heinz König
Fachbereich Mathematik
Universität des Saarlandes
6600 Saarbrücken/BRD

AMS Subject Classifications (1970): 46J10, 46J15

ISBN 3-540-08252-2 Springer-Verlag Berlin · Heidelberg · New York
ISBN 0-387-08252-2 Springer-Verlag New York · Heidelberg · Berlin

Printed in Germany
Printing and binding: Beltz Offsetdruck, Hemsbach/Bergstr.
2141/3140-543210

Preface

The present work wants to be the systematic presentation of a functional-analytic theory. It is an abstract version of those parts of classical analytic function theory which can be circumscribed by boundary value theory and Hardy spaces H^p. The fascination of the field comes from the fact that famous classical theorems of typical complex-analytic flavor appear as instant outflows of an abstract theory the tools of which are standard real-analytic methods such as elementary functional analysis and measure theory. The abstract theory started about twenty years ago in papers of Arens and Singer, Gleason, Helson and Lowdenslager, Bochner, Bishop, Wermer,... and went through several steps of abstraction (Dirichlet algebras, logmodular algebras,...). It never ceased to radiate back and illuminate the concrete classical theory. We present the ultimate step of abstraction which has been under work for about ten years.

The central concept is the abstract Hardy algebra situation. It is comprehensive as well as pure and simple and permits to build up a coherent theory of remarkable and pleasant width and depth. We attempt to present a systematic account in Chapters IV-IX. The abstract Hardy algebra situation can be looked upon as a local section of the abstract function algebra situation. To achieve the localization is the main business of the abstract F.and M.Riesz theorem and of the resultant Gleason part decomposition procedure. These are central themes in Chapters II-III devoted to the abstract function algebra situation. Chapter I presents the concrete unit disk situation in such a spirit as to prepare the abstract concepts. Chapter X is devoted to standard applications of the abstract theory to polynomial and rational approximation in the complex plane and is the most conventional part of the book.

In comparison with the respective parts of the earlier treatises on uniform algebras, the most comprehensive of which is GAMELIN [1969], the present work contains numerous new results. In concepts and systematization it is shaped after the work of König. Most of the chapters contain substantial new material. A prime point is the systematic use of the associated algebra $H^{\#}$. The most important individual new result is perhaps the approximation theorem VI.4.1. Let us also quote Section VI.5 on the Marcel Riesz estimation for the abstract conjugation after fundamental results of Pichorides in the unit disk situation. For more details we refer to the Introductions and Notes to the individual chapters.

In its overall structure and in certain parts the present work resembles the lectures on function algebras which König held in 1967/68 at the California Institute of Technology in Pasadena/California, and which in part had been distributed in a provisional form. He wants to express his warmest thanks to Wim Luxemburg and Charles DePrima who were his hosts in those days, and likewise to Gunter Lumer to whom he owes the participation in the Function Algebra Seminar at the University of Washington in Seattle/Washington in 1970. Above all he sends his deepest thanks to Galen Seever and to Kôzô Yabuta for most valuable and pleasant cooperation, and he wants to include his former student Klaus Barbey who started to participate with the elaboration of 1970/71 lecture notes which formed the next step in the evolution of the present text.

In conclusion we want to express our sincere thanks to Michael Neumann for his active interest and valuable work in connection with a common seminar, to Ulla Faust and Gisela Schirmbeck who typed the final text with impressive care and thoughtfulness, to Horst Loch who read most of the proofs with distinctive care, and to Karla May and Gerd Rodê for their kind assistance.

Contents

Chapter I

Boundary Value Theory

for Harmonic and Holomorphic Functions in the Unit Disk

The present chapter describes the concrete situation which forms the basic model for the subsequent abstract theory. It leads up to the point where the abstract theory can be put into action. The abstract theory will then illuminate the reasons for which the individual classical theorems are valid.

1. Harmonic Functions

For G an open subset of the complex plane $\mathbb{C}$ let Harm(G) denote the class of harmonic functions $G\rightarrow\mathbb{C}$, $\mathrm{Harm}^{\infty}(G)$ the class of bounded functions in Harm(G), and CHarm(G) the class of those functions in Harm(G) which admit continuous extensions $\overline{G}\rightarrow\mathbb{C}$. Here $\overline{G}$ means the closure of G relative to the Riemann sphere so that $\infty\in\overline{G}$ if G is unbounded. Furthermore let Hol(G), $\mathrm{Hol}^{\infty}(G)$ and CHol(G) denote the respective classes of holomorphic functions $G\rightarrow\mathbb{C}$.

The boundary value theory for the above function classes for unit disk $D=\{z\in\mathbb{C}:|z|<1\}$ and unit circle $S=\{s\in\mathbb{C}:|s|=1\}$ is dominated by the Poisson kernel $P:D\times S\rightarrow\mathbb{R}$. It is defined to be

$$P(z,s)=\mathrm{Re}\,\frac{s+z}{s-z}=\frac{s}{s-z}+\frac{s\bar{z}}{1-s\bar{z}}=\frac{1-|z|^2}{|s-z|^2}\qquad\forall\ z\in D\text{ and }s\in S,$$

$$P(Re^{iu},e^{iv})=\frac{1-R^2}{1-2R\cos(u-v)+R^2}\qquad\forall\ 0\leqq R<1\text{ and real }u,v.$$

We list some immediate properties. i) P is continuous on $D\times S$ and

$$0<\frac{1-|z|}{1+|z|}\leqq P(z,s)\leqq\frac{1+|z|}{1-|z|}\qquad\forall\ z\in D\text{ and }s\in S.$$

ii) $P(R,e^{iv})\leqq P(R,e^{iu})$ for real u,v with $|u|\leqq|v|\leqq\pi$ and $0\leqq R<1$.

iii) $P(R,e^{it})\rightarrow 0$ for $R\uparrow 1$ pointwise in $0<|t|\leqq\pi$ and hence uniformly in $\delta\leqq|t|\leqq\pi$ for each $\delta>0$.

iv) $P(z\alpha,s\alpha)=P(z,s)$ for $z\in D$ and $s,\alpha\in S$.

v) $P(R\alpha,\beta)=P(R\beta,\alpha)$ for $\alpha,\beta\in S$ and $0\leq R<1$.

And we recall from elementary analytic function theory the basic representation theorem. Here λ denotes one-dimensional Lebesgue measure on S with the normalization $\lambda(S)=1$.

1.1 REPRESENTATION THEOREM: For $f\in Hol(D)$ we have

$$f(Rz) = \int_S \frac{s}{s-z} f(Rs)d\lambda(s) = iImf(0) + \int_S \frac{s+z}{s-z} Ref(Rs)d\lambda(s)$$

$$= \int_S P(z,s)f(Rs)d\lambda(s) \qquad \forall\ z\in D \text{ and } 0\leq R<1.$$

Hence for $f\in Harm(D)$ we have

$$f(Rz) = \int_S P(z,s)f(Rs)d\lambda(s) \qquad \forall\ z\in D \text{ and } 0\leq R<1.$$

In particular we have $\int_S P(z,s)d\lambda(s) =1$ for $z\in D$.

We turn to the boundary behaviour of the functions in Harm(D). For $f:D\to\mathbb{C}$ and $0\leq R<1$ we put $f_R:f_R(s)=f(Rs)$ $\forall s\in S$.

1.2 REMARK: Let $1\leq p\leq\infty$. For $f\in Harm(D)$ then

$$\|f_R\|_{L^p(\lambda)} := \begin{cases} \left(\int_S |f(Rs)|^p d\lambda(s)\right)^{\frac{1}{p}} & \text{for } 1\leq p<\infty \\ \operatorname*{Max}_{s\in S} |f(Rs)| & \text{for } p=\infty \end{cases}$$

is monotone increasing in $0\leq R<1$.

Proof: For $0\leq r<R<1$ we obtain from 1.1

$$f(rz) = f(R\tfrac{r}{R}z) = \int_S P(\tfrac{r}{R}z,s)f(Rs)d\lambda(s) \qquad \forall\ z\in S.$$

From this the cases $p=1$, $1<p<\infty$ and $p=\infty$ require separate treatment. The cases $p=1$ and $p=\infty$ are almost obvious, so we restrict ourselves to $1<p<\infty$. For $1<q<\infty$ the conjugate exponent we obtain

$$\int_S |f(rz)|^p d\lambda(z) \leq \int_S \left(\int_S P(\tfrac{r}{R}z,s)|f(Rs)|d\lambda(s)\right)^p d\lambda(z)$$

$$= \int_S \left(\int_S P(\tfrac{r}{R}z,s)^{1/q}\left(P(\tfrac{r}{R}z,s)^{1/p}|f(Rs)|\right)d\lambda(s)\right)^p d\lambda(z)$$

$$\leq \int_S \left(\int_S P(\tfrac{r}{R}z,s)d\lambda(s)\right)^{p/q}\left(\int_S P(\tfrac{r}{R}z,s)|f(Rs)|^p d\lambda(s)\right)d\lambda(z)$$

$$= \int_S \left(\int_S P(\tfrac{r}{R}z,s)|f(Rs)|^p d\lambda(s)\right)d\lambda(z) = \int_S |f(Rs)|^p d\lambda(s),$$

where v) above has been applied. QED.

For $1\leq p\leq\infty$ we define $Harm^p(D)$ to consist of the functions $f\in Harm(D)$ with

$$N_p f := \lim_{R\uparrow 1} \|f_R\|_{L^p(\lambda)} = \sup_{0\leq R<1} \|f_R\|_{L^p(\lambda)} < \infty.$$

For $p=\infty$ this coincides with the earlier definition. From

$$N_1 f \leq N_p f \leq N_\infty f \qquad \text{for } f\in Harm(D)$$

we see that $Harm^\infty(D) \subset Harm^p(D) \subset Harm^1(D)$. We formulate the boundary behaviour of the functions in $Harm^p(D)$ in the subsequent propositions. Here ca(S) denotes the class of complex-valued Baire measures on S.

<u>1.3</u> <u>PROPOSITION</u>: i) For $\theta\in ca(S)$ define the function

$$\langle\theta\rangle : \langle\theta\rangle(z) = \int_S P(z,s)d\theta(s) \qquad \forall\ z\in D.$$

Then $\langle\theta\rangle\in Harm^1(D)$ and $N_1\langle\theta\rangle = \|\theta\|$:=total variation of θ. Furthermore for $R\uparrow 1$ we have convergence $\langle\theta\rangle_R\lambda\to\theta$ in the weak* topology $\sigma(ca(S),C(S)) = \sigma(C(S)',C(S))$.

ii) Let $1\leq p\leq\infty$. For $F\in L^p(\lambda)$ consider the function

$$f = \langle F\lambda\rangle : f(z) = \int_S P(z,s)F(s)d\lambda(s) \qquad \forall\ z\in D.$$

Then $f\in Harm^p(D)$ and $N_p f = \|F\|_{L^p(\lambda)}$. Furthermore for $R\uparrow 1$ we have convergence $f_R\to F$ in $L^p(\lambda)$-norm if $1\leq p<\infty$ and in the weak* topology $\sigma(L^\infty(\lambda),L^1(\lambda)) = \sigma(L^1(\lambda)',L^1(\lambda))$ if $p=\infty$.

iii) For $F\in C(S)$ we have $f=\langle F\lambda\rangle\in CHarm(D)$. Furthermore for $R\uparrow 1$ we have convergence $f_R\to F$ uniformly on S.

Proof: 1) $\langle\theta\rangle\in Harm(D)$ for $\theta\in ca(S)$ is obvious since for real-valued θ the definition represents $\langle\theta\rangle$ as the real part of a function in Hol(D). 2) In order to prove iii) it suffices to show the uniform convergence $f_R\to F$ for $R\uparrow 1$. For $0\leq R<1$ and $z\in S$ we have

$$f_R(z)-F(z) = \int_S P(Rz,s)\big(F(s)-F(z)\big)d\lambda(s) = \int_S P(R,\tfrac{s}{z})\big(F(s)-F(z)\big)d\lambda(s)$$
$$= \frac{1}{2\pi}\int_{-\pi}^{\pi} P(R,e^{it})\big(F(ze^{it})-F(z)\big)dt,$$

and hence for $0<\delta<\pi$ after subdivision into $|t|\leqq\delta$ and $\delta\leqq|t|\leqq\pi$

$$|f_R(z)-F(z)| \leqq \frac{1}{2\pi}\int_{-\delta}^{\delta} P(R,e^{it})\omega(\delta)dt + 2\,||F||\,P(R,e^{i\delta}) \leqq \omega(\delta) + 2\,||F||\,P(R,e^{i\delta}),$$

where ω is the modulus of continuity of the function $F\in C(S)$. Therefore

$$\limsup_{R\uparrow 1} ||f_R - F|| \leqq \omega(\delta) \qquad \text{for each } 0<\delta<\pi,$$

so that $||f_R-F||\to 0$ for $R\uparrow 1$.

3) We next prove i). For $f=\langle\theta\rangle\in\mathrm{Harm}(D)$ and $0\leqq R<1$ we have

$$\int_S |f(Rz)|d\lambda(z) \leqq \int_S\Big(\int_S P(Rz,s)d|\theta|(s)\Big)d\lambda(z)$$
$$=\int_S\Big(\int_S P(Rs,z)d\lambda(z)\Big)d|\theta|(s) = ||\theta||,$$

therefore $f\in\mathrm{Harm}^1(D)$ and $N_1 f\leqq ||\theta||$. The weak* convergence to be shown means that $\int_S Hf_R d\lambda \to \int_S Hd\theta$ for $R\uparrow 1$ for each $H\in C(S)$. But this is true since for $h=\langle H\lambda\rangle$ we know from iii) that

$$\int_S Hf_R d\lambda = \int_S\Big(\int_S H(z)P(Rz,s)d\theta(s)\Big)d\lambda(z)$$
$$= \int_S\Big(\int_S H(z)P(Rs,z)d\lambda(z)\Big)d\theta(s) = \int_S h_R(s)d\theta(s) \to \int_S Hd\theta \text{ for } R\uparrow 1.$$

And then $|\int_S Hf_R d\lambda|\leqq ||H||\int_S|f_R|d\lambda\leqq ||H||N_1 f$ for $0\leqq R<1$ implies that $|\int_S Hd\theta|\leqq ||H||N_1 f$ for each $H\in C(S)$. This means $||\theta||\leqq N_1 f$, so that we obtain $N_1 f= ||\theta||$.

4) In order to prove ii) we obtain

$$||f_R||_{L^p(\lambda)} \leqq ||F||_{L^p(\lambda)} \qquad \text{for } 0\leqq R<1$$

as in the proof of 1.2. Thus $f\in\mathrm{Harm}^p(D)$ and $N_p f\leqq ||F||_{L^p(\lambda)}$. Then in the case $1\leqq p<\infty$ we use the fact that $C(S)$ is dense in $L^p(\lambda)$. Thus for $H\in C(S)$

and $h=\langle H\lambda\rangle$ we have

$$\|f_R-F\|_{L^p(\lambda)} \leq \|(f-h)_R\|_{L^p(\lambda)} + \|F-H\|_{L^p(\lambda)} + \|h_R-H\|_{L^p(\lambda)}$$

$$\leq 2\|F-H\|_{L^p(\lambda)} + \|h_R-H\| \qquad \text{for } 0 \leqq R<1,$$

$$\limsup_{R\uparrow 1} \|f_R-F\|_{L^p(\lambda)} \leqq 2\|F-H\|_{L^p(\lambda)} \qquad \text{in view of iii),}$$

and hence $f_R \to F$ in $L^p(\lambda)$-norm. From this it follows that

$$\|F\|_{L^p(\lambda)} = \lim_{R\uparrow 1} \|f_R\|_{L^p(\lambda)} = N_p f.$$

In the case $p=\infty$ the weak* convergence to be shown means that $\int_S Hf_R d\lambda \to \int_S HFd\lambda$ for $R\uparrow 1$ for each $H\in L^1(\lambda)$. But this is true since for $h=\langle H\lambda\rangle \in \in Harm^1(D)$ we know that $h_R \to H$ in $L^1(\lambda)$-norm and hence

$$\int_S Hf_R d\lambda = \int_S h_R F d\lambda \to \int_S HFd\lambda \qquad \text{for } R\uparrow 1.$$

And then

$$\left|\int_S Hf_R d\lambda\right| \leqq \|H\|_{L^1(\lambda)} \|f_R\|_{L^\infty(\lambda)} \leqq \|H\|_{L^1(\lambda)} N_\infty f \qquad \text{for } 0 \leqq R<1$$

implies that $\left|\int_S HFd\lambda\right| \leqq \|H\|_{L^1(\lambda)} N_\infty f$ for each $H\in L^1(\lambda)$. This means $\|F\|_{L^\infty(\lambda)} \leqq N_\infty f$, so that we obtain $N_\infty f = \|F\|_{L^\infty(\lambda)}$. QED.

1.4 PROPOSITION: i) For each $f\in Harm^1(D)$ there exists a unique $\theta \in ca(S)$ with $f=\langle\theta\rangle$.

ii) Let $1<p\leqq\infty$. For each $f\in Harm^p(D)$ there exists a unique $F\in L^p(\lambda)$ with $f=\langle F\lambda\rangle$.

iii) For each $f\in CHarm(D)$ there exists a unique $F\in C(S)$ with $f=\langle F\lambda\rangle$.

Proof: i) The measures $f_R\lambda\in ca(S)$ fulfill

$$\|f_R\lambda\| = \|f_R\|_{L^1(\lambda)} \leqq N_1 f < \infty \quad \forall\ 0\leqq R<1.$$

Let $\theta \in ca(S)$ be a weak* limit point of these measures for $R\uparrow 1$. Thus for each $H\in C(S)$ there is a sequence $R(n)\uparrow 1$ with $\int_S Hf_{R(n)}d\lambda \to \int_S Hd\theta$. Now from 1.1 we have

$$f(Rz) = \int_S P(z,s) f_R(s) d\lambda(s) \qquad \forall\ z\in D \text{ and } 0\leqq R<1.$$

For fixed $z\in D$ we take $H=P(z,\cdot)$ and a suitable sequence $R=R(n)\uparrow 1$ to obtain

$$f(z) = \int_S P(z,s)d\theta(s) = \langle\theta\rangle(z) \quad \forall\ z\in D.$$

ii) For $1\leqq q<\infty$ the conjugate exponent we have $L^p(\lambda)=L^q(\lambda)'$. In view of $\|f_R\|_{L^p(\lambda)}\leqq N_p f<\infty$ for $0\leqq R<1$ there exists a weak* limit point $F\in L^p(\lambda)$ of the f_R for $R\uparrow 1$. Then we obtain $f=\langle F\lambda\rangle$ as in the proof of i). The proof of iii) is similar but simpler since the compactness argument is not needed. The uniqueness assertions are immediate from 1.3. QED.

1.5 COROLLARY (HERGLOTZ): The formula $f=\langle\theta\rangle$ defines a bijection between the nonnegative functions $f\in Harm(D)$ and the nonnegative measures $\theta\in Pos(S)$.

Proof: For $f\in Harm(D)$ nonnegative we have

$$f(0) = \int_S f(Rs)d\lambda(s) = \|f_R\|_{L^1(\lambda)} \qquad \text{for } 0\leqq R<1,$$

so that $f\in Harm^1(D)$ and $N_1 f=f(0)$. Then the assertions follow from 1.3 and 1.4. QED.

2. Pointwise Convergence: The Fatou Theorem and its Converse

Let $f\in Harm^1(D)$. We ask for the pointwise convergence behaviour of the functions f_R for $R\uparrow 1$. The answer is the famous Fatou theorem. Besides of this abelian-type theorem we prove its tauberian converse due to Loomis. As usual the converse is not true unless an extra tauberian condition is satisfied: here we have to assume that $f\geqq 0$. The Loomis theorem will in Section V.5 be valuable for the abstract theory.

It is convenient for us to work with functions of bounded variation. The Baire measures $\theta\in ca(S)$ are in bijective correspondence with the functions of bounded variation $\Theta:[-\pi,\pi]\to\mathbb{C}$ with the normalization

$$\Theta(t) = \frac{1}{2}(\Theta(t+)+\Theta(t-)) \text{ for } |t|<\pi,\ \Theta(\pi)-\Theta(\pi-) = \Theta(-\pi+)-\Theta(-\pi),$$

and $\Theta(0)=0$: the correspondence is

$$\int_S F d\theta = \int_{-\pi}^{\pi} F(e^{it})d\Theta(t) \qquad \forall\ F\in C(S).$$

We can extend $\Theta:[-\pi,\pi]\to\mathbb{C}$ to a unique function $\Theta:\mathbb{R}\to\mathbb{C}$ with the periodicity property $\Theta(t+2\pi)-\Theta(t)=\text{const}\ \forall$ real t, which then is a function of local bounded variation with the normalization $\Theta(t)=\frac{1}{2}(\Theta(t+)+\Theta(t-))\ \forall$ real t and $\Theta(0)=0$. The above correspondence reads

$$\int_S F d\theta = \int_{\tau-\pi}^{\tau+\pi} F(e^{it})d\Theta(t) \quad \forall\ F\in C(S) \text{ and each real } \tau.$$

Equivalent is

$$\theta(\{e^{it}:u<t<v\}) = \Theta(v-)-\Theta(u+)\ \forall \text{ real } u<v\leqq u+2\pi.$$

In particular θ is nonnegative $\in \text{Pos}(S)$ iff Θ is real-valued and monotone increasing.

2.1 FATOU THEOREM: Let $f\in \text{Harm}^1(D)$ with corresponding $\theta\in ca(S)$ and $\Theta:\mathbb{R}\to\mathbb{C}$. Let $\alpha\in\mathbb{R}$ with

$$\frac{\Theta(\alpha+t)-\Theta(\alpha-t)}{2t} \to \frac{A}{2\pi} \quad \text{for } t\downarrow 0.$$

Then $f(Re^{i\alpha})\to A$ for $R\uparrow 1$.

From the above we see that

$$\frac{\Theta(\alpha+t-)-\Theta(\alpha-t+)}{2t} = \frac{1}{2t}\theta(\{e^{iu}:\alpha-t<u<\alpha+t\})$$

$$= \frac{1}{2\pi}\frac{\theta(\{e^{iu}:\alpha-t<u<\alpha+t\})}{\lambda(\{e^{iu}:\alpha-t<u<\alpha+t\})} \quad \forall\ 0<t\leqq\pi,$$

and this tends $\to \frac{1}{2\pi}\frac{d\theta}{d\lambda}(e^{i\alpha})$ for $t\downarrow 0$ for λ-almost all $e^{i\alpha}\in S$. Thus we have the subsequent corollary.

2.2 COROLLARY: Let $f\in \text{Harm}^1(D)$ with corresponding $\theta\in ca(S)$. Then the radial limit $\lim_{R\uparrow 1} f(Rs)$ exists for λ-almost all $s\in S$, and the limit function is $=\frac{d\theta}{d\lambda}\in L^1(\lambda)$.

The result can be extended from radial limits to non-tangential limits. We shall come back to this point in Section 4.

2.3 LOOMIS THEOREM: Let $f\in \text{Harm}^1(D)$ be nonnegative with corresponding $\theta\in ca(S)$ and $\Theta:\mathbb{R}\to\mathbb{C}$. Let $\alpha\in\mathbb{R}$ with $f(Re^{i\alpha})\to A$ for $R\uparrow 1$. Then

$$\frac{\Theta(\alpha+t)-\Theta(\alpha-t)}{2t} \to \frac{A}{2\pi} \qquad \text{for } t\downarrow 0.$$

The proofs have to be based on the relation

$$f(z) = \int_{-\pi}^{\pi} P(z,e^{it})d\Theta(t) = \int_{-\pi}^{\pi} \text{Re}\,\frac{e^{it}+z}{e^{it}-z}\,d\Theta(t) \qquad \forall\ z\in D.$$

An obvious transformation allows us to assume that $\alpha=0$. But then it is convenient to transfer the problem from the unit disk D to the halfplane $\Delta=\{s\in\mathbb{C}:\text{Re}\, s>0\}$ via an appropriate fractional-linear map. Let us write up the transition.

The function $h:h(s)=\frac{1-s}{1+s}$ for $s\in\mathbb{C}$ maps $\Delta\to D$ (and is equal to its inverse). It maps $i\mathbb{R}\to S$ with

$$h(ix) = e^{it} \Leftrightarrow x = -\tan\frac{t}{2} \quad \forall\ x\in\mathbb{R} \text{ and } |t|<\pi.$$

Under this map a normalized function of bounded variation $\Theta:[-\pi,\pi]\to\mathbb{C}$ corresponds to a pair which consists of a normalized function of bounded variation $\phi:\mathbb{R}\to\mathbb{C}$ and a complex number c: the correspondence is

$$\Theta(t) = -\phi(x) \qquad \text{for corresponding } x\in\mathbb{R} \text{ and } |t|<\pi,$$

$$\Theta(\pi)-\Theta(\pi-) = \Theta(-\pi+)-\Theta(-\pi) = \frac{c}{2}.$$

In particular Θ is real-valued and monotone increasing iff ϕ is so and c is $\geqq 0$. Let now $s\in\Delta$ and $z=h(s)\in D$. We find

$$\frac{e^{it}+z}{e^{it}-z} = \frac{1-ixs}{s-ix} \qquad \text{for corresponding } x\in\mathbb{R} \text{ and } |t|<\pi.$$

Therefore we have

$$f(z) = \int_{-\pi}^{\pi} P(z,e^{it})d\Theta(t) = c\text{Re}\frac{1-z}{1+z} + \lim_{\delta\downarrow 0}\int_{-\pi+\delta}^{\pi-\delta} P(z,e^{it})d\Theta(t)$$

$$= c\text{Re}\, s - \lim_{\delta\downarrow 0}\int_{-\tan\frac{-\pi+\delta}{2}}^{-\tan\frac{\pi-\delta}{2}} \text{Re}\,\frac{1-ixs}{s-ix}\,d\phi(x)$$

$$= c\text{Re}\, s + \int_{-\infty}^{\infty} \text{Re}\,\frac{1-ixs}{s-ix}\,d\phi(x).$$

In particular for $0<s<\infty$ and $-1<z=h(s)<1$ we have

$$f(z) = cs + \int_{-\infty}^{\infty} s\,\frac{1+x^2}{s^2+x^2}d\phi(x) = cs + \int_0^{\infty} s\,\frac{1+x^2}{s^2+x^2}d\varphi(x),$$

with $\varphi:\varphi(x)=\phi(x)-\phi(-x)$ for $x\geqq 0$ a function of bounded variation with the normalization $\varphi(x)=\frac{1}{2}(\varphi(x+)+\varphi(x-))$ $\forall x>0$ and $\varphi(0)=0$, and φ real-valued and monotone increasing if ϕ is so. It follows that $f(z)\to A$ for $z\uparrow 1$ is equivalent to

$$\int_0^{\infty} s\,\frac{1+x^2}{s^2+x^2}\,d\varphi(x)\to A \text{ for } s\downarrow 0 \quad\text{and hence to } \int_0^{\infty}\frac{s}{s^2+x^2}\,d\varphi(x)\to A \text{ for } s\downarrow 0.$$

On the other hand we have

$$\frac{\varphi(x)}{x}=\frac{\phi(x)-\phi(-x)}{x}=\frac{\Theta(t)-\Theta(-t)}{\tan\frac{t}{2}} \text{ for corresponding } x>0 \text{ and } 0<t<\pi,$$

so that

$$\frac{\Theta(t)-\Theta(-t)}{2t}\to\frac{A}{2\pi} \text{ for } t\downarrow 0 \text{ is equivalent to } \frac{\varphi(x)}{x}\to\frac{2}{\pi}A \text{ for } x\downarrow 0.$$

We can therefore reformulate our assertions as follows.

2.4 FATOU THEOREM: Let $\varphi:[0,\infty[\to\mathbb{C}$ be of bounded variation with $\varphi(0)=0$ and

$$F:\ F(s) = \frac{2}{\pi}\int_0^{\infty}\frac{s}{s^2+x^2}\,d\varphi(x) \quad\text{for } s>0.$$

Then $\frac{\varphi(x)}{x}\to a$ for $x\downarrow 0$ implies that $F(s)\to a$ for $s\downarrow 0$.

2.5 LOOMIS THEOREM: Let $\varphi:[0,\infty[\to\mathbb{R}$ be monotone increasing and bounded with $\varphi(0)=0$ (and F as above). Then $F(s)\to a$ for $s\downarrow 0$ implies that $\frac{\varphi(x)}{x}\to a$ for $x\downarrow 0$.

Proof of 2.4: This is a typical proof of an abelian theorem. Fix $\delta>0$. Then for $s>0$ we have

$$F(s) - \frac{2a}{\pi}\arctan\frac{\delta}{s} = \frac{2}{\pi}\int_0^{\delta}\frac{s}{s^2+x^2}\,d(\varphi(x)-ax) + \frac{2}{\pi}\int_{\delta}^{\infty}\frac{s}{s^2+x^2}\,d\varphi(x)$$

$$= \frac{2}{\pi}\frac{s}{s^2+\delta^2}(\varphi(\delta)-a\delta) + \frac{2}{\pi}\int_0^{\delta}\left(\frac{\varphi(x)}{x}-a\right)\frac{2sx^2}{(s^2+x^2)^2}dx + \frac{2}{\pi}\int_{\delta}^{\infty}\frac{s}{s^2+x^2}d\varphi(x),$$

$$\left|F(s) - \frac{2a}{\pi}\arctan\frac{\delta}{s}\right| \leqq \frac{2}{\pi}\frac{s}{s^2+\delta^2}|\varphi(\delta) - a\delta|$$

$$+ 2\operatorname{Sup}\left\{\left|\frac{\varphi(x)}{x}-a\right| : 0<x\leqq\delta\right\} + \frac{2}{\pi}\frac{s}{s^2+\delta^2}\operatorname{Var}(\varphi)$$

and hence $\limsup_{s\downarrow 0} |F(s)-a| \leq 2 \operatorname{Sup}\{|\frac{\varphi(x)}{x}-a| : 0<x\leq\delta\}$.

For $\delta\downarrow 0$ the assertion follows. QED.

The proof of 2.5 will be based on the subsequent uniqueness remark.

2.6 UNIQUENESS REMARK: Let $H:[0,\infty[\to\mathbb{R}$ be a bounded Baire function. If

$$\int_0^\infty \frac{H(x)}{s^2+x^2}dx = 0 \qquad \text{for all } 0<s<1,$$

then $H(x)=0$ for Lebesgue-almost all $x>0$.

Proof of 2.6: The formula

$$\phi : \phi(x) = \int_0^x \frac{H(|t|)}{1+t^2}dt \qquad \forall\ x\in\mathbb{R}$$

defines an odd continous function of bounded variation $\phi:\mathbb{R}\to\mathbb{R}$. We have

$$\int_{-\infty}^\infty \frac{1-ixs}{s-ix}\, d\phi(x) = \int_{-\infty}^\infty \frac{s(1+x^2)+ix(1-s^2)}{s^2+x^2}\, \frac{H(|x|)}{1+x^2}\, dx = 2s\int_0^\infty \frac{H(x)}{s^2+x^2}\, dx = 0$$

for $0<s<1$ and hence for all $s\in\Delta$. Under the transition $\Delta\to D$ described above we obtain

$$\int_{-\pi}^{\pi} \frac{e^{it}+z}{e^{it}-z}\, d\Theta(t) = 0 \quad \forall\ z\in D \quad \text{and hence} \int_{-\pi}^{\pi} P(z,e^{it})\, d\Theta(t) = 0 \qquad \forall\ z\in D,$$

where $\Theta:[-\pi,\pi]\to R$ is the corresponding odd continuous function of bounded variation. From 1.3.i) we conclude that $\Theta=0$ and hence $\phi=0$ and hence the assertion. QED.

Alternative proof of 2.6: This is an ab-ovo proof based on the Stone-Weierstraß theorem. Let $A\subset \operatorname{Re} C([0,\infty])$ be the closed linear subspace spanned by the constants and by the functions

$$h_\alpha : h_\alpha(x) = \frac{1}{\alpha^2+x^2} \qquad \forall\ 0\leq x\leq\infty \quad \text{with } 0<\alpha<1.$$

The identity $h_\alpha - h_\beta = (\beta^2-\alpha^2)h_\alpha h_\beta$ shows that A is an algebra. Therefore $A=\operatorname{Re}C([0,\infty])$ in view of Stone-Weierstraß. Now

$$\int_0^\infty h_\alpha(x)\,\frac{H(x)}{s^2+x^2}\, dx = \frac{1}{\alpha^2-s^2}\left(\int_0^\infty \frac{H(x)}{s^2+x^2}\, dx - \int_0^\infty \frac{H(x)}{\alpha^2+x^2}\, dx\right) = 0$$

for all $0<\alpha,s<1$ with $\alpha\neq s$ and hence

$$\int_0^\infty h(x)\,\frac{H(x)}{s^2+x^2}\,dx = 0 \qquad \forall\ h\in A = \mathrm{ReC}([0,\infty])\ \text{and}\ 0<s<1.$$

From this the assertion is obvious. QED.

Proof of 2.5:i) There exists an $M>0$ such that $0\leq\varphi(x)\leq Mx$ for all $x\geq 0$. In fact we have

$$F(s) \geq \frac{2}{\pi}\int_0^s \frac{s}{s^2+x^2}\,d\varphi(x) \geq \frac{2}{\pi}\int_0^s \frac{1}{2s}\,d\varphi(x) = \frac{\varphi(s)}{\pi s} \quad \text{for } s>0,$$

which in view of the assumptions implies the result. ii) Let us form for $t>0$ the function $\varphi_t:\varphi_t(x)=\frac{1}{t}\varphi(tx)$ for $x\geq 0$. Then $\varphi_t:[0,\infty[\to\mathbb{R}$ is likewise monotone increasing and bounded with $\varphi_t(0)=0$, and also $0\leq\varphi_t(x)\leq Mx$ for all $x\geq 0$. Now after partial integration

$$F(s) = \frac{2}{\pi}\int_0^\infty \frac{2sx}{(s^2+x^2)^2}\,\varphi(x)\,dx \quad \text{for } s>0.$$

It follows that

$$F(ts) = \frac{2}{\pi}\int_0^\infty \frac{2sx}{(s^2+x^2)^2}\,\varphi_t(x)\,dx \quad \text{for } s>0 \text{ and } t>0.$$

iii) In order to prove $\frac{\varphi(x)}{x}\to a$ for $x\downarrow 0$ we consider a sequence $t(n)\downarrow 0$ for which the values $\frac{1}{t(n)}\varphi(t(n))=\varphi_{t(n)}(1)$ converge to some limit c. In view of i) c must be finite. Then we have to show that $c=a$. In view of ii) the functions $\varphi_{t(n)}$ $(n=1,2,\ldots)$ are equibounded on each bounded subinterval of $[0,\infty[$. Therefore the Helly selection theorem (for which we refer to WIDDER [1946] Chapter I.16) applied successively to $[0,N]$ $(N=1,2\ldots)$ and a subsequent diagonal selection lead to a subsequence which converges pointwise on $[0,\infty[$ to a function $\theta:[0,\infty[\to\mathbb{R}$ which is likewise monotone increasing and satisfies $0\leq\theta(x)\leq Mx$ for all $x\geq 0$. We keep the notation $\varphi_{t(n)}$ for the subsequence in question. Now from

$$F(t(n)s) = \frac{2}{\pi}\int_0^\infty \frac{2s}{s^2+x^2}\,\frac{x^2}{s^2+x^2}\,\frac{1}{x}\,\varphi_{t(n)}(x)\,dx \quad \text{for } s>0$$

and from the assumption $F(s)\to a$ for $s\downarrow 0$ we deduce via dominated convergence that

$$a = \frac{2}{\pi}\int_0^\infty \frac{2sx^2}{(s^2+x^2)^2}\,\frac{\theta(x)}{x}\,dx = \frac{2}{\pi}\int_0^\infty \frac{2sx}{(s^2+x^2)^2}\,\theta(x)\,dx \quad \text{for } s>0.$$

The operation $\frac{1}{t}\int_0^t \ldots ds$ for $t>0$ leads to

$$a = \frac{2}{\pi}\int_0^\infty \frac{1}{t}\Big(\frac{1}{x}-\frac{x}{t^2+x^2}\Big)\,\theta(x)\,dx = \frac{2}{\pi}\int_0^\infty \frac{t}{t^2+x^2}\,\frac{\theta(x)}{x}\,dx \quad \text{for } t>0,$$

$$\frac{2}{\pi}\int_0^\infty \frac{t}{t^2+x^2}\left(\frac{\theta(x)}{x}-a\right)dx = 0 \quad \text{for } t>0.$$

Thus from 2.6 we obtain $\theta(x)=ax$ for Lebesgue-almost all $x>0$. Hence $\theta(x\pm)=ax$ for all $x>o$, and since θ is monotone we have in fact $\theta(x)=ax$ for all $x>0$. In particular $c=\lim_{n\to\infty} \varphi_{t(n)}(1)=\theta(1)=a$. QED.

3. Holomorphic Functions

We introduce the function classes $\mathrm{Hol}^p(D):=\{f\in \mathrm{Hol}(D):N_p f<\infty\} = \mathrm{Hol}(D)\cap \mathrm{Harm}^p(D)$ for $1\leqq p\leqq\infty$. For $p=\infty$ this coincides with the earlier definition. We had also defined $\mathrm{CHol}(D):=\mathrm{Hol}(D)\cap \mathrm{CHarm}(D)$. On the side of the boundary values we introduce

$$\mathrm{an}(D) := \{\theta\in \mathrm{ca}(S) : \langle\theta\rangle \in \mathrm{Hol}(D)\},$$

the class of analytic measures, and the function classes

$$H^p(D) := \{F\in L^p(\lambda) : \langle F\lambda\rangle \in \mathrm{Hol}(D)\} \qquad \text{for } 1\leqq p\leqq\infty,$$

$$A(D) := \{F\in C(S) : \langle F\lambda\rangle \in \mathrm{Hol}(D)\}.$$

Then 1.3 and 1.4 contain the bijective correspondences

$$\begin{array}{ccccccccc}
\mathrm{CHol}(D) & \subset & \mathrm{Hol}^\infty(D) & \subset & \mathrm{Hol}^p(D) & \subset & \{\langle F\lambda\rangle : F\in H^1(D)\} & \subset & \mathrm{Hol}^1(D) \\
\updownarrow & & \updownarrow & & \updownarrow & & \updownarrow & & \updownarrow \\
A(D) & \subset & H^\infty(D) & \subset & H^p(D) & \subset & H^1(D) \cong H^1(D)\lambda & \subset & \mathrm{an}(D)
\end{array}$$

with $1<p<\infty$. It will be a main theorem that $H^1(D)\lambda=\mathrm{an}(D)$ (F.and M.RIESZ).

A new aspect relative to the harmonic function situation is multiplicativity: $\mathrm{Hol}(D)$ is an algebra. Therefore $A(D)$ and $H^\infty(D)$ are algebras, and the Hölder inequality leads to obvious multiplicative relations between the $H^p(D)$ for $1\leqq p\leqq\infty$. In particular the unexpected notion of a multiplicative measure enters the scene: The obvious fact that $F,G\in A(D)$ implies $FG\in A(D)$ and $\langle(FG)\lambda\rangle = \langle F\lambda\rangle\langle G\lambda\rangle$ means that

$$\int_S F(s)G(s)P(z,s)d\lambda(s) = \left(\int_S F(s)P(z,s)d\lambda(s)\right)\left(\int_S G(s)P(z,s)d\lambda(s)\right) \quad \forall\ z\in D,$$

which means that for each $z \in D$ the measure $P(z,\cdot)\lambda$ (and in particular for $z=0$ the measure λ itself) is multiplicative on $A(D)$. The same is true for $H^\infty(D)$.

3.1 REMARK: Let $\theta \in an(D)$ and $F \in A(D)$. Then $F\theta \in an(D)$ and $<F\theta> = <F\lambda><\theta>$.

Proof: Let $h=<\theta>$ and $f=<F\lambda>$. For $0 \leq R < 1$ then $Fh_R \in A(D)$ and

$$f(z)h(Rz) = \int_S P(z,s)F(s)\, h_R(s)d\lambda(s) \quad \forall\ z \in D.$$

For $R \uparrow 1$ it follows that

$$f(z)h(z) = \int_S P(z,s)F(s)d\theta(s) \qquad \forall\ z \in D. \qquad \text{QED.}$$

3.2 REMARK: For $\theta \in ca(S)$ the subsequent properties are equivalent.

i) $\theta \in an(D)$, that is $<\theta> \in Hol(D)$.

ii) $\int_S s^n d\theta(s) = 0$ $(n=1,2,\ldots)$.

iii) $\int_S Hd\theta = H(0)\int_S d\theta$ for all polynomials H (in one complex variable).

iv) $\int_S Hd\theta = h(0)\int_S d\theta$ for all $H \in A(D)$ with $h(0) = <H\lambda>(0) = \int_S Hd\lambda$.

In this case we have the Cauchy formula

$$<\theta>(z) = \int_S \frac{s}{s-z}\, d\theta(s) \qquad \forall\ z \in D.$$

Proof: i) $\Rightarrow$ iv) is 3.1 for $z=0$. iv) $\Rightarrow$ iii) is obvious since $A(D)$ contains the polynomials. iii) $\Rightarrow$ ii) is trivial. ii) $\Rightarrow$ i) and the last assertion follow from

$$<\theta>(z) = \int_S P(z,s)d\theta(s) = \int_S \left(\frac{s}{s-z} + \frac{s\bar{z}}{1-s\bar{z}}\right)d\theta(s)$$

$$= \int_S \frac{s}{s-z}d\theta(s) + \sum_{n=1}^{\infty} \bar{z}^n \int_S s^n d\theta(s) \qquad \forall\ z \in D. \quad \text{QED.}$$

3.3 PROPOSITION: i) $A(D) \subset C(S)$ is the supnorm closure of the subalgebra of the polynomials. Furthermore $\operatorname{Re}A(D) \subset \operatorname{Re}C(S)$ is supnorm dense in $\operatorname{Re}C(S)$ (the DIRICHLET property).

ii) Let $1 \leq p < \infty$. Then $H^p(D) \subset L^p(\lambda)$ is the $L^p(\lambda)$-norm closure of $A(D)$ (and hence of the subalgebra of the polynomials).

iii) $H^\infty(D) \subset L^\infty(\lambda)$ is the weak* closure of $A(D)$ (and hence of the subalgebra of the polynomials). We have more: For each $F\in H^{\infty}(D)$ there exist functions $F_n\in A(D)$ with $\|F_n\| \leq \|F\|_{L^{\infty}(\lambda)}$ and $F_n\to F$ pointwise in the $L^{\infty}(\lambda)$ sense (and hence the F_n can be chosen to be polynomials).

Proof: i) It is clear that $A(D)\subset C(S)$ is supnorm closed. And in view of 3.2 the measures $\theta\in ca(S)$ which annihilate $A(D)$ are the same as those which annihilate the polynomials. The density of the subalgebra of the polynomials also has a simple direct proof: For $F\in A(D)$ and $f=<F\lambda>$ we have $\|f_R-F\|\to 0$ for $R\uparrow 1$, and in view of the Taylor series expansion each f_R for $0\leq R<1$ is the uniform limit of polynomials. In order to prove the Dirichlet property let $\theta\in ca(S)$ annihilate $\mathrm{Re}A(D)$ and hence F and $\bar{F}$ for all $F\in A(D)$. It follows that $\int_S s^n d\theta(s)=0$ $\forall n\in\mathbb{Z}$ and hence $\theta=0$ in view of the Weierstraß theorem.

ii)iii) In view of 3.2 $H^p(D)$ consists of the $F\in L^p(\lambda)$ with $\int_S s^n F(s)d\lambda(s) = 0 (n=1,2,\ldots)$ and is therefore closed in the respective sense. For $F\in H^p(D)$ and $f=<F\lambda>$ we have $f_R\in A(D)$ for $0\leq R<1$, and $f_R\to F$ for $R\uparrow 1$ in $L^p(\lambda)$-norm if $1\leq p<\infty$, whereas $\|f_R\| \leq \|F\|_{L^{\infty}(\lambda)}$ and $f_R\to F$ pointwise in the $L^{\infty}(\lambda)$ sense if $p=\infty$ after 2.1. QED.

We conclude this section with the formulation of the most fundamental classical theorems. We present almost no proofs. All these theorems will be put into broad and natural context and appear as simplest special cases in the abstract theories of Chapters II and III.

<u>3.4</u> <u>F. and M.RIESZ THEOREM</u>: Each analytic measure is absolutely continuous with respect to λ, that is $\mathrm{an}(D) = H^1(D)\lambda$.

In view of the future abstraction we deduce this theorem from a certain modified version.

<u>3.5</u> <u>MODIFIED F.and M.RIESZ THEOREM</u>: Assume that $\theta\in ca(S)$ annihilates $A(D)$, that is $\int_S Fd\theta=0$ $\forall F\in A(D)$. If $\theta=\alpha+\beta$ with λ-continous α and λ-singular β, then α and β likewise annihilate $A(D)$.

<u>3.6</u> <u>LEMMA</u>: Let $\theta\in \mathrm{an}(D)$ be $\neq 0$. Then there exists an $n(n=0,1,2\ldots)$ such that

$$\frac{1}{Z^n}\theta \in \mathrm{an}(D) \text{ and } \int_S \frac{1}{Z^n}d\theta \neq 0,$$

where $Z:Z(s)=s$ is the identity function.

Proof of 3.6: From 3.2 we know that $\int_S s^n d\theta(s)=0\ \forall n\geq 1$. Thus in view of the Weierstraß theorem we have $\int_S s^{-n} d\theta(s)\neq 0$ for some $n\geq 0$ and hence for a smallest $n\geq 0$. Then 3.2 shows that this $n\geq 0$ fulfills the assertion. QED.

Proof of 3.5 ⇒ 3.4: i) Let $\theta\in an(D)$ and $\theta=\alpha+\beta$ with λ-continuous α and λ-singular β. Then $\theta-\theta(S)\lambda=(\alpha-\theta(S)\lambda)+\beta$ annihilates $A(D)$ so that from 3.5 we see that β likewise annihilates $A(D)$. ii) In particular i) shows that a λ-singular analytic measure must annihilate $A(D)$, that is has an integral=0. But if now the above β were $\neq 0$, then 3.6 would lead to a λ-singular analytic measure with integral $\neq 0$. This contradiction shows that β must be $=0$. QED.

3.7 <u>JENSEN INEQUALITY</u>: Let $F\in A(D)$ and $f=\langle F\lambda\rangle$. Then

$$\log|f(z)| \leq \int_S P(z,s)\log|F(s)|d\lambda(s) \qquad \forall\ z\in D.$$

Hence the same is true for $F\in H^1(D)$.

3.8 <u>COROLLARY</u>: If $F\in H^1(D)$ is $\neq 0$ then $\log|F|\in L^1(\lambda)$.

Proof of $A(D)\Rightarrow H^1(D)$ in 3.7: Let $F\in H^1(D)$. For $0\leq R<1$ we have

$$\log|f(Rz)| \leq \int_S P(z,s)\log|f_R(s)|d\lambda(s)$$

$$\leq \int_S P(z,s)\log(|f_R(s)|+\varepsilon)d\lambda(s) \qquad \forall\ z\in D \text{ and } \varepsilon>0.$$

Now $\|f_R-F\|_{L^1(\lambda)}\to 0$ for $R\uparrow 1$ so that for a suitable sequence $R(n)\uparrow 1$ we have $f_{R(n)}\to F$ pointwise and under an $L^1(\lambda)$-majorant. Then the same is true for $\log(|f_{R(n)}|+\varepsilon)\to\log(|F|+\varepsilon)$. It follows that

$$\log|f(z)| \leq \int_S P(z,s)\log(|F(s)|+\varepsilon)d\lambda(s) \qquad \forall\ z\in D \text{ and } \varepsilon>0,$$

and for $\varepsilon\downarrow 0$ the result follows from Beppo Levi. Proof of 3.8: If $F\neq 0$ then $f(z)\neq 0$ for some $z\in D$. Thus the assertion is obvious. QED.

For the last theorem we introduce the functionals

$$D^p: D^p(\sigma) = \text{Inf}\{\int_S |F|^p d\sigma : F\in A(D) \text{ with } f(0)=1\} \qquad \forall\ \sigma\in Pos(S),$$

where $1 \leq p < \infty$. Thus $0 \leq D^p(\sigma) \leq \sigma(S)$.

<u>3.9</u> <u>SZEGÖ-KOLMOGOROV-KREIN THEOREM</u>: For $1 \leq p < \infty$ we have

$$D^p(\sigma) = \exp(\int_S (\log\frac{d\sigma}{d\lambda})d\lambda) \qquad \forall\ \sigma \in Pos(S).$$

In particular

$$D^p(F\lambda) = \exp(\int_S (\log F)d\lambda) \qquad \forall\ 0 \leq F \in L^1(\lambda),$$

which is the Szegö theorem.

4. The Function Classes $Hol^{\#}(D)$ and $H^{\#}(D)$

In the future abstract theory the function class which corresponds to the class $H^{\#}(D)$ to be defined in the present section will be far more important than the function classes which correspond to the $H^p(D)$ for finite $p \geq 1$. As before our main concern will be the transition from D to S.

For G an open subset of $\mathbb{C}$ we define a function $f: G \to \mathbb{C}$ to be of class $Hol^{\#}(G)$ iff there exists a sequence of functions $f_n \in Hol^{\infty}(G)$ with $|f_n| \leq 1$ on G, $f_n \to 1$ pointwise on G (and hence uniformly on each compact subset of G), and $f_n f \in Hol^{\infty}(G)$ for all $n \geq 1$. We list some immediate consequences. i) $Hol^{\infty}(G) \subset Hol^{\#}(G) \subset Hol(G)$, and $Hol^{\#}(G)$ is an algebra. ii) $Hol^{\#}(G)$ contains the class $Hol^{+}(G)$ of the functions $f \in Hol(G)$ with $\mathrm{Re}\, f \geq 0$. In fact, we can take $f_n := \frac{n}{n+f}$. iii) If $U, V \subset \mathbb{C}$ are open and $\Theta: U \to V$ a holomorphic map, then $f \in Hol^{\#}(V)$ implies that $f \circ \Theta \in Hol^{\#}(U)$.

Let us now turn to the unit disk situation. We define a function $F \in L(\lambda)$ to be of class $H^{\#}(D)$ iff there exists a sequence of functions $F_n \in H^{\infty}(D)$ with $|F_n| \leq 1$, $F_n \to 1$ pointwise (as usual in the $L(\lambda)$-sense), and $F_n F \in H^{\infty}(D)$ for all $n \geq 1$. Then $H^{\infty}(D) \subset H^{\#}(D)$, and $H^{\#}(D)$ is an algebra.

<u>4.1</u> <u>PROPOSITION</u>: For $f \in Hol^{\#}(D)$ the radial limit

$$F(s) := \lim_{R \uparrow 1} f_R(s) \qquad \text{exists for } \lambda\text{-almost all } s \in S,$$

and produces an element $F\in H^{\#}(D)$. The map $f\mapsto F$ thus defined is a bijection $Hol^{\#}(D)\to H^{\#}(D)$.

Proof: i) Let $f\in Hol^{\#}(D)$, and take functions $f_n\in Hol^{\infty}(D)$ as required in the definition with $g_n:=f_nf\in Hol^{\infty}(D)$. Let $F_n,G_n\in H^{\infty}(D)$ the respective boundary functions. Then $|F_n|\leqq 1$ and

$$\int_S |F_n-1|^2 d\lambda \leq 2-2\mathrm{Re}\int_S F_n d\lambda = 2\big(1-\mathrm{Re}F_n(0)\big)\to 0.$$

Therefore after transition to a suitable subsequence we can assume that $F_n\to 1$ pointwise. ii) We choose a Baire set $N\subset S$ with $\lambda(N)=0$ such that in each point $s\in S-N$ 1) radial convergence $f_n(Rs)\to F_n(s)$ and $g_n(Rs)\to G_n(s)$ for $R\uparrow 1$ takes place for each $n\geqq 1$, and 2) the representatives $s\to F_n(s)$ of the $F_n\in H^{\infty}(D)$ thus obtained on $S-N$ fulfills $F_n(s)\to 1$ for $n\to\infty$. Consider now the equation $f_n(Rs)f(Rs)=g_n(Rs)$ for $s\in S-N$ and $0\leqq R<1$. For fixed $s\in S-N$ choose an $n\geqq 1$ with $F_n(s)\neq 0$. Then $f_n(Rs)\neq 0$ for R sufficiently close to 1. Thus the limit $F(s):=\lim_{R\uparrow 1} f(Rs)$ exists in each point $s\in S-N$. The element $F\in L(\lambda)$ thus produced fulfills $F_nF=G_n$ for all $n\geqq 1$. Therefore $F\in H^{\#}(D)$. iii) In case $F=0$ we have $G_n=0$ and hence $g_n=0$ for all $n\geqq 1$. In view of $f_n\to 1$ this implies that $f=0$. Therefore the above map $f\mapsto F$ is injective. iv) Let us now start with a function $F\in H^{\#}(D)$, and take functions $F_n\in H^{\infty}(D)$ as required in the definition with $G_n:=F_nF\in H^{\infty}(D)$. And put $f_n=\langle F_n\lambda\rangle$, $g_n=\langle G_n\lambda\rangle\in Hol^{\infty}(D)$. Then $|f_n|\leqq 1$ and $f_n\to 1$ on D. Now $f_lg_n=f_ng_l$ for all $l,n\geqq 1$ since the difference is in $Hol^{\infty}(D)$ and possesses the boundary function $F_lG_n-F_nG_l=0$. Therefore $f_n\to 1$ on D implies that there exists a function $f:D\to\mathbb{C}$ such that $f_nf=g_n$ for all $n\geqq 1$. This implies that $f\in Hol^{\#}(D)$. Let now $\tilde{F}\in H^{\#}(D)$ be the radial boundary function produced by f. Then $F_n\tilde{F}=G_n=F_nF$ for all $n\geqq 1$. In view of $F_n\to 1$ this implies that $\tilde{F}=F$. Thus the map $f\mapsto F$ under consideration is surjective. QED.

<u>4.2</u> <u>COROLLARY</u>: Let $f\in Hol^{\#}(D)$. Then for λ-almost all $s\in S$ the angular limit

$$F(s):=\lim_{\substack{z\to s\\ z\in\omega(s,\alpha)}} f(z) \text{ on } \omega(s,\alpha):=\{z\in D:-\mathrm{Re}(z-s)\bar{s}\geq |z-s|\cos\alpha\}$$

exists for all $0<\alpha<\frac{\pi}{2}$ (and of course each time is equal to the radial limit obtained in 4.1).

Proof: i) An immediate adaptation of parts i) and ii) of the above proof shows that we can restrict ourselves to the case $f \in Hol^{\infty}(D)$. Then fix a point $s \in S$ such that the radial limit $\lim_{R \uparrow 1} f(Rs) =: c$ exists. We shall prove that $f(z) \to c$ for $z \to s$ on $\omega(s,\alpha)$ for each $0 < \alpha < \frac{\pi}{2}$. We can of course assume $s=1$. Then it is convenient to transfer the problem from D to the halfplane Δ via the fractional-linear map $h: h(z) = \frac{1-z}{1+z}$ $\forall z \in \mathbb{C}$. After the transition we have a function $f \in Hol^{\infty}(\Delta)$ such that $f(z) \to c$ for $z \downarrow 0$, and we have to prove that $f(z) \to c$ for $z \to 0$ on $\omega(\alpha) := \{z \in \Delta : \operatorname{Re} z \geqq |z| \cos\alpha\}$ for each $0 < \alpha < \frac{\pi}{2}$. This will be done with the aid of the Vitali theorem (see CARATHEODORY [1950] Vol.I p.186). ii) The sequence of the functions $f_n : f_n(z) = f(\frac{z}{n})$ $\forall z \in \Delta$ is equibounded in Δ and satisfies $f_n(z) \to c$ for $n \to \infty$ in each point $z > 0$. Therefore in view of Vitali $f_n \to c$ uniformly on each compact subset of Δ. Thus for fixed $0 < \alpha < \frac{\pi}{2}$ we have a sequence of $\varepsilon_n \downarrow 0$ such that

$$|f_n(z) - c| = |f(\tfrac{z}{n}) - c| \leqq \varepsilon_n \quad \forall\ z \in \omega(\alpha) \text{ with } \tfrac{1}{2} \leqq |z| \leqq 1 \text{ and } n \geqq 1.$$

This implies that $|f(z) - c| \leqq \varepsilon_n$ $\forall z \in \omega(\alpha)$ with $|z| \leqq \frac{1}{n}$ and $n \geqq 1$. QED.

The next results contain important information on the richness of the classes $Hol^{\#}(D)$ and $H^{\#}(D)$. Note that 4.5-4.7 depend upon the Jensen inequality 3.7 and 3.8. So the proofs of these results as well as those of 3.4-3.5 and 3.7-3.9 will not be complete until the end of Chapter II.

<u>4.3</u> <u>LEMMA</u>: Assume that $f \in Hol^{+}(D)$. From 1.5 we have $\theta \in Pos(S)$ such that $\operatorname{Re} f = \langle \theta \rangle$ and hence

$$f(z) = i \operatorname{Im} f(0) + \int_S \frac{s+z}{s-z} d\theta(s) \qquad \forall\ z \in D.$$

Then $e^f \in Hol^{\#}(D)$ iff θ is absolutely continuous with respect to λ.

Proof: i) Assume that θ is λ-continuous and $\theta = F\lambda$ with $0 \leqq F \in L^1(\lambda)$. Put $F_n := \operatorname{Min}(F, n)$ and

$$f_n : f_n(z) = i \operatorname{Im} f(0) + \int_S \frac{s+z}{s-z} F_n(s) d\lambda(s) \qquad \forall\ z \in D \text{ and } n \geqq 1.$$

Then $f_n \in Hol(D)$ with $f_n \to f$ and $\operatorname{Re} f_n = \langle F_n \lambda \rangle \in Harm^{\infty}(D)$ with $\operatorname{Re} f_n \leqq \operatorname{Re} f$.

Thus $h_n:=\exp(f_n-f)\in Hol^\infty(D)$ with $|h_n|\leqq 1$ and $h_n\to 1$ and $h_n e^f=\exp(f_n)\in Hol^\infty(D)$ $\forall n\geqq 1$. Therefore $e^f \in Hol^\#(D)$. ii) Let $\theta=\alpha+\beta$ with λ-continuous α and λ-singular β. Then $\alpha,\beta\geqq 0$. And $f=u+v$ with

$$u:u(z) = i\,\mathrm{Im}\,f(0) + \int_S \frac{s+z}{s-z}d\alpha(s),$$

$$v:v(z) = \int_S \frac{s+z}{s-z}d\beta(s) \qquad \forall\ z\in D.$$

Assume that $e^f\in Hol^\#(D)$. Then $e^v=e^{f-u}\in Hol^\#(D)$ since $|e^{-u}|\leqq 1$. Take functions $f_n\in Hol^\infty(D)$ with $|f_n|\leqq 1$ and $f_n\to 1$ and $g_n:=f_n e^v\in Hol^\infty(D)$ $\forall n\geqq 1$, and let $F_n,G_n\in H^\infty(D)$ be the respective boundary functions. Now $\mathrm{Re}\,v(Rs)\to 0$ for $R\uparrow 1$ for λ-almost all $s\in S$ after 2.2. Thus $|g_n|=|f_n|e^{\mathrm{Re}\,v}$ implies that $|G_n|=|F_n|\leqq 1$ and hence $|g_n|=|f_n|e^{\mathrm{Re}\,v}\leqq 1$ and hence $e^{\mathrm{Re}\,v}\leqq 1$ since $f_n\to 1$. Therefore $\mathrm{Re}\,v=0$ and hence $\beta=0$. QED.

4.4 COROLLARY: Let $F\in\mathrm{Re}\,L^1(\lambda)$ and

$$f:f(z) = \int_S \frac{s+z}{s-z}F(s)d\lambda(s) \quad \forall\ z\in D.$$

Then $e^f\in Hol^\#(D)$, and is in fact an invertible element of the algebra $Hol^\#(D)$.

4.5 COROLLARY: We have $H^1(D)\subset H^\#(D)$.

Proof: Let $F\in H^1(D)$ be $\neq 0$ and $f=\langle F\lambda\rangle\in Hol^1(D)$. From 3.8 we know that $\log|F|\in L^1(\lambda)$. Thus for

$$h:h(z) = \int_S \frac{s+z}{s-z}\log|F(s)|d\lambda(s) \quad \forall\ z\in D$$

we have $e^h\in Hol^\#(D)$ in view of 4.4. Now the Jensen inequality 3.7 reads $|f|\leqq|e^h|$. Therefore $f\in Hol^\#(D)$ and hence $F\in H^\#(D)$. QED.

4.6 JENSEN INEQUALITY: Let $f\in Hol^\#(D)$ with boundary function $F\in H^\#(D)$. In case $f\neq 0$ we have $\log|F|\in L^1(\lambda)$ and

$$\log|f(z)| \leqq \int_S P(z,s)\log|F(s)|d\lambda(s) \quad \forall\ z\in D.$$

Proof: Take functions $f_n\in Hol^\infty(D)$ with $|f_n|\leqq 1$ and $f_n\to 1$ and $g_n:=f_n f\in$ $\in Hol^\infty(D)$ $\forall n\geqq 1$, and let $F_n,G_n\in H^\infty(D)$ be the respective boundary functions.

Then $G_n = F_n F$. We can of course assume $f_n \neq 0$ and hence $g_n \neq 0$. Thus $\log|F_n|$, $\log|G_n| \in L^1(\lambda)$ from 3.8 and hence $\log|F| \in L^1(\lambda)$. And from 3.7 we have for $z \in D$

$$\log|g_n(z)| \leq \int_S P(z,s)\log|G_n(s)|d\lambda(s) \leq \int_S P(z,s)\log|F(s)|d\lambda(s)$$

in view of $|F_n| \leq 1$, from which the result follows for $n \to \infty$. QED.

4.7 PROPOSITION: $(\mathrm{Hol}^{\#}(D))^{\times}$:= the set of invertible elements of the algebra $\mathrm{Hol}^{\#}(D)$ consists of the functions $h = ce^f$, where

$$f: f(z) = \int_S \frac{s+z}{s-z} F(s)d\lambda(s) \quad \forall\ z \in D \quad \text{with } F \in \mathrm{Re}\, L^1(\lambda),$$

and $c \in \mathbb{C}$ with $|c| = 1$. Thus $|h| = \exp(\langle F\lambda \rangle)$. The boundary function $H \in (H^{\#}(D))^{\times}$ which corresponds to h satisfies $|H| = e^F$.

Proof: We have to prove that each $h \in (\mathrm{Hol}^{\#}(D))^{\times}$ is of the above form which has already been exhibited in 4.4. From 4.6 we obtain $F := \log|H| \in L^1(\lambda)$ and

$$\log|h(z)| = \int_S P(z,s)\log|H(s)|d\lambda(s) = \int_S P(z,s)F(s)d\lambda(s) = \mathrm{Re}\, f(z) \ \forall\ z \in D,$$

with $f \in \mathrm{Hol}(D)$ as defined above. Thus $|he^{-f}| = 1$ and hence $h = ce^f$ with c a constant of modulus $|c| = 1$. QED.

To end the section let us consider the subclass $\mathrm{Hol}^{+}(D) \subset \mathrm{Hol}^{\#}(D)$. Let $H^{+}(D) \subset H^{\#}(D)$ consist of the boundary functions of the functions in $\mathrm{Hol}^{+}(D)$. We ask for a direct characterization of $H^{+}(D)$. Of course $H^{+}(D) \subset \{F \in H^{\#}(D) : \mathrm{Re}\, F \geq 0\}$, but here $\neq$ holds true. In fact, the function $f: f(z) = \frac{1-z}{1+z} \forall z \in D$ is in $\mathrm{Hol}^{+}(D)$ and its boundary function $F = \frac{1-Z}{1+Z} \in H^{+}(D)$ has $\mathrm{Re}F = 0$. So $-F \in H^{\#}(D)$ has $\mathrm{Re}(-F) \geq 0$ as well, but it corresponds to the function $-f \in \mathrm{Hol}^{\#}(D)$ which is $\notin \mathrm{Hol}^{+}(D)$ so that $-F \notin H^{+}(D)$.

To explain the phenomenon recall from 4.3 that to $f \in \mathrm{Hol}^{+}(D)$ there exists a measure $\theta \in \mathrm{Pos}(S)$ such that $\mathrm{Re}\, f = \langle\theta\rangle$ and hence

$$f(z) = i\mathrm{Im}f(0) + \int_S \frac{s+z}{s-z} d\theta(s) \qquad \forall\ z \in D.$$

From 2.2 we have $\mathrm{Re}F = \frac{d\theta}{d\lambda}$ for the boundary function $F \in H^{+}(D)$. Thus the real part $\mathrm{Re}F$ suffers from an essential lack of information and in particular does not determine the entire F up to an additive constant. So the $\neq$ above becomes quite clear. But we can prove a positive result.

4.8 PROPOSITION: For $F \in L(\lambda)$ with $\operatorname{Re} F \geq 0$ the subsequent properties are equivalent.

i) $F \in H^+(D)$.

ii) $\frac{1}{F+s} \in H^\infty(D)$ for all $s \in \Delta :=$ the open halfplane $\operatorname{Re} s > 0$.

iii) $\frac{1}{F+s} \in H^\infty(D)$ for some $s \in \Delta$.

Proof: i)$\Rightarrow$ii) and ii)$\Rightarrow$iii) are obvious. iii)$\Rightarrow$i) We have $G := \frac{\bar{s}-F}{s+F} = \frac{\bar{s}+s}{s+F} - 1 \in H^\infty(D)$ with $|G| \leq 1$ and hence $g := \langle G\lambda \rangle \in \operatorname{Hol}^\infty(D)$ with $|g| \leq 1$, and g is not the constant -1. Therefore $f := \frac{\bar{s}-sg}{1+g} \in \operatorname{Hol}^+(D)$, and from $g = \frac{\bar{s}-f}{s+f}$ we see that f has the boundary function F. Thus $F \in H^+(D)$. QED.

Notes

The concrete theory in the unit disk is presented in the spirit of functional analysis in the beautiful books of HOFFMAN [1962a] and DUREN [1970]. In HOFFMAN [1962a] also the initial step of abstraction (the Dirichlet algebra situation) is within consideration. The older treatises of NEVANLINNA [1953] and PRIWALOW [1956] are more in the spiri of classical analysis. In these Notes we restrict ourselves to some additional remarks.

The Loomis theorem 2.3 is in LOOMIS [1943]. We also refer to ALLEN and KERR [1953] and GEHRING [1957]. Our proof is from KÖNIG [1960] and is much simpler than the former ones. The proof of 4.2 is from DUREN [1970] p.6. The function class $\operatorname{Hol}^\#(G)$ for open $G \subset \mathbb{C}$ is introduced in KÖNIG [1970a]. For the unit disk it coincides with the Smirnov class named D in PRIWALOW [1956] and N^+ in DUREN [1970]. But the definition here is quite different, and in the form of $H^\#(D)$ it is the one which can be transferred to the abstract theory. Also $\operatorname{Hol}^+(D)$ will be transferred in the form of $H^+(D)$.

Chapter II

Function Algebras: The Bounded-Measurable Situation

In the present chapter we fix a measurable space (X,Σ) and a complex subalgebra $A\subset B(X,\Sigma)$ which contains the constants. It is natural then to consider the annihilator

$$A^{\perp} := \{\theta\in ca(X,\Sigma): \int f d\theta = 0 \ \forall f\in A\},$$

the subspace $\subset ca(X,\Sigma)$ of the measures which annihilate A. Furthermore define the spectrum $\Sigma(A)$ of A to consist of those nonzero multiplicative linear functionals $\varphi:A\to\mathbb{C}$ which are continuous in the weak topology $\omega := \sigma(B(X,\Sigma),ca(X,\Sigma))$, that is which can be represented in the form $\varphi(f) = \int f d\theta \ \forall f\in A$ for some $\theta\in ca(X,\Sigma)$. The functionals $\varphi\in\Sigma(A)$ are of course supnorm continuous and hence of norm $\|\varphi\|=1$, but the converse need not be true. It will soon become clear that each $\varphi\in\Sigma(A)$ can be represented even by positive measures which in view of $\varphi(1)=1$ must then be probability measures $\theta \in Prob(X,\Sigma)$. Thus the set

$$M(\varphi) = M(A,\varphi) := \{\theta\in Prob(X,\Sigma):\varphi(f) = \int f d\theta \ \forall f\in A\} \text{ will be } \neq \emptyset.$$

Then the union $M(A)$ of the $M(A,\varphi)$ for all $\varphi\in\Sigma(A)$ consists of all $\theta\in Prob(X,\Sigma)$ which are multiplicative on A.

The chapter is devoted to two fundamental theorems, both for a fixed $\varphi\in\Sigma(A)$: The abstract F. and M.Riesz theorem, and the abstract Szegö-Kolmogorov-Krein theorem with a subsequent universal version of the Jensen inequality. The F. and M.Riesz theorem leads to the Gleason-Part decompositions of $\Sigma(A)$ and of $ca(X,\Sigma)$ and $A^{\perp}$. The latter of these is decisive for the direction into which the abstract theory had to be developed.

1. Szegö Functional and Fundamental Lemma

In the present section we fix $\varphi\in\Sigma(A)$. For a measure $\sigma\in Pos(X,\Sigma)$ and $1\leq p<\infty$ it is natural to ask whether the functional $\varphi:A\to\mathbb{C}$ is well-defined on the subspace $A \bmod \sigma \subset L^{p}(\sigma)$ and is $L^{p}(\sigma)$-norm continuous there, that is whether

$$\|\varphi\|_{L^p(\sigma)} := \operatorname{Sup}\{|\varphi(f)| : f\in A \text{ with } \|f\|_{L^p(\sigma)} \leq 1\} \quad \text{is } < \infty \text{ or } = \infty?$$

Instead of this norm which can be $=\infty$ it is advisable to introduce the Szegö functional

$$D^p : D^p(\sigma) = \operatorname{Inf}\{\int |f|^p d\sigma : f\in A \text{ with } \varphi(f)=1\} \qquad \forall \sigma\in \operatorname{Pos}(X,\Sigma).$$

Then $0 \leq D^p(\sigma) \leq \sigma(X)$. One verifies that

$$\|\varphi\|_{L^p(\sigma)} = (D^p(\sigma))^{-\frac{1}{p}} \qquad \forall\ \sigma\in\operatorname{Pos}(X,\Sigma) \text{ and } 1\leq p<\infty,$$

so that the $L^p(\sigma)$-norm in question is $=\infty$ iff $D^p(\sigma) = 0$. The limit case $D^p(\sigma) = \sigma(X)$ also deserves special attention. For this we can assume that $\sigma(X) = 1$.

1.1 REMARK: Let $\sigma\in\operatorname{Prob}(X,\Sigma)$ and $1\leq p<\infty$. Then $D^p(\sigma) = 1$ iff $\sigma\in M(\varphi)$.

Proof: i) Assume that $\sigma\in M(\varphi)$. For $f\in A$ with $\varphi(f)=1$ then $\int|f|^p d\sigma \geq (\int|f|d\sigma)^p \geq |\int f d\sigma|^p = 1$, so that $D^p(\sigma) \geq 1$ and hence $D^p(\sigma)=1$. ii) Assume that $D^p(\sigma)=1$. Let us fix $f\in A$ with $\varphi(f)=0$. For $t>0$ then $\int|1+tf|^p d\sigma \geq 1$ or $\int\frac{1}{t}(|1+tf|^p-1)d\sigma \geq 0$. For $t\downarrow 0$ we obtain

$$\frac{1}{t}(|1+tf|^p-1) = \frac{1}{t}\,\frac{(|1+tf|^2)^p-1}{|1+tf|^p+1} = \frac{1}{t}\,\frac{(1+2t\operatorname{Re}f+t^2|f|^2)^p-1}{|1+tf|^p+1}$$

$$= \frac{2p\operatorname{Re}f + \text{terms in } t,t^2,\dots}{|1+tf|^p+1} \longrightarrow p\operatorname{Re}f\ .$$

And the convergence is majorized by a constant since

$$||1+z|^p-1| = |p\int_1^{|1+z|} t^{p-1}dt| \leq p(1+|z|)^{p-1}|z| \quad \forall z\in\mathbb{C}\ ,$$

$$\left|\frac{1}{t}(|1+tf|^p-1)\right| \leq p(1+|f|)^{p-1}|f| \qquad \forall 0<t\leq 1\ .$$

Thus it follows that $\int\operatorname{Re}f\,d\sigma \geq 0$. Then consideration of cf for $c\in\mathbb{C}$ shows that $\int f d\sigma = 0$ for $f\in A$ with $\varphi(f)=0$. Therefore $\varphi(f) - \int f d\sigma = \int(\varphi(f)-f)d\sigma = 0$ for all $f\in A$, so that $\sigma\in M(\varphi)$. QED.

The next lemma will sometimes be useful in that it permits to reduce the case $1 \leq p < \infty$ to $1 < p < \infty$.

1.2 LEMMA: For fixed $\sigma \in Pos(X,\Sigma)$ the function $p \mapsto D^p(\sigma)$ is continuous from the right in $1 \leq p < \infty$.

Proof: Fix $1 \leq p < \infty$. For $f \in A$ with $\varphi(f)=1$ then $D^s(\sigma) \leq \int |f|^s d\sigma$ and hence $\limsup_{s \to p} D^s(\sigma) \leq \int |f|^p d\sigma$. Therefore $\limsup_{s \to p} D^s(\sigma) \leq D^p(\sigma)$. On the other hand for $1 \leq p < s < \infty$ we obtain from Hölder

$$\int |f|^p d\sigma \leq (\int |f|^s d\sigma)^{\frac{p}{s}} (\sigma(X))^{1-\frac{p}{s}} \quad \forall f \in A,$$

$$D^p(\sigma) \leq (D^s(\sigma))^{\frac{p}{s}} (\sigma(X))^{1-\frac{p}{s}},$$

and hence $D^p(\sigma) \leq \liminf_{s \downarrow p} D^s(\sigma)$. QED.

1.3 LEMMA: Let $\sigma \in Pos(X,\Sigma)$ and $1 < p,q < \infty$ with $\frac{1}{p} + \frac{1}{q} = 1$. For $0 \leq P \in L^p(\sigma)$ and $0 \leq Q \in L^q(\sigma)$ then

$$\int PQ d\sigma = \|P\|_{L^p(\sigma)} \|Q\|_{L^q(\sigma)}$$

iff $P^p, Q^q \in L^1(\sigma)$ are linearly dependent.

Proof: i) We quote a simple calculus fact: For $u,v \geq 0$ we have

$$uv \leq t^p \frac{u^p}{p} + \frac{1}{t^q} \frac{v^q}{q} \qquad \forall 0 < t < \infty,$$

with equality iff $t^p u^p = t^{-q} v^q$. ii) We can assume that $a := \|P\|_{L^p(\sigma)}$ and $b := \|Q\|_{L^q(\sigma)}$ be >0. Choose $t>0$ with $t^p a^p = t^{-q} b^q$. Then integrate

$$PQ \leq t^p \frac{P^p}{p} + \frac{1}{t^q} \frac{Q^q}{q} \quad \text{to obtain} \quad \int PQ d\sigma \leq t^p \frac{a^p}{p} + \frac{1}{t^q} \frac{b^q}{q} = ab.$$

Thus $\int PQ d\sigma = ab$ iff the inequality we integrated was an equality, that is iff $t^p P^p = t^{-q} Q^q$ after i). QED.

1.4 FUNDAMENTAL LEMMA: Let $\sigma \in Pos(X,\Sigma)$ and $1 < p,q < \infty$ with $\frac{1}{p} + \frac{1}{q} = 1$. Assume that $D^p(\sigma) > 0$. Then there exist functions $P \in L^p(\sigma)$ and $Q \in L^q(\sigma)$ such that

i) P is in the $L^p(\sigma)$-norm closure of A mod σ,

ii) $\varphi(f) = \|\varphi\|_{L^p(\sigma)} \int fQd\sigma$ for all $f\in A$,

iii) $|P|^p = PQ = |Q|^q =: F\in L^1(\sigma)$ and $F\sigma\in M(\varphi)$.

Proof: 1) On the linear subspace $T := \{h\in L^q(\sigma): \int uhd\sigma = \varphi(u)\int hd\sigma \ \forall u\in A\} \subset L^q(\sigma)$ we consider the linear functional $\theta: \theta(h) = \int hd\sigma \ \forall h\in T$. Then

$$|\theta(h)| = |\int uhd\sigma| \leq \|u\|_{L^p(\sigma)} \|h\|_{L^q(\sigma)} \quad \forall\ h\in T \text{ and } u\in A \text{ with } \varphi(u)=1,$$

so that $\|\theta\| \leq C := (D^p(\sigma))^{\frac{1}{p}} = (\|\varphi\|_{L^p(\sigma)})^{-1}$. From Hahn-Banach and $L^p(\sigma) = (L^q(\sigma))'$ we obtain a function $P\in L^p(\sigma)$ with $\int |P|^p d\sigma \leq 1$ and $\theta(h) = C\int hPd\sigma$ for all $h\in T$. 2) This function P is in the $L^p(\sigma)$-closure of A mod σ. Otherwise there were an $f\in L^q(\sigma)$ with $\int ufd\sigma = 0 \ \forall u\in A$ and with $\int Pfd\sigma \neq 0$. But then $f\in T$ and $0 = \int fd\sigma = \theta(f) = C\int fPd\sigma$ which is a contradiction.

3) We know that $|\varphi(u)| \leq \frac{1}{C}\|u\|_{L^p(\sigma)} \ \forall u\in A$. Thus from Hahn-Banach and $L^q(\sigma) = (L^p(\sigma))'$ we obtain a function $Q\in L^q(\sigma)$ with $\int |Q|^q d\sigma \leq 1$ and $\varphi(u) = \frac{1}{C}\int uQd\sigma \ \forall u\in A$. 4) We put $F := |Q|^q \in L^1(\sigma)$ and claim that $F\sigma\in M(\varphi)$. In view of 1.1 and $\int Fd\sigma \leq 1$ it suffices to show that $D^q(F\sigma) \geq 1$. But we have

$$1 = \varphi(uv) = \frac{1}{C}\int uvQd\sigma \leq \frac{1}{C}(\int |u|^p d\sigma)^{\frac{1}{p}}(\int |v|^q Fd\sigma)^{\frac{1}{q}} \quad \forall\ u,v\in A \text{ with } \varphi(u)=\varphi(v)=1,$$

and hence $1 \leq \int |v|^q Fd\sigma \ \forall v\in A$ with $\varphi(v)=1$, so that in fact $D^q(F\sigma) \geq 1$.

5) From 3) we have $\int uQd\sigma = C\varphi(u) \ \forall u\in A$. In particular $\int Qd\sigma = C$. Thus $Q\in T$. From 1) therefore $C = \int Qd\sigma = \theta(Q) = C\int PQd\sigma$. Thus

$$1 = \int PQd\sigma \leq \int |PQ|d\sigma \leq \|P\|_{L^p(\sigma)} \|Q\|_{L^q(\sigma)} \leq 1,$$

so that all these are equalities. We obtain $PQ = |PQ|$ and $\int |P|^p d\sigma = \int |Q|^q d\sigma = 1$, and from 1.3 that $|Q|^q = c|P|^p$ for some real c which of course then must be $=1$. Thus $|P|^p = PQ = |Q|^q = F$. QED.

1.5 THEOREM: Let $\sigma\in Pos(X,\Sigma)$ with $D^p(\sigma)>0$ for some $1\leq p<\infty$. Then there exists an $m\in M(\varphi)$ which is $<<\sigma$.

Proof: Combine 1.4 with 1.2. QED.

1.6 COROLLARY: Let $\theta \in ca(X,\Sigma)$ with $\varphi(f) = \int f d\theta \ \forall f \in A$. Then there exists an $m \in M(\varphi)$ which is $<< \theta$. In particular $M(\varphi) \neq \emptyset$.

Proof: We have $\int |f| d|\theta| \geq |\int f d\theta| \geq 1 \ \forall \ f \in A$ with $\varphi(f) = 1$ and hence $D^1(|\theta|) > 0$. QED.

The next consequence of the fundamental lemma will be the root of the abstract F. and M.Riesz theorem in Section 3.

1.7 THEOREM: Assume that $\sigma,\tau \in Pos(X,\Sigma)$ are such that τ is singular to all $m \in M(\varphi)$ which are $<< \sigma+\tau$. Then $D^p(\sigma+\tau) = D^p(\sigma) \quad \forall 1 \leq p < \infty$.

Proof: In view of 1.2 we can assume that $1<p<\infty$. We have to prove that $D^p(\sigma+\tau) \leq D^p(\sigma)$ and can therefore assume that $C^p := D^p(\sigma+\tau) > 0$. Let $Q \in L^q(\sigma+\tau)$ be the function obtained in 1.4. Then $|Q|^q(\sigma+\tau) \in M(\varphi)$ and hence is singular to τ. Thus $Q = 0$ modulo τ. For $f \in A$ with $\varphi(f)=1$ thus 1.4.ii) implies that

$$C = C\varphi(f) = \int fQd(\sigma+\tau) = \int fQd\sigma \leq (\int |f|^p d\sigma)^{\frac{1}{p}} (\int |Q|^q d\sigma)^{\frac{1}{q}} = (\int |f|^p d\sigma)^{\frac{1}{p}} .$$

Therefore $C^p \leq D^p(\sigma)$. QED.

2. Measure Theory: Prebands and Bands

2.1 PROPOSITION: For a nonvoid subset $K \subset ca(X,\Sigma)$ the subsequent properties are equivalent.

i) For each countable subset $T \subset K$ there exists an $m \in K$ such that $\tau << m \ \forall \tau \in T$.

ii) Each $\theta \in ca(X,\Sigma)$ admits a decomposition $\theta = \alpha+\beta$ with $\alpha \in ca(X,\Sigma)$ absolutely continuous to some $m \in K$ and $\beta \in ca(X,\Sigma)$ singular to all $m \in K$.

In this situation for each $\theta \in ca(X,\Sigma)$ the decomposition $\theta = \alpha+\beta$ in ii) is unique. Furthermore for each $\theta \in ca(X,\Sigma)$ there exist $m \in K$ such that $\|\theta_m\|$ is maximal, and we have $\alpha = \theta_m$ and $\beta = \theta_{\hat{m}}$ for all these $m \in K$.

Proof: We start with ii) $\Rightarrow$ i). Let $T = \{m(l):l=1,2,\ldots\} \subset K$ be a countable subset. Decompose

$$\theta := \sum_{l=1}^{\infty} \frac{1}{2^l} \frac{|m(l)|}{1+\|m(l)\|} \in Pos(X,\Sigma) \text{ into } \theta = \alpha+\beta \text{ after ii).}$$

Then $\alpha << m$ for some $m \in K$. Since β is singular to m it follows that $\alpha,\beta \geqq 0$. Now β is singular to each $m(l)$ and hence singular to θ. Since $0 \leqq \beta \leqq \theta$ it follows that $\beta = 0$. Therefore $\theta=\alpha<<m$ so that $m(l)<<m$ $(l=1,2,\ldots)$.

For the rest of the proof we assume i). 1) For each $\theta \in ca(X,\Sigma)$ there exist $m \in K$ such that $\|\theta_m\|$ is maximal. To see this let $c := \mathrm{Sup}\{\|\theta_m\|: m \in K\} \leqq \|\theta\|$ and take $m(l) \in K$ $(l=1,2,\ldots)$ with $\|\theta_{m(l)}\| \to c$ for $l \to \infty$. And let $m \in K$ with $m(l)<<m$ $\forall l \geqq 1$. Then from

$$\theta = \theta_m + \theta_{\hat{m}} = \theta_{m(l)} + \theta_{\hat{m}(l)}$$

we have $\theta_{\hat{m}(l)} = (\theta_m - \theta_{m(l)}) + \theta_{\hat{m}}$, where the first term is $<<m$ and the second is singular to m. It follows that $\|\theta_{\hat{m}(l)}\| = \|\theta_m - \theta_{m(l)}\| + \|\theta_{\hat{m}}\|$ and hence

$$\|\theta_m\| - \|\theta_{m(l)}\| = \|\theta_{\hat{m}(l)}\| - \|\theta_{\hat{m}}\| = \|\theta_m - \theta_{m(l)}\| \geqq 0.$$

Therefore $\|\theta_m\| \geqq c$ so that $\|\theta_m\| = c$. 2) Let $\theta \in ca(X,\Sigma)$ and $m \in K$ such that $\|\theta_m\|$ is maximal. Then $\theta_{\hat{m}}$ is singular to all $\sigma \in K$. To see this fix $\sigma \in K$ and take $\tau \in K$ with $\sigma,m << \tau$. We obtain as in 1)

$$\theta = \theta_m + \theta_{\hat{m}} = \theta_\tau + \theta_{\hat{\tau}},$$
$$\theta_{\hat{m}} = (\theta_\tau - \theta_m) + \theta_{\hat{\tau}}, \quad \|\theta_{\hat{m}}\| = \|\theta_\tau - \theta_m\| + \|\theta_{\hat{\tau}}\|,$$
$$\|\theta_\tau\| - \|\theta_m\| = \|\theta_{\hat{m}}\| - \|\theta_{\hat{\tau}}\| = \|\theta_\tau - \theta_m\|,$$

and hence $\theta_\tau = \theta_m$ or $\theta_{\hat{\tau}} = \theta_{\hat{m}}$. Thus $\theta_{\hat{m}}$ lives on some $|\tau|$-null set which is also a $|\sigma|$-null set so that $\theta_{\hat{m}}$ is singular to σ. 3) With 1) and 2) we have in particular proved for each $\theta \in ca(X,\Sigma)$ the decomposition $\theta = \alpha+\beta$ as claimed in ii). What remains is the uniqueness of this decomposition. Consider

$$K^{\vee} := \{\alpha \in ca(X,\Sigma):\alpha << \text{some } m \in K\},$$
$$K^{\wedge} := \{\beta \in ca(X,\Sigma):\beta << \text{singular to all } m \in K\}.$$

It is obvious that $K^\wedge$ is a linear subspace $\subset ca(X,\Sigma)$. And i) shows at once that $K^\vee$ is a linear subspace as well. $K^\vee \cap K^\wedge = \{0\}$ is immediate from the definitions. Therefore $ca(X,\Sigma) = K^\vee + K^\wedge$ means in fact the direct sum decomposition $ca(X,\Sigma) = K^\vee \oplus K^\wedge$. QED.

A nonvoid subset $K \subset ca(X,\Sigma)$ with the above property i) will be called a preband. The direct sum decomposition $ca(X,\Sigma) = K^\vee \oplus K^\wedge$ into the linear subspaces

$$K^\vee := \{\alpha \in ca(X,\Sigma) : \alpha << \text{some } m \in K\} \quad \text{and}$$

$$K^\wedge := \{\beta \in ca(X,\Sigma) : \beta \quad \text{singular to all } m \in K\}$$

will be called the preband decomposition with respect to K and will be written $\theta = \theta_K + \theta_{\hat{K}}$ for $\theta \in ca(X,\Sigma)$.

2.2 EXAMPLES: i) The simplest prebands are $K=\{m\}$ for fixed $m \in ca(X,\Sigma)$. Here $\theta = \theta_K + \theta_{\hat{K}}$ for $\theta \in ca(X,\Sigma)$ is the usual Lebesgue decomposition $\theta = \theta_m + \theta_{\hat{m}}$.

ii) Consider a fixed $\alpha \in ca(X,\Sigma)$ and a preband $K \subset ca(X,\Sigma)$ such that $\sigma << \alpha\ \forall \sigma \in K$. Then for $m \in K$ we claim: $\|\alpha_m\|$ is maximal $\Leftrightarrow \sigma << m\ \forall \sigma \in K$. In particular there exist $m \in K$ such that $\sigma << m\ \forall \sigma \in K$. For such an $m \in K$ it is clear that

$$\|\theta_m\| \text{ is maximal and } \theta_K = \theta_m\ ,\ \theta_{\hat{K}} = \theta_{\hat{m}} \quad \forall \theta \in ca(X,\Sigma).$$

Proof $\Leftarrow$: For $\sigma \in K$ we have $\sigma << m$ and hence $\|\alpha_m\| - \|\alpha_\sigma\| = \|\alpha_m - \alpha_\sigma\| \geqq 0$ as in the proof of 2.1. Proof $\Rightarrow$: Fix $\sigma \in K$ and take $\tau \in K$ with $\sigma, m << \tau$. As above we see $\|\alpha_\tau\| - \|\alpha_m\| = \|\alpha_\tau - \alpha_m\|$ and hence $\alpha_\tau = \alpha_m$. Now choose $T \in \Sigma$ such that $|\tau|(X-T) = 0$ and

$$\alpha_\tau(B) = \alpha_m(B) = \alpha(B \cap T) \quad \forall B \in \Sigma.$$

We take a $B \in \Sigma$ with $|m|(B) = 0$ and have to prove that $|\sigma|(B) = 0$, for which it suffices to prove that $|\tau|(B) = 0$. Now $|\alpha|(B \cap T) = |\alpha_m|(B) = 0$ since $|m|(B) = 0$, and hence $|\tau|(B \cap T) = 0$ since $\tau << \alpha$. This combined with $|\tau|(X-T) = 0$ shows that in fact $|\tau|(B) = 0$. QED.

iii) In the algebra situation considered in the present chapter for each $\varphi \in \Sigma(A)$ the set $M(\varphi) = M(A,\varphi) \subset \text{Prob}(X,\Sigma)$ is a preband. In fact, for $m_l \in M(\varphi)$ $(l=1,2,\ldots)$ we have $m := \sum_{l=1}^{\infty} 2^{-l} m_l \in M(\varphi)$ and $m_l << m\ \forall l \geqq 1$.

A linear subspace $B \subset ca(X,\Sigma)$ is called a band iff it satisfies the conditions

i) if $\beta \in B$ and $\alpha \ll \beta$ then $\alpha \in B$,

ii) if $0 \leqq \sigma_n \in B$ and $\sigma_n \uparrow \sigma$ then $\sigma \in B$.

Under the assumption of i) condition ii) is equivalent to

ii′) B is norm-closed.

In fact, the implication ii′) ⇒ ii) is obvious since $0 \leqq \sigma_n \uparrow \sigma$ implies that $\|\sigma - \sigma_n\| = \sigma(X) - \sigma_n(X) \to 0$. For ii) ⇒ ii′) observe that $\sigma_n \in B$ $(n=1,2,\ldots)$ implies that

$$\theta := \sum_{n=1}^{\infty} 2^{-n} \frac{|\sigma_n|}{1+\|\sigma_n\|} \in B .$$

Therefore in case $\|\sigma - \sigma_n\| \to 0$ we see from $\sigma_n \ll \theta$ $\forall n \geqq 1$ that $\sigma \ll \theta$ and hence that $\sigma \in B$ after i).

2.3 REMARK: i) For a nonvoid subset $K \subset ca(X,\Sigma)$ define $K^\vee$, $K^\wedge \subset ca(X,\Sigma)$ as above. Then $K^\wedge$ is a band called the singular band to K. But $K^\vee$ need not even be a linear subspace.

ii) Let $B \subset ca(X,\Sigma)$ be a band. Then B is a preband and $B^\vee = B$. Thus $ca(X,\Sigma) = B \oplus B^\wedge$. This is the F.Riesz band decomposition theorem.

iii) For $K \subset ca(X,\Sigma)$ nonvoid $K^{\wedge\wedge}$ is a band and in fact the smallest band which contains K. It is called the band generated by K. Furthermore $K \subset K^\vee \subset K^{\wedge\wedge}$, and we have $K^\vee = K^{\wedge\wedge}$ iff K is a preband.

Proof: 1) i) and ii) are obvious. 2) For nonvoid $K \subset ca(X,\Sigma)$ the inclusions $K \subset K^\vee \subset K^{\wedge\wedge}$ are obvious as well. For a preband K combine $ca(X,\Sigma) = K^\vee \oplus K^\wedge$ with $K^\vee \subset K^{\wedge\wedge}$ and $K^\vee \cap K^{\wedge\wedge} = \{0\}$ to obtain $K^\vee = K^{\wedge\wedge}$. On the other hand $K^\vee = K^{\wedge\wedge}$ implies that $ca(X,\Sigma) = K^\wedge \oplus K^{\wedge\wedge} = K^\wedge \oplus K^\vee$ so that K must be a preband in view of 2.1. 3) For nonvoid $K \subset ca(X,\Sigma)$ and $B \subset ca(X,\Sigma)$ a band we have $K \subset B \Rightarrow K^\wedge \supset B^\wedge \Rightarrow K^{\wedge\wedge} \subset B^{\wedge\wedge} = B^\vee = B$. Thus $K^{\wedge\wedge}$ is in fact the smallest band which contains K. QED.

2.4 EXAMPLE: The simplest bands are $B = \{m\}^\vee = \{\theta \in ca(X,\Sigma) : \theta \ll m\}$ for fixed $m \in ca(X,\Sigma)$.

We conclude the section with some results which will be needed for the Gleason-Part decompositions in Section 4. For a preband $K \subset ca(X,\Sigma)$ let $\langle K \rangle : ca(X,\Sigma) \to K^\vee$ denote the projection $\theta \mapsto \theta_K$ in the direct sum decomposition $ca(X,\Sigma) = K^{\vee} \oplus K^{\wedge}$. Then of course $\langle K\rangle = \langle K^{\vee}\rangle$.

2.5 REMARK: Let $E,F \subset ca(X,\Sigma)$ be bands. Then

i) $E\cap F$ is a band and $\langle E\cap F\rangle = \langle E\rangle\langle F\rangle = \langle F\rangle\langle E\rangle$.

ii) $E+F$ is a band and $\langle E+F\rangle = \langle E\rangle + \langle F\rangle - \langle E\cap F\rangle$.

iii) $(E\cap F)^{\wedge} = E^{\wedge} + F^{\wedge}$.

iv) $(E+F)^{\wedge} = E^{\wedge} \cap F^{\wedge}$.

Proof: 1) It is obvious that $E\cap F$ is a band. Now observe that $\langle E\rangle\theta = \theta_E \ll \theta \quad \forall\theta\in ca(X,\Sigma)$. Therefore $\theta\in F$ implies that $\langle E\rangle\theta \in F$. For $\theta \in ca(X,\Sigma)$ we thus have

$$\theta = \langle E\rangle\theta + \langle E^{\wedge}\rangle\theta = \langle E\rangle\langle F\rangle\theta + \langle E\rangle\langle F^{\wedge}\rangle\theta + \langle E^{\wedge}\rangle\theta,$$

where $\langle E\rangle\langle F\rangle\theta \in E\cap F$ and $\langle E\rangle\langle F^{\wedge}\rangle\theta + \langle E^{\wedge}\rangle\theta \in F^{\wedge} + E^{\wedge} \subset (E\cap F)^{\wedge}$. Compare this with $\theta = \langle E\cap F\rangle\theta + \langle (E\cap F)^{\wedge}\rangle\theta$. It follows that $\langle E\cap F\rangle = \langle E\rangle\langle F\rangle$ and $(E\cap F)^{\wedge} = E^{\wedge} + F^{\wedge}$ with $\langle (E\cap F)^{\wedge}\rangle = \langle E\rangle\langle F^{\wedge}\rangle + \langle E^{\wedge}\rangle$. This proves i) and iii). 2) It follows that $(E^{\wedge} \cap F^{\wedge})^{\wedge} = E^{\wedge\wedge} + F^{\wedge\wedge} = E+F$. Therefore $E+F$ is a band, and we have $(E+F)^{\wedge} = E^{\wedge} \cap F^{\wedge}$. Also from 1) we obtain $\langle E+F\rangle = \langle (E^{\wedge} \cap F^{\wedge})^{\wedge}\rangle = \langle E^{\wedge}\rangle\langle F\rangle + \langle E\rangle = \langle E\rangle + \langle F\rangle - \langle E\rangle\langle F\rangle$. This proves ii) and iv). QED.

2.6 REMARK: For bands $E,F \in ca(X,\Sigma)$ the subsequent properties are equivalent. i) $E\cap F = \{0\}$. ii) $E\subset F^{\wedge}$. iii) $F\subset E^{\wedge}$.

Proof: It suffices to prove i)$\Rightarrow$ii). For $\theta\in E$ we obtain $\langle F\rangle\theta = \langle F\rangle\langle E\rangle\theta = \langle F\cap E\rangle\theta = 0$ and hence $\theta = \langle F^{\wedge}\rangle\theta \in F^{\wedge}$. QED.

2.7 REMARK: Let $(B_s)_{s\in I}$ be a family of bands $\subset ca(X,\Sigma)$ with $B_s\cap B_t = \{0\}$ for all $s\neq t$ in I. Put $B := (\bigcup_{s\in I} B_s)^{\wedge\wedge}$. Then $B^{\wedge} = \bigcap_{s\in I} B_s^{\wedge}$. And we have

$$\sum_{s\in I} \|\langle B_s\rangle\theta\| < \infty \quad \text{and} \quad \sum_{s\in I} \langle B_s\rangle\theta = \langle B\rangle\theta \qquad \forall\theta\in ca(X,\Sigma).$$

Symbolically this will be written $\bigoplus_{s\in I} B_s = B$.

Proof: It is trivial that

$$B^{\wedge} = (\bigcup_{s\in I} B_s)^{\wedge} = \bigcap_{s\in I} B_s^{\wedge}.$$

Now $B_s \subset B_t^\wedge$ for $s \neq t$ from 2.6. For pairwise different $s(1),\ldots,s(n) \in I$ we obtain from 2.5

$$\begin{aligned} Id &= (\langle B_{s(1)}\rangle + \langle B^\wedge_{s(1)}\rangle) \cdots (\langle B_{s(n)}\rangle + \langle B^\wedge_{s(n)}\rangle) \\ &= \langle \bigcap_{l=1}^{n} B^\wedge_{s(l)}\rangle + \sum_{l=1}^{n} \langle B_{s(l)}\rangle . \end{aligned}$$

It follows that

$$\|\theta\| = \|\langle \bigcap_{l=1}^{n} B^\wedge_{s(l)}\rangle\theta\| + \sum_{l=1}^{n} \|\langle B_{s(l)}\rangle\theta\| \qquad \forall \theta \in ca(X,\Sigma).$$

Therefore $\sum_{s\in I} \|\langle B_s\rangle\theta\| \leq \|\theta\| < \infty$, so that $\sum_{s\in I} \langle B_s\rangle\theta$ is absolutely summable and hence summable since $ca(X,\Sigma)$ is norm-complete. Of course $\alpha := \sum_{s\in I} \langle B_s\rangle\theta$ is $\in B$. And from

$$\langle B_t\rangle\alpha = \langle B_t\rangle \sum_{s\in I} \langle B_s\rangle\theta = \sum_{s\in I} \langle B_t \cap B_s\rangle\theta = \langle B_t\rangle\theta \qquad \forall t \in I ,$$

which is true since nontrivial band projections are continuous linear operators of norm = 1, we see that $\theta-\alpha \in B_t^\wedge$ $\forall t\in I$ and hence $\theta-\alpha \in B^\wedge$. It follows that $\alpha = \langle B\rangle\theta$. QED.

3. The abstract F. and M.Riesz Theorem

We return to the algebra situation to be considered in the present chapter. Let us fix $\varphi \in \Sigma(A)$. Then we have the preband $M(\varphi)$ and the band $M(\varphi)^\vee \subset ca(X,\Sigma)$. And it is obvious that for any band $B \subset ca(X,\Sigma)$ the intersection $B \cap M(\varphi)$ is a preband as well, provided of course that $B \cap M(\varphi) \neq \emptyset$.

3.1 ABSTRACT F. and M.RIESZ THEOREM: Let $B \subset ca(X,\Sigma)$ be a band with $B \cap M(\varphi) \neq \emptyset$. Then $\theta \in B \cap A^\perp$ implies that $\theta_{B\cap M(\varphi)} \in A^\perp$.

The most prominent special case is $B = ca(X,\Sigma)$.

3.2 ABSTRACT F. and M.RIESZ THEOREM: $\theta \in A^\perp$ implies that $\theta_{M(\varphi)} \in A^\perp$. Thus the band projection $\langle M(\varphi)\rangle = \langle M(\varphi)^\vee\rangle : ca(X,\Sigma) \to M(\varphi)^\vee$ maps $A^\perp \to A^\perp$.

It is natural to introduce

$$an(\varphi) = an(A,\Sigma) := \{\theta \in ca(X,\Sigma): \int f d\theta = \varphi(f)\int d\theta \ \forall f \in A\},$$

the class of analytic measures for φ. Thus $A^{\perp}$ consists of the $\theta \in an(\varphi)$ with $\int d\theta = 0$. Then let us formulate another special case of the theorem.

3.3 COROLLARY: Assume that $\theta \in an(\varphi)$ with $\int d\theta \neq 0$. Then there exist $m \in M(\varphi)$ with $m \ll \theta$ such that $\sigma \ll m$ $\forall \sigma \in M(\varphi)$ with $\sigma \ll \theta$. For such an $m \in M(\varphi)$ we have $\theta_m \in an(\varphi)$ with $\int d\theta_m = \int d\theta$ (and θ_m is in fact independent from the particular m).

Proof of 3.1 ⇒ 3.3: The band $B := \{\theta\}^{\vee} = \{\sigma \in ca(X,\Sigma): \sigma \ll \theta\}$ fulfills $B \cap M(\varphi) \neq \emptyset$ in view of 1.6. Thus $B \cap M(\varphi)$ is a preband upon which 2.2.ii) can be applied. It follows that there exist measures $m \in B \cap M(\varphi)$ as claimed, and that for these m we have $\theta_{B\cap M(\varphi)} = \theta_m$. Now $\theta - (\int d\theta) m \in B \cap A^{\perp}$. Thus from 3.1 we obtain $\theta_m - (\int d\theta) m \in A^{\perp}$ which means that $\theta_m \in an(\varphi)$ and $\int d\theta_m = \int d\theta$. QED.

The proof of 3.1 will be based on 1.7 via the subsequent technical lemma.

3.4 LEMMA: Let $m \in M(\varphi)$ and $\sigma, \tau \in Pos(X,\Sigma)$ with $\sigma \ll m$ and $D^2(m+t\sigma+\tau) = D^2(m+t\sigma)$ $\forall t>0$. Then there exists a sequence of functions $u_n \in A$ with $\varphi(u_n)=1$ such that $u_n \to 1$ in $L^2(m+\sigma)$ and $u_n \to 0$ in $L^2(\tau)$.

Proof of 3.4: We choose $u_t \in A$ with $1 = \varphi(u_t) = \int u_t dm$ and

$$\int |u_t|^2 d(m+t\sigma+\tau) < D^2(m+t\sigma) + t^2 \quad \forall t>0.$$

Then 1) $\int |u_t|^2 d(m+t\sigma) \geq D^2(m+t\sigma)$ implies that $\int |u_t|^2 d\tau < t^2$. Thus $u_t \to 0$ in $L^2(\tau)$ for $t \downarrow 0$. 2) We have

$$\int |u_t|^2 d(m+t\sigma) < D^2(m+t\sigma) + t^2 \leq 1 + t\sigma(X) + t^2,$$
$$\int |u_t - 1|^2 dm < 1 + t\sigma(X) + t^2 - 2\mathrm{Re}\int u_t dm + 1 = t\sigma(X) + t^2,$$

and hence $u_t \to 1$ in $L^2(m)$ for $t \downarrow 0$. 3) From the above and $\int |u_t|^2 dm \geq (\int |u_t| dm)^2 \geq 1$ we have $\int |u_t|^2 d\sigma \leq \sigma(X) + t$ and hence

$$\int |u_t - 1|^2 d\sigma \leq \sigma(X) + t - 2\mathrm{Re}\int u_t d\sigma + \sigma(X) = 2\Big(\sigma(X) - \mathrm{Re}\int u_t d\sigma\Big) + t.$$

Thus $u_t \to 1$ in $L^2(\sigma)$ for $t\downarrow 0$ and hence the result if we have proved that $\int u_t d\sigma \to \sigma(X)$ for $t\downarrow 0$. Assume that this is false. Then there exist $t(n)\downarrow 0$ such that $\int u_{t(n)} d\sigma \to \lambda \neq \sigma(X)$. Our estimations show that $\int |u_t|^2 d(m+\sigma)$ is bounded for $0<t\leq 1$. Therefore in view of weak sequential compactness there exists $P \in L^2(m+\sigma)$ such that after selection of a subsequence we have $\int u_{t(n)} h d(m+\sigma) \to \int Phd(m+\sigma)$ for all $h \in L^2(m+\sigma)$. In particular $\int u_{t(n)} hdm \to \int Phdm$ for all $h \in L^2(m)$. Hence P mod m = 1 after 2) and therefore P=1 in view of $\sigma << m$. It follows that $\int u_{t(n)} d\sigma \to \int d\sigma$ and hence a contradiction. QED.

Proof of 3.1: Let $\theta \in B \cap A^\perp$. Since $B \cap M(\varphi)$ is a preband we have from 2.1 the decomposition $\theta = \theta_{B\cap M(\varphi)} + \theta_{\hat{B\cap M(\varphi)}} =: \alpha+\beta$ with $\alpha <<$ some $m \in B \cap M(\varphi)$. Let now $t>0$ and $\xi \in M(\varphi)$ with $\xi << m+t|\alpha|+|\beta|$. Then $\xi \in B$ in view of $m,\theta \in B$. Thus $\xi \in B \cap M(\varphi)$ and hence is singular to $|\beta|$. Therefore from 1.7 we obtain $D^2(m+t|\alpha|+|\beta|) = D^2(m+t|\alpha|)$ for all $t>0$. Hence after 3.4 there exists a sequence of functions $u_n \in A$ with $u_n \to 1$ in $L^2(|\alpha|)$ and $u_n \to 0$ in $L^2(|\beta|)$. Now in view of $\theta \in A^\perp$ we have

$$0 = \int uu_n d\theta = \int uu_n d\alpha + \int uu_n d\beta \qquad \forall u \in A .$$

For $n\to\infty$ it follows that $0 = \int ud\alpha$ $\forall u\in A$ and hence $\theta_{B\cap M(\varphi)} = \alpha \in A^\perp$. QED.

We conclude the section with the discussion of important special situations.

<u>3.5</u> <u>COROLLARY</u>: For a fixed $m \in M(\varphi)$ the subsequent properties are equivalent. i) $\theta \in A^\perp$ implies that $\theta_m \in A^\perp$. ii) $\sigma << m$ $\forall \sigma \in M(\varphi)$ (in which case m is called dominant). In particular: if $M(\varphi)=\{m\}$ then $\theta \in A^\perp$ implies that $\theta_m \in A^\perp$.

Proof: ii) $\Rightarrow$ i) We have $M(\varphi)^\vee = \{m\}^\vee$ and hence $\theta_{M(\varphi)} = \theta_m$ $\forall \theta \in ca(X,\Sigma)$, so that the result follows from 3.2. i) $\Rightarrow$ ii) For $\sigma \in M(\varphi)$ we have $\sigma - m \in A^\perp$ and hence $\sigma_m - m \in A^\perp$. In particular $\int d\sigma_m = 1$ so that $\sigma = \sigma_m << m$. QED.

<u>3.6</u> <u>RETURN to the UNIT DISK</u>: We consider the algebra $A(D) \subset C(S) \subset B(S,\text{Baire})$ defined in Section I.3. The connection with the abstract theory is effected in the subsequent simple remarks.

i) In view of the Dirichlet property I.3.3.i) the annihilator $A(D)^\perp \subset ca(S)$ contains no nonzero real-valued measures. Therefore each $\varphi \in \Sigma(A(D))$ is represented by a unique probability measure: $M(\varphi) = \{m(\varphi)\}$.

ii) The spectrum $\Sigma(A(D))$ has some obvious members: the point evaluations in the points of $D \cup S$. For $a \in S$ the point evaluation $\varphi_a : \varphi_a(F) = F(a)$ $\forall F \in A(D)$ has the representing measure $m(\varphi_a) = \delta_a :=$ Dirac measure for a. And for $a \in D$ the point evaluation

$$\varphi_a: \varphi_a(F) = f(a) = \langle F\lambda \rangle(a) = \int_S F(s)P(a,s)d\lambda(s) \quad \forall F \in A(D)$$

has the representing measure $m(\varphi_a) = P(a,\cdot)\lambda$. In particular $m(\varphi_o) = \lambda$.

iii) The point evaluations φ_a $\forall a \in D \cup S$ exhaust the whole of $\Sigma(A(D))$. In fact, if $\varphi \in \Sigma(A(D))$ then $a := \varphi(Z)$ is a complex number of modulus $|a| \leq 1$, and we have $\varphi(F) = F(a) = \varphi_a(F)$ for all polynomials F and hence $\varphi(F) = \varphi_a(F)$ $\forall F \in A(D)$ after I.3.3.i). Thus $\varphi = \varphi_a$.

In view of remarks i) and ii) the concrete Modified F. and M.Riesz theorem I.3.5 is an immediate consequence of the above 3.5.

4. Gleason Parts

In the present section the whole of $\Sigma(A)$ is under consideration. We start with a basic remark. A band $B \subset ca(X,\Sigma)$ is called reducing iff $\theta \in A^{\perp}$ implies that $\langle B \rangle \theta = \theta_B \in A^{\perp}$. Thus 3.2 says that $M(\varphi)^{\vee}$ is a reducing band for each $\varphi \in \Sigma(A)$.

4.1 REMARK: Let $B \subset ca(X,\Sigma)$ be a reducing band. Then for each $\varphi \in \Sigma(A)$ we have either $M(\varphi) \subset B$ or $M(\varphi) \subset B^{\wedge}$, and hence either $M(\varphi)^{\vee} \subset B$ or $M(\varphi)^{\vee} \subset B^{\wedge}$.

Proof: i) We fix $m \in M(\varphi)$ and prove that either $m \in B$ or $m \in B^{\wedge}$. We have $m = m_B + m_{\hat{B}}$ with positive m_B , $m_{\hat{B}}$. Now $(u-\varphi(u))m \in A^{\perp}$ $\forall u \in A$, since this means $\int v(u-\varphi(u))dm = \int vudm - \varphi(u)\int vdm = 0$ $\forall v \in A$. It follows that $(u-\varphi(u))m_B \in A^{\perp}$ and in particular $\int udm_B - \varphi(u)\int dm_B = 0$. Thus $m_B - m_B(X)m \in A^{\perp}$, and so once more from the assumption $m_B - m_B(X)m_B = m_{\hat{B}}(X)m_B \in A^{\perp}$. In particular $m_{\hat{B}}(X)m_B(X) = 0$ so that either $m_{\hat{B}} = 0$ or $m_B = 0$. ii) Now one $m \in M(\varphi)$ in B forces all other $\sigma \in M(\varphi)$ to be $\in B$, since otherwise there were some $\sigma \in M(\varphi)$ in $B^{\wedge}$ and then $\frac{1}{2}(m+\sigma)$ neither $\in B$ nor $\in B^{\wedge}$. QED.

4.2 THEOREM: For $\varphi,\psi \in \Sigma(A)$ the bands $M(\varphi)^{\vee}$, $M(\psi)^{\vee}$ are either identical or mutually singular (which after 2.6 is equivalent to $M(\varphi)^{\vee} \cap M(\psi)^{\vee} = \{0\}$).

Proof: From 4.1 in connection with 3.2 we have either $M(\varphi)^\vee \subset M(\psi)^\vee$ or $M(\varphi)^\vee \subset M(\psi)^\wedge$, and likewise $M(\psi)^\vee \subset M(\varphi)^\vee$ or $M(\psi)^\vee \subset M(\varphi)^\wedge$. The combination $M(\varphi)^\vee \subset M(\psi)^\vee$ and $M(\psi)^\vee \subset M(\varphi)^\wedge$ cannot occur. Therefore either $M(\varphi)^\vee = M(\psi)^\vee$ or $M(\varphi)^\vee \subset M(\psi)^\wedge$. QED.

It is natural to define $\varphi,\psi \in \Sigma(A)$ to be equivalent iff $M(\varphi)^\vee = M(\psi)^\vee$. This means that each $\sigma \in M(\varphi)$ is << some $\tau \in M(\psi)$, and vice versa. In view of 4.2 the alternative is that each $\sigma \in M(\varphi)$ is singular to each $\tau \in M(\psi)$.

We have thus defined an equivalence relation on $\Sigma(A)$. Under this relation $\Sigma(A)$ decomposes into equivalence classes which will be called the Gleason parts for A. Let $Gl(\varphi)=Gl(A,\varphi)$ denote the Gleason part which contains $\varphi \in \Sigma(A)$, and let $\Gamma(A)$ denote the collection of all Gleason parts for A.

To each Gleason part $P \in \Gamma(A)$ we can associate the band $b(P) := M(\varphi)^\vee$ for any $\varphi \in P$. Then our above results can be reformulated as follows.

<u>4.3</u> <u>REFORMULATION</u>: i) For each $P \in \Gamma(A)$ the band $b(P)$ is reducing.

ii) For different $P,Q \in \Gamma(A)$ the bands $b(P), b(Q)$ are mutually singular (which after 2.6 is equivalent to $b(P) \cap b(Q) = \{0\}$).

iii) If $B \subset ca(X,\Sigma)$ is a reducing band then for each $P \in \Gamma(A)$ either $b(P) \subset B$ or $b(P) \subset B^\wedge$.

Now we apply 2.7 to the family of bands $(b(P))_{P \in \Gamma(A)}$. Then we are led to

$$B := \Big(\bigcup_{P \in \Gamma(A)} b(P)\Big)^{\wedge\wedge} = \Big(\bigcup_{\varphi \in \Sigma(A)} M(\varphi)^\vee\Big)^{\wedge\wedge} = \Big(\bigcup_{\varphi \in \Sigma(A)} M(\varphi)\Big)^{\wedge\wedge} = M(A)^{\wedge\wedge},$$

which thus is the band generated by the set $M(A)$ of all $m \in Prob(X,\Sigma)$ which are multiplicative on A. And $B^\wedge = M(A)^\wedge$ consists of the $\theta \in ca(X,\Sigma)$ which are singular to all these $m \in M(A)$. The measures $\theta \in M(A)^\wedge$ will be called completely singular.

<u>4.4</u> <u>THEOREM</u>: We have

$$M(A)^{\wedge\wedge} = \bigoplus_{P \in \Gamma(A)} b(P) \text{ and hence } ca(X,\Sigma) = \bigoplus_{P \in \Gamma(A)} b(P) \oplus M(A)^\wedge,$$

in the sense that for each $\theta \in ca(X,\Sigma)$ we have

$$\sum_{P \in \Gamma(A)} \| \langle b(P) \rangle \theta \| \leq \|\theta\| < \infty ,$$

$$\langle M(A)^{\wedge\wedge}\rangle\theta = \sum_{P\in\Gamma(A)} \langle b(P)\rangle\theta \text{ and hence } \theta = \sum_{P\in\Gamma(A)} \langle b(P)\rangle\theta + \langle M(A)^{\wedge}\rangle\theta.$$

In particular $\theta\in A^{\perp}$ iff $\langle b(P)\rangle\theta\in A^{\perp}$ $\forall P\in\Gamma(A)$ and $\langle M(A)^{\wedge}\rangle\theta\in A^{\perp}$.

Here the last sentence expresses what must be considered to be the decisive effect of the abstract F.and M.Riesz theorem: It reduces arbitrary $\theta\in A^{\perp}$ to very special ones, namely to those which are either contained in some band $b(P)=M(\varphi)^{\vee}$ and hence are $\ll$ some $m\in M(A)$, or else are completely singular $\in M(A)^{\wedge}$. The existence of such a reduction is the basic reason for the consideration of the Hardy algebra situation which starts with Chapter IV.

<u>4.5</u> <u>THEOREM</u>: A function $h\in B(X,\Sigma)$ is in $A^{\perp\perp}$ (which after the bipolar theorem is the closure of A in the topology $\omega=\sigma(B(X,\Sigma),ca(X,\Sigma))$ iff $\int h d\theta = 0$ for all $\theta\in A^{\perp}$ which are either $\ll$ some $m\in M(A)$ or completely singular $\in M(A)^{\wedge}$.

<u>4.6</u> <u>RETURN to the UNIT DISK</u>: In 3.6.iii) we have seen that $\Sigma(A(D)) = \{\varphi_a : a\in D\cup S\}$. From ii) the Gleason parts for $A(D)$ are immediate: these are $\{\varphi_z : z\in D\}$ and the $\{\varphi_a\}$ for the points $a\in S$. It is usual to identify $\varphi_a \equiv a$ for $a\in D\cup S$ (see III.1 for the connection with the Gelfand theory of Banach algebras). Then $\Sigma(A(D)) = D\cup S$. And the Gleason parts for $A(D)$ are the interior D and the one-point parts $\{a\}$ of the points $a\in S$.

5. The abstract Szegö-Kolmogorov-Krein Theorem

The present section is devoted to the other fundamental theorem of the chapter. While the abstract F. and M.Riesz theorem has the main purpose to pave the road to the Hardy algebra theory which starts with Chapter IV, the abstract Szegö-Kolmogorov-Krein theorem will be introduced into that theory and in fact is bound to become its core.

We retain the algebra situation of the chapter and fix $\varphi\in\Sigma(A)$. The abstract Szegö-Kolmogorov-Krein theorem will be deduced from the fundamental lemma 1.4. We need two simple lemmata which are unrelated to the algebra situation.

<u>5.1</u> <u>GEOMETRIC MEAN LEMMA</u>: Assume that $m\in Prob(X,\Sigma)$ and $0\leq f\in L^1(m)$ and define

$$\theta: \theta(t) = (\int f^t dm)^{\frac{1}{t}} \quad \text{for } 0<t\leq 1.$$

Then i) θ is monotone increasing and in fact strictly increasing except when f is constant, and ii) $\theta(0+) = \exp(\int \log f\, dm) =: I(f)$ (which is called the geometric mean of f).

Proof: 1) For $0<s<t\leq 1$ we obtain from Hölder that $\theta(s)\leq\theta(t)$ and deduce from 1.3 that f must be constant if $\theta(s)=\theta(t)$. 2) In order to prove $\theta(0+)\geq I(f)$ we can assume that $\int \log f\, dm > -\infty$. From $e^x \geq 1+x \quad \forall x\in\mathbb{R}$ we obtain $f^t = \exp(t \log f) \geq 1 + t \log f$ for $t>0$, hence $\int f^t dm \geq$ $\geq 1 + t\int \log f\, dm > 0$ and $\theta(t) \geq (1 + t\int \log f\, dm)^{1/t}$ for $t>0$ sufficiently small. Thus $\theta(0+)\geq I(f)$. 3) The proof that $\theta(0+)\leq I(f)$ can in view of Beppo Levi be carried out for $\max(f,\varepsilon)$ with $\varepsilon>0$ instead of f, so that we can assume that $f \geq$ some $\varepsilon>0$ and hence even $f\geq 1$. Then $\log f \geq 0$. We fix $0<\delta\leq 1$ and obtain for $0<t\leq\delta$

$$f^t = \exp(t \log f) = 1 + t \log f + \sum_{l=2}^{\infty} \frac{t^l}{l!}(\log f)^l$$

$$\leq 1 + t \log f + t\delta \sum_{l=2}^{\infty} \frac{1}{l!}(\log f)^l \leq 1 + t(\log f + \delta f),$$

$$\theta(t) \leq \left(1 + t(\int \log f\, dm + \delta\int f dm)\right)^{\frac{1}{t}},$$

and hence $\theta(0+)\leq I(f)\exp(\delta\int f dm)$. For $\delta\downarrow 0$ the results follows. QED.

5.2 EMBRYONIC MEAN VALUE THEOREM: Assume that $-\infty<a<b<\infty$ and $\theta:[a,b]\to\mathbb{R}$. Then there exists a point $\tau\in[a,b]$ such that

$$\frac{\theta(b)-\theta(a)}{b-a} \leq \text{der sup } \theta(\tau) := \limsup_{t\to\tau} \frac{\theta(t)-\theta(\tau)}{t-\tau}.$$

Proof: i) For $a\leq u<c<v\leq b$ we have

$$\frac{\theta(v)-\theta(u)}{v-u} = \frac{\theta(v)-\theta(c)}{v-c}\,\frac{v-c}{v-u} + \frac{\theta(c)-\theta(u)}{c-u}\,\frac{c-u}{v-u}$$

$$\leq \max\left(\frac{\theta(v)-\theta(c)}{v-c}, \frac{\theta(u)-\theta(c)}{u-c}\right).$$

ii) From i) we obtain a chain of intervals $[a,b]=[a_o,b_o]\supset\ldots\supset[a_n,b_n]\supset\ldots$ with $b_n-a_n = 2^{-n}(b-a)$ and

$$\frac{\theta(b)-\theta(a)}{b-a} = \frac{\theta(b_o)-\theta(a_o)}{b_o-a_o} \leq \dots \leq \frac{\theta(b_n)-\theta(a_n)}{b_n-a_n} \leq \dots$$

iii) Let $\tau \in [a_n, b_n]$ $\forall n \geq 1$. Then for each n we have, whatever $a_n < \tau < b_n$ or $\tau = a_n$ or $\tau = b_n$, from i) that

$$\frac{\theta(b_n)-\theta(a_n)}{b_n-a_n} \leq \frac{\theta(t_n)-\theta(\tau)}{t_n-\tau} \quad \text{with } t_n = a_n \text{ or } = b_n \text{ and } t_n \neq \tau.$$

For $n \to \infty$ the result follows. QED.

5.3 ABSTRACT SZEGÖ-KOLMOGOROV-KREIN THEOREM: Assume that $\sigma \in Pos(X,\Sigma)$ and $0 \leq F \in L^1(\sigma)$ and $1 \leq p < \infty$ such that $D^p(\sigma)$ and $D^p(F\sigma)$ are not both $= 0$. Then there exist measures $m \in M(\varphi)$ with $m << \sigma$ such that $\int \log F dm$ exists in the extended sense, that is such that $\int (\log F)^{\pm} dm$ are not both $= \infty$. And

$$\text{Inf}\{\exp(\int \log F dm) : \text{all these } m\} \leq \frac{D^p(F\sigma)}{D^p(\sigma)} \leq$$

$$\leq \text{Sup}\{\exp(\int \log F dm) : \text{all these } m\}.$$

The main step will be the proof of the subsequent special case.

5.4 SPECIAL CASE: Assume that $\sigma \in Pos(X,\Sigma)$ and $1 < p < \infty$ with $D^p(\sigma) > 0$, and $F \in L^1(\sigma)$ with $F \geq$ some $\varepsilon > 0$. Then there exists an $m \in M(\varphi)$ with $m << \sigma$ such that $\log F \in L^1(m)$ and

$$\exp(\int \log F dm) \leq \frac{D^p(F\sigma)}{D^p(\sigma)}.$$

Proof of 5.4: We can assume that $F \geq 1$. For $0 \leq t \leq 1$ then $1 \leq F^t \leq F$ and hence $D^p(F^t\sigma) \geq D^p(\sigma) > 0$. Therefore from 1.4 we obtain functions $P_t \in L^p(F^t\sigma)$ and $Q_t \in L^q(F^t\sigma)$ with

i) $\text{Inf}\{\int |P_t - u|^p F^t d\sigma : u \in A\} = 0$,

ii) $\int u Q_t F^t d\sigma = \varphi(u)(D^p(F^t\sigma))^{\frac{1}{p}}$ $\forall u \in A$,

iii) $|P_t|^p = P_t Q_t = |Q_t|^q =: F_t$ with $F_t F^t \sigma \in M(\varphi)$.

Let us define the function $\theta : \theta(t) = -\log D^p(F^t\sigma)$ for $0 \leq t \leq 1$. Our aim is to prove that

(*) $$\operatorname{der}\sup \theta(t) \leq -\int (\log F) F_t F^t d\sigma \qquad \forall\, 0 \leq t \leq 1.$$

Then 5.2 furnishes a point $0 \leq \tau \leq 1$ such that $m := F_\tau F^\tau \sigma \in M(\varphi)$ fulfills the assertion. The proof of (*) requires several steps.

1) Fix $0 \leq s < t \leq 1$. In view of i) there exist functions $u_n \in A$ with

$$\int |P_t - u_n|^p F^t d\sigma = \int |P_t F^{\frac{t}{p}} - u_n F^{\frac{t}{p}}|^p d\sigma \to 0.$$

From ii) we obtain

$$\int u_n Q_t F^t d\sigma = \varphi(u_n) \exp\left(-\frac{1}{p}\theta(t)\right) \text{ and hence } \varphi(u_n) \to \exp\left(\frac{1}{p}\theta(t)\right),$$

$$\varphi(u_n) \exp\left(-\frac{1}{p}\theta(s)\right) = \int u_n Q_s F^s d\sigma = \int \left(u_n F^{\frac{t}{p}}\right)\left(\frac{1}{F}\right)^{\frac{t-s}{p}}\left(Q_s F^{\frac{s}{q}}\right) d\sigma,$$

and therefore

$$\exp \frac{1}{p}(\theta(t) - \theta(s)) = \int \left(P_t F^{\frac{t}{p}}\right)\left(\frac{1}{F}\right)^{\frac{t-s}{p}}\left(Q_s F^{\frac{s}{q}}\right) d\sigma.$$

Here we apply the Hölder inequality in two different ways, in that we combine the middle factor once with the first one and once with the last one. It follows that

$$\exp \frac{\theta(t) - \theta(s)}{t-s} \leq \left(\int \left(\frac{1}{F}\right)^{t-s} F_t F^t d\sigma\right)^{\frac{1}{t-s}},$$

$$\exp \frac{\theta(t) - \theta(s)}{t-s} \leq \left(\int \left(\frac{1}{F}\right)^{\frac{q}{p}(t-s)} F_s F^s d\sigma\right)^{\frac{p}{q}\frac{1}{t-s}}.$$

2) In the first of these inequalities we fix $0 < t \leq 1$ and let $s \uparrow t$. Then from 5.1 we have

$$\limsup_{s \uparrow t} \frac{\theta(s) - \theta(t)}{s-t} \leq \int \log\left(\frac{1}{F}\right) F_t F^t d\sigma \qquad \text{for } 0 < t \leq 1.$$

And in the other one we fix $0 \leq s < 1$ and let $t \downarrow s$. Then from 5.1 we have

$$\limsup_{t \downarrow s} \frac{\theta(t) - \theta(s)}{t-s} \leq \int \log\left(\frac{1}{F}\right) F_s F^s d\sigma \qquad \text{for } 0 \leq s < 1.$$

Now combine these results to obtain (*). QED.

Proof of 5.3: In view of 1.2 we can assume that $1 < p < \infty$.
i) Assume that $D^p(\sigma) > 0$. Then we can apply 5.4 to the function $F + \varepsilon$ for

$\varepsilon>0$ and thus obtain an $m_\varepsilon \in M(\varphi)$ with $m_\varepsilon \ll \sigma$. From $\int(\log F)^+ dm_\varepsilon \leqq \int(\log(F+\varepsilon))^+ dm_\varepsilon < \infty$ we see that the assertion on the existence of appropriate $m \in M(\varphi)$ in 5.3 is fulfilled, so that the Inf in question is well-defined. And it follows that

$$\mathrm{Inf} \leqq \int \log F\, dm_\varepsilon \leqq \int \log(F+\varepsilon)\, dm_\varepsilon \leqq \frac{D^p((F+\varepsilon)\sigma)}{D^p(\sigma)} \leqq \frac{\int |u|^p (F+\varepsilon)\, d\sigma}{D^p(\sigma)},$$

$$\mathrm{Inf} \leqq \frac{\int |u|^p F d\sigma}{D^p(\sigma)} \ \forall\ u \in A \ \text{ with } \varphi(u)=1, \text{ and hence } \mathrm{Inf} \leqq \frac{D^p(F\sigma)}{D^p(\sigma)}.$$

ii) Next assume that $D^p(F\sigma)>0$. Choose a fixed function $f: x \mapsto f(x) \geqq 0$ which represents $F \in L^1(\sigma)$ and define

$$g : g(x) = \begin{cases} \dfrac{1}{f(x)} & \text{if } f(x) \neq 0 \\ 0 & \text{if } f(x)=0 \end{cases}.$$

Put $\tau := F\sigma \in \mathrm{Pos}(X,\Sigma)$. Then $0 \leqq G := g \bmod \tau \in L^1(\tau)$ and $G\tau = g\tau = gf\sigma \leqq \sigma$. We can apply i) to τ and G. Thus there exist measures $m \in M(\varphi)$ with $m \ll \tau = F\sigma$ such that $\int \log G\, dm$ exists in the extended sense, and

$$\mathrm{Inf}\left\{\exp\left(\int \log G\, dm\right) : \text{all these } m\right\} \leqq \frac{D^p(G\tau)}{D^p(\tau)} \leqq \frac{D^p(\sigma)}{D^p(F\sigma)}.$$

Now $g(x)f(x)=1$ whenever $f(x) \neq 0$. Thus $gf=1$ modulo $F\sigma$ and hence modulo $m \ll F\sigma$, so that $GF=1$ and $\log G + \log F = 0$ in $L(F\sigma)$ and hence in $L(m)$. Let us reformulate: There exist measures $m \in M(\varphi)$ with $m \ll F\sigma$ such that $\int \log F dm$ exists in the extended sense, and

$$\mathrm{Inf}\{\exp(-\int \log F\, dm) : \text{all these } m\} \leqq \frac{D^p(\sigma)}{D^p(F\sigma)},$$

$$\frac{D^p(F\sigma)}{D^p(\sigma)} \leqq \mathrm{Sup}\{\exp(\int \log F\, dm) : \text{all these } m\}.$$

Now observe that the last inequality remains true when we allow m to be taken from the larger set of the $m \in M(\varphi)$ with $m \ll \sigma$ such that $\int \log F\, dm$ exists in the extended sense. iii) From i) and ii) we obtain the entire 5.3. In fact, the assertion on the existence of appropriate $m \in M(\varphi)$ is fulfilled in any case, so that the Inf and Sup in question are well-defined. Now the Inf inequality is proved in i) for $D^p(\sigma)>0$ and is trivial in case $D^p(\sigma)=0$ since then $D^p(F\sigma)>0$. And the Sup inequality is proved in ii) for $D^p(F\sigma)>0$ and is trivial in case $D^p(F\sigma)=0$ since then $D^p(\sigma)>0$. QED.

The special case $\sigma\in M(\varphi)$ leads to an estimation of $D^p(F\sigma)$ for $0\leqq F\in L^1(\sigma)$, that is for the positive measures << some $\sigma\in M(\varphi)$. We proceed to derive an estimation of $D^p(\alpha)$ for all $\alpha\in Pos(X,\Sigma)$.

5.5 THEOREM: Let $\alpha\in Pos(X,\Sigma)$ and $1\leqq p<\infty$. Assume that $m\in M(\varphi)$ is such that $\sigma<<m$ $\forall\sigma\in M(\varphi)$ with $\sigma<<\alpha$. Then

$$\mathrm{Inf}\left\{\exp\left(\int\log\frac{d\alpha}{dm}\,d\sigma\right) : \sigma\in M(\varphi) \text{ with } \sigma<<m \text{ and } \exists\right\}\leqq D^p(\alpha) = D^p(\alpha_m)\leqq$$

$$\leqq \mathrm{Sup}\left\{\exp\left(\int\log\frac{d\alpha}{dm}\,d\sigma\right) : \sigma\in M(\varphi) \text{ with } \sigma<<m \text{ and } \exists\right\},$$

where $\exists$ of course means that $\int\log\frac{d\alpha}{dm}\,d\sigma$ is to exist in the extended sense. If $D^p(\alpha)>0$ then there exist $m\in M(\varphi)$ with $m<<\alpha$ such that $\sigma<<m$ $\forall\sigma\in M(\varphi)$ with $\sigma<<\alpha$ as required.

Proof: From 1.7 we have $D^p(\alpha)=D^p(\alpha_m)$. Thus the above inequalities follow from 5.3 in view of $D^p(m)=1$. In the case $D^p(\alpha)>0$ the band $B:=\{\alpha\}^{\vee}=\{\sigma\in ca(X,\Sigma):\sigma<<\alpha\}$ fulfills $B\cap M(\varphi)\neq\emptyset$ so that $B\cap M(\varphi)$ is a preband upon which 2.2.ii) can be applied. It follows that there exist measures $m\in B\cap M(\varphi)$ as claimed. QED.

5.6 UNIVERSAL JENSEN INEQUALITY: Let $\theta\in an(\varphi)$. Assume that $m\in M(\varphi)$ is such that $\sigma<<m$ $\forall\sigma\in M(\varphi)$ with $\sigma<<\theta$. Then

$$\log\left|\int d\theta\right| \leqq \mathrm{Sup}\left\{\int\log\left|\frac{d\theta}{dm}\right|d\sigma : \sigma\in M(\varphi) \text{ with } \sigma<<m \text{ and } \exists\right\}.$$

If $\int d\theta\neq 0$ then there exist $m\in M(\varphi)$ with $m<<\theta$ such that $\sigma<<m$ $\forall\sigma\in M(\varphi)$ with $\sigma<<\theta$ as required.

Proof: We have $\int u\,d\theta = \varphi(u)\int d\theta$ $\forall u\in A$ and hence $D^1(|\theta|)\geqq\left|\int d\theta\right|$. Now 5.5 can be applied. QED.

5.7 COROLLARY: For $u\in A$ and $m\in M(\varphi)$ we have

$$\log|\varphi(u)| \leqq \mathrm{Sup}\left\{\int\log|u|\,d\sigma : \sigma\in M(\varphi) \text{ with } \sigma<<m\right\}.$$

Proof: Take $\theta := um \in an(\varphi)$ in 5.6. QED.

The corollary 5.7 looks at least as nice as 5.6. But we shall see that it is much less powerful.

5.8 THEOREM: Let $m\in M(\varphi)$ and $1\leq p<\infty$. Then the subsequent properties are equivalent.

i) $M(\varphi)=\{m\}$.

ii) $D^p(\sigma) = \exp\left(\int \log \frac{d\sigma}{dm}\, dm\right)$ for all $\sigma\in Pos(X,\Sigma)$.

iii) $D^p(\sigma) \leq \exp\left(\int \log \frac{d\sigma}{dm}\, dm\right)$ for all $\sigma\in Pos(X,\Sigma)$.

iv) $\log|\int d\theta| \leq \int \log\left|\frac{d\theta}{dm}\right| dm$ for all $\theta\in an(\varphi)$.

In this case we see from 5.7 that $\log|\varphi(u)| \leq \int \log|u|dm$ for all $u\in A$.

Proof: i) $\Rightarrow$ ii) is immediate from 5.5 and ii) $\Rightarrow$ iii) is trivial. iii) $\Rightarrow$ i) Let $\sigma\in M(\varphi)$. Then $1 = D^p(\sigma) \leq \exp\left(\int \log \frac{d\sigma}{dm}\, dm\right) \leq \int \frac{d\sigma}{dm}\, dm = \sigma_m(X) \leq \sigma(X) = 1$ from 5.1. Thus $\sigma_m=\sigma$, and $\frac{d\sigma}{dm}$ = const once more from 5.1. Therefore $\frac{d\sigma}{dm} = 1$ and $\sigma=\sigma_m=m$. i) $\Rightarrow$ iv) is immediate from 5.6. iv) $\Rightarrow$ i) Let $\theta\in M(\varphi)\subset an(\varphi)$ and deduce $\theta=m$ as in the proof of iii) $\Rightarrow$ i). QED.

5.9 RETURN to the UNIT DISK: Consider the point evaluations $\varphi_a \in \Sigma(A(D))$ in the points $a\in D$. From 3.6.ii) we know that $M(\varphi_a)=\{P(a,.)\lambda\}$. Therefore 5.8 contains the classical Szegö-Kolmogorov-Krein theorem I.3.9 (for a=0) as well as the Jensen inequality I.3.7. Thus we have closed all gaps we left in Chapter I.

Notes

In HOFFMAN [1962a] Chapter 4 the concrete theory in the unit disk and the departure into abstraction are presented in such a way that the central character of the F. and M.Riesz and Szegö-Kolmogorov-Krein theorems becomes quite evident. The treatises on the abstract theory which appeared around 1970 all contain more or less comprehensive versions of the abstract F. and M. Riesz and Szegö-Kolmogorov-Krein theorems: BROWDER [1969], GAMELIN [1969], LEIBOWITZ [1970], and STOUT [1971]. But none of them covers the full extent of either theorem - in part due the restriction to uniform algebras, the more special compact-continuous situation of the subsequent Chapter III.

The abstract F. and M.Riesz theorem 3.2 evolved in the steps HELSON-LOWDENSLAGER [1958] and BOCHNER [1959], HOFFMAN [1962b], FORELLI [1963], AHERN [1965] where 3.5 was proved, GLICKSBERG [1967], and KÖNIG-SEEVER [1969]. All these papers except the last one are restricted to the compact-continuous situation. Let us announce at this point that Section III.2 will contain another independent proof of 3.2 in the compact-continuous situation which uses the particular methods of that situation, that is the F.Riesz representation theorem on Baire measures on compact Hausdorff spaces. Also the fundamental reduction procedure 4.4-4.5 was not implied except in special situations before KÖNIG-SEEVER [1969]. The GLICKSBERG [1967] version had to be supplemented by the RAINWATER [1969] theorem that the different band decompositions involved are in fact identical, and before this had been discovered its application required certain extra efforts such as the simultaneous measure theory for a system of measures due to LUMER [1968], or as in GARNETT-GLICKSBERG [1967]. The basic idea of the reduction prodecure is due (in the Dirichlet algebra situation) to GLICKSBERG-WERMER [1963]. The extended version 3.1 of the abstract F. and M.Riesz theorem is close to the results of Brian Cole - unpublished but summarized in GLICKSBERG [1972] Section 3 - and SEEVER [1973] in an approach in which the band decompositions are performed via idempotents in extended algebras of second dual type. But this approach requires rather complicated new machinery. Thus it is of interest to know that the full result 3.1 can be obtained by the simple method of KÖNIG-SEEVER [1969].

Gleason parts were defined in GLEASON [1957]. The equivalence of that definition with the one used in Section 4 (which is true in the compact-continuous situation) is contained in a fundamental result of BISHOP [1964]. We shall come back to the Gleason part theme in Sections III.3-4.

The abstract Szegö-Kolmogorov-Krein theorem evolved in the steps HELSON-LOWDENSLAGER [1958] and BOCHNER [1959], HOFFMAN [1962b], LUMER [1964], and HOFFMAN-ROSSI [1965] and KÖNIG [1965]- as far as the frame of uniqueness situations is concerned where the theorem retains the form of an equation. 5.8 is a purified version of the main result of the important note LUMER [1964]. Beyond this frame there are special cases in AHERN-SARASON [1967a], GAMELIN-LUMER [1968], LUMER [1968], and GAMELIN [1969]. The full theorem is due to KÖNIG [1967b][1967c].

Chapter III

Function Algebras: The Compact-Continuous Situation

In the present chapter we abandon the bounded-measurable situation of Chapter II and restrict ourselves to the more special compact-continuous situation: We fix a compact Hausdorff topological space X and a complex subalgebra $A \subset C(X)$ which contains the constants and is closed in the supremum norm (= supnorm). Then we are in the situation of Chapter II with $\Sigma :=$ the Baire sets of X. But in the present compact-continuous situation the F.Riesz representation theorem enables us to use the powerful versions of the Hahn-Banach theorem described in the Appendix to obtain independent and natural proofs for some of the fundamental facts of the theory: in particular for the existence of representative measures and of so-called Jensen measures and for the abstract F. and M.Riesz theorem.

Furthermore the chapter presents another approach to the Gleason part theme via the Gleason and Harnack metrics. It produces a decomposition of $\Sigma(A)$ which is a priori different form the one obtained in Section II.4. In the compact-continuous situation the two decompositions are seen to be identical.

1. Representative Measures and Jensen Measures

The completeness of A implies by a well-known Banach algebra theorem that each nonzero multiplicative linear functional $\varphi: A \to \mathbb{C}$ is norm continuous (with $\|\varphi\| = \varphi(1) = 1$) and hence continuous in the weak topology $\sigma(C(X),(C(X))') = \sigma(C(X), ca(X)) = \omega|C(X)$. Thus $\Sigma(A)$ as defined in the Introduction to Chapter II is the set of all nonzero multiplicative linear functionals on A, that is the spectrum of A in the Banach algebra sense.

In the present section we fix $\varphi \in \Sigma(A)$. Recall $M(\varphi) = M(A,\varphi) \subset Prob(X)$ the set of representative measures for φ. A measure $\sigma \in Pos(X)$ is defined to be a Jensen measure for φ iff $\log|\varphi(u)| \leqq \int \log|u| d\sigma$ for all $u \in A$. Let $MJ(\varphi) = MJ(A,\varphi)$ denote the set of all Jensen measures for φ. Exponentiation shows that $MJ(\varphi) \subset M(\varphi)$: For $u \in A$ we have $e^u \in A$ with

$\varphi(e^u) = e^{\varphi(u)}$ so that $\sigma \in MJ(\varphi)$ implies that $\mathrm{Re}\varphi(u) = \log|e^{\varphi(u)}| = \log|\varphi(e^u)| \leq \int \log|e^u| d\sigma = \int \mathrm{Re}\, u d\sigma$, from which it follows that $\varphi(u) = \int u d\sigma$ for all $u \in A$. There is a very short proof for the existence of Jensen measures and hence for the existence of representative measures (which is known from II.1.6).

1.1 PROPOSITION: $MJ(\varphi) \neq \emptyset$. We claim a bit more: If $K \subset X$ is compact $\neq \emptyset$ such that $\|u\| = \|u|K\|$ $\forall u \in A$ then there exists $\sigma \in MJ(\varphi)$ with $\sigma(K)=1$.

Proof: Follows from A.2.4 applied to $S := \{\log|u| - \log|\varphi(u)| : u \in A$ with $\varphi(u) \neq 0\} \subset USC(X)$. QED.

The sets $MJ(\varphi) \subset M(\varphi)$ are convex and weak*closed and hence weak*compact subsets of $Prob(X) \subset ca(X) = (C(X))'$. The latter fact is one of the prominent peculiarities of the compact-continuous situation.

1.2 REMARK: In the case $M(\varphi) = \{m\}$ of a unique representative measure m it follows from 1.1 that m must be a Jensen measure (this is also known from II.5.8). If A has the Dirichlet property, that is if $\mathrm{Re}\, A \subset \mathrm{Re}C(X)$ is supnorm dense in $\mathrm{Re}C(X)$, then it is obvious that each $\varphi \in \Sigma(A)$ has a unique representative measure. Applied to the unit disk situation we thus obtain a fast and unconventional proof of the classical Jensen inequality I.3.7.

We introduce the important functional $\phi : USC(X) \to [-\infty, \infty[$. It is defined to be

$$\phi(f) = \mathrm{Inf}\{\mathrm{Re}\varphi(u) : u \in A \text{ with } \mathrm{Re}\, u \geq f\} \qquad \forall f \in USC(X).$$

We list some immediate properties. i) $-\infty \leq \mathrm{Inf}\, f \leq \phi(f) \leq \mathrm{Max}\, f$ $\forall f \in USC(X)$. The upper estimation is obvious. The lower estimation follows from the lower estimation in $\mathrm{Min}\,\mathrm{Re}\, u \leq \mathrm{Re}\varphi(u) \leq \mathrm{Max}\,\mathrm{Re}\, u$ $\forall u \in A$ which in turn is immediate via exponentiation. In particular ϕ is finite-valued on $\mathrm{Re}C(X)$. ii) ϕ is sublinear. iii) ϕ is isotone, that is $f \leq g$ implies that $\phi(f) \leq \phi(g)$. iv) $\phi(\mathrm{Re}\, u) = \mathrm{Re}\varphi(u)$ for all $u \in A$.

1.3 REMARK: Let $\sigma \in (\mathrm{Re}C(X))^*$. Then $\sigma(f) \leq \phi(f)$ $\forall f \in \mathrm{Re}C(X) \Leftrightarrow \sigma \in M(\varphi) \subset Prob(X) \subset (\mathrm{Re}C(X))'$ (via the F.Riesz representation theorem).

Proof: In order to prove $\Rightarrow$ note that $\sigma(f) \leq \phi(f) \leq \mathrm{Max}\, f$ $\forall f \in \mathrm{Re}C(X)$ implies that $\sigma \in Prob(X) \subset (\mathrm{Re}C(X))'$ after A.2.2. The other details are obvious. QED.

We can combine 1.3 with the primitive Hahn-Banach version that each sublinear functional on a real vector space dominates some linear functional, to obtain another proof for the existence of representative measures. But combination with the Hahn-Banach version A.1.1 yields a much more informative result.

1.4 THEOREM: $\phi(f) = \mathrm{Max}\{\int f d\sigma : \sigma \in M(\varphi)\}$ for all $f \in USC(X)$.

Proof: In view of 1.3 we have to prove that for each $f \in USC(X)$ there exists $\sigma \in M(\varphi)$ such that $\phi(f) = \int f d\sigma$. We apply A.1.1 to the convex set $\{g \in ReC(X) : g \geqq f\} \subset ReC(X)$ to obtain a $\sigma \in M(\varphi)$ such that $\mathrm{Inf}\{\phi(g) : g \in ReC(X)$ with $g \geqq f\} = \mathrm{Inf}\{\sigma(g) : g \in ReC(X)$ with $g \geqq f\}$. But here the first member is $= \phi(f)$ as is immediate from the definition while the second member is $= \int f d\sigma$ per definitionem. QED.

We conclude with a useful modification lemma which will be applied in the alternative proof of the abstract F.and M.Riesz theorem in the next section.

1.5 LEMMA: Assume that $0 \leqq f_n \in USC(X)$ with $\phi(f_n) \to 0$. Then there exists a sequence of functions $v_n \in A$ with $|v_n| \leqq 1$ and $\phi(|1-v_n|) \to 0$ such that $\| v_n f_n \| \to 0$.

Proof: Choose numbers $t_n > 0$ with $\phi(f_n) < t_n^2$ and $t_n \to 0$ and then functions $u_n \in A$ with $\mathrm{Re}\, u_n \geqq f$ and $\varphi(u_n) = \mathrm{Re}\varphi(u_n) < t_n^2$ after the definition of ϕ. Then $v_n := \exp(-u_n/t_n) \in A$ satisfies

$$|v_n| = \exp(-\mathrm{Re}\, u_n/t_n) \leqq \exp(-f_n/t_n) \leqq 1 ,$$

$$|v_n f_n| \leqq t_n (f_n/t_n) \exp(-f_n/t_n) \leqq t_n \frac{1}{e} .$$

It remains to show that $\phi(|1-v_n|) \to 0$. But

$$\varphi(v_n) = \exp(-\varphi(u_n)/t_n) > \exp(-t_n) \geqq 1-t_n ,$$

$$\left(\int |1-v_n| d\sigma\right)^2 \leqq \int |1-v_n|^2 d\sigma \leqq 2 - 2\mathrm{Re}\varphi(v_n) < 2t_n \qquad \forall \sigma \in M(\varphi),$$

and hence $\phi(|1-v_n|) \leqq (2t_n)^{1/2}$ after 1.4. QED.

2. Return to the abstract F.and M.Riesz Theorem

The crucial point in the alternative proof of the abstract F.and M.Riesz theorem is the subsequent lemma. It will be useful for other purposes as well.

2.1 LEMMA: Let $S,T\subset Pos(X)$ be convex weak*compact $\neq \emptyset$ such that each $\sigma\in S$ is singular to each $\tau\in T$. Then there exists a sequence of functions $u_n\in C(X)$ with $|u_n|\leq 1$ such that

$$\sigma([u_n\to 1])=\sigma(X) \quad \forall\sigma\in S \quad \text{and} \quad \tau([u_n\to 0])=\tau(X) \quad \forall\tau\in T.$$

If in particular $S=M(A,\varphi)$ for some $\varphi\in\Sigma(A)$ then we can achieve that $u_n\in A$.

Proof: i) Define on $ReC(X)$ the sublinear functional

$$\Theta_S: \Theta_S(f) = \underset{\sigma\in S}{Max} \int f d\sigma \qquad \forall f\in ReC(X).$$

The bipolar theorem A.1.7 shows that the linear functionals $\in(ReC(X))^*$ which are $\leq\Theta_S$ are precisely the members of S. ii) Consider on the product vector space $V:=ReC(X)\times ReC(X)$ the sublinear functional

$$\Theta: \Theta(f,g) = \Theta_S(f) + \Theta_T(g) \qquad \forall(f,g)\in V.$$

The linear functionals $\psi\in V^*$ have the form $\psi(f,g)=\sigma(f)+\tau(g)$ $\forall(f,g)\in V$ with $\sigma,\tau\in(ReC(X))^*$. And i) shows that $\psi\leq\Theta$ is equivalent to $\sigma\in S$ and $\tau\in T$. iii) We apply the Hahn-Banach version A.1.1 to Θ and to the convex set $\{(f,g)\in V: f,g\geq 0 \text{ and } f+g=1\}\subset V$. It follows that there exist $\sigma\in S$ and $\tau\in T$ such that

$$I := Inf\{\Theta_S(f)+\Theta_T(g): 0\leq f,g\in ReC(X) \text{ with } f+g=1\}$$
$$= Inf\{\sigma(f)+\tau(g): 0\leq f,g\in ReC(X) \text{ with } f+g=1\}.$$

We assert that $I=0$. In fact, there exists a Baire set $B\subset X$ such that $\sigma(B)=0$ and $\tau(B)=\tau(X)$, and hence to each $\varepsilon>0$ a compact G_δ-set $K\subset B$ such that $\tau(K)>\tau(X)-\varepsilon$. We can find functions $f_n\in ReC(X)$ with $0\leq f_n\leq 1$ and $f_n\downarrow\chi_K$. Then $\sigma(f_n)\to 0$ and $\tau(1-f_n)\to\tau(X)-\tau(K)<\varepsilon$. It follows that $I\leq\varepsilon$ for all $\varepsilon>0$ and hence $I=0$. iv) We complete the proof of the first assertion. From iii) we obtain functions $u_n \in ReC(X)$ with $0 \leq u_n \leq 1$ such that $\theta_S(1-u_n) < n^{-2}$ and $\theta_T(u_n) < n^{-2}$. Then the Beppo Levi theorem tells us that $[u_n \to 1]$ has full measure for all $\sigma \in S$ and that $[u_n \to 0]$ has full measure for all $\tau \in T$. v) It remains to prove the second assertion. In the case $S = M(A,\varphi)$ for some $\varphi \in \Sigma(A)$ we have $\Theta_S = \phi$ after 1.4. We apply the modification lemma 1.5 to the functions $1-u_n$ constructed in iv) to obtain functions $v_n \in A$ with $|v_n| \leq 1$ and $\phi(|1-v_n|) \to 0$ such that $\|v_n(1-u_n)\| \to 0$. We can assume that $\phi(|1-v_n|) < n^{-2}$. Then $[v_n \to 1]$ has full measure for all $\sigma \in S = M(A,\varphi)$ while $[u_n \to 0] \subset [v_n \to 0]$ shows that $[v_n \to 0]$ has full measure for all $\tau \in T$. QED.

Before we come to the abstract F.and M.Riesz theorem we mention another related consequence of 2.1: an improvement of the preband decomposition after II.2.1 in an important particular case.

2.2 REMARK: Let $M \subset Pos(X)$ be convex weak*compact $\neq \emptyset$. Then M is a preband in the sense of II.2.1. Thus each $\theta \in ca(X)$ admits a unique decomposition $\theta = \theta_M + \theta_{\hat{M}}$ with $\theta_M \ll$ some $m \in M$ and $\theta_{\hat{M}}$ singular to all $m \in M$. Now 2.1 applied to M and $|\theta_{\hat{M}}|$ yields a much sharper statement on $\theta_{\hat{M}}$: it follows that $\theta_{\hat{M}}$ lives on some Baire set $B \subset X$ which is a common null set for all $m \in M$.

We conclude with the alternative proof of the abstract F.and M.Riesz theorem.

2.3 ABSTRACT F.and M.RIESZ THEOREM: Let $\varphi \in \Sigma(A)$. Then $\theta \in A^{\perp}$ implies that $\theta_{M(\varphi)} \in A^{\perp}$.

Proof: We apply 2.1 to $M(\varphi)$ and $|\theta_{\hat{M}(\varphi)}|$ to obtain functions $u_n \in A$ with $|u_n| \leq 1$ such that $[u_n \to 1]$ has full measure for all $m \in M(\varphi)$ and $[u_n \to 0]$ has full measure for $|\theta_{\hat{M}(\varphi)}|$. For $u \in A$ then

$$0 = \int u u_n d\theta = \int u u_n d\theta_{M(\varphi)} + \int u u_n d\theta_{\hat{M}(\varphi)} \to \int u d\theta_{M(\varphi)} ,$$

so that $\int u d\theta_{M(\varphi)} = 0$. QED.

3. The Gleason and Harnack Metrics

We present another approach to the Gleason part theme via the Gleason and Harnack metrics. The present section contains the essentials modulo the F.Riesz representation theorem. It is a surprise that they can be established under very little structure: We fix a nonvoid set X and a complex subalgebra $A \subset B(X)$ which contains the constants. The Gleason and Harnack functions are defined to be

$$G: G(u,v) = \operatorname{Sup}\{|f(u)-f(v)| : f \in A \text{ with } \|f\| \leq 1\},$$

$$H: H(u,v) = \operatorname{Sup}\left\{\frac{F(v)}{F(u)} : F \in \operatorname{Re}A \text{ with } F > 0\right\} \quad \forall u,v \in X,$$

where of course $\|.\|$ denotes the supnorm on $B(X)$.

3.1 REMARK: i) G is a semi-metric on X and a metric iff A separates the points of X. We have $0 \leq G(u,v) \leq 2 \quad \forall u,v \in X$. And the relation $G(u,v) < 2$ is symmetric. ii) We have $1 \leq H(u,v) \leq \infty \quad \forall u,v \in X$. And in view of $H(u,w) \leq H(u,v)H(v,w) \quad \forall u,v,w \in X$ the relation $H(u,v) < \infty$ is transitive.

At this point it is far from evident that the relation $G(u,v) < 2$ is transitive and that the relation $H(u,v) < \infty$ is symmetric. The next theorem shows in particular that the two relations are in fact identical. Thus $G(u,v) < 2$ or $H(u,v) < \infty$ is an equivalence relation on X.

3.2 THEOREM: We have $H = \left(\frac{2+G}{2-G}\right)^2$.

It will be convenient to consider besides G and H the family of the functions S_t for $0 \leq t < 1$ defined to be

$$S_t(u,v) = \operatorname{Sup}\{|f(v)| : f \in A \text{ with } \|f\| \leq 1 \text{ and } |f(u)| \leq t\} \quad \forall u,v \in X.$$

Note that $t \leq S_t(u,v) \leq 1$.

3.3 THEOREM: We have $\dfrac{1+S_t}{1-S_t} = \dfrac{1+t}{1-t} H$ for each $0 \leq t < 1$.

3.4 COROLLARY: For $u,v \in X$ and $0 \leq t < 1$ we have the equivalence $G(u,v) < 2 \Leftrightarrow H(u,v) < \infty \Leftrightarrow S_t(u,v) < 1$. In particular the relation $G(u,v) < 2$ is an equivalence relation on X. Under this relation X decomposes into equivalence classes which will be called the Gleason-Harnack parts for A.

3.5 COROLLARY: The functions G and log H as well as S_o are semi-metrics on X (log H in the extended sense $\leq \infty$) which are equivalent, that is produce the same uniform structure on X.

In the proofs of the above results we can assume that A is supnorm closed. For we have the subsequent remark which admits an obvious routine proof.

3.6 REMARK: The functions $\bar{G}$, $\bar{H}$, $\bar{S}_t$ computed for the supnorm closure $\bar{A}$ of A coincide with the respective functions G, H, S_t for A.

Furthermore we make essential use of the fractional linear transformations which map the unit disk D onto itself. These are the functions

$$z \mapsto \langle a,z\rangle := \frac{a-z}{1-\bar{a}z} \qquad \text{for fixed } a\in D,$$

multiplied with constant factors of modulus one. We list some simple facts which will be needed in the sequel.

3.7 REMARK: Assume that $a,b\in D$ and $z\in D\cup S$. Then

i) $\langle a,\langle a,z\rangle\rangle = z$.

ii) $|\langle a,\langle b,z\rangle\rangle| = |\langle\langle b,a\rangle,z\rangle|$. For z=0 in particular $|\langle a,b\rangle| = |\langle b,a\rangle|$.

iii) $\lambda\langle a,b\rangle = \langle\lambda a,\lambda b\rangle$ for all $\lambda\in S$.

iv) $\dfrac{|a|-|b|}{1-|a||b|} \leq |\langle a,b\rangle| \leq \dfrac{|a|+|b|}{1+|a||b|}$.

Proofs of 3.2-3.5: 1) Define

$$c_t(u,v) = \frac{S_t(u,v)-t}{1-tS_t(u,v)} \qquad \text{for } u,v\in X \text{ and } 0\leq t<1.$$

We claim that

$$|\langle f(u),f(v)\rangle| \leq c_t(u,v) \qquad \forall\ f\in A \text{ with } \|f\|<1,$$

$$|f(u)| \leq \frac{c_t(u,v)+|f(v)|}{1+c_t(u,v)|f(v)|} \qquad \forall\ f\in A \text{ with } \|f\|\leq 1.$$

To prove the first inequality write $\langle f(u),f(v)\rangle =: z = |z|\varepsilon$ with $|\varepsilon|=1$. Geometric series expansion shows that $g := \langle\langle f(u),-t\varepsilon\rangle,f\rangle \in A$. And $|g| = |\langle -t\varepsilon,\langle f(u),f\rangle\rangle|$ from 3.7.ii). Thus from $|g(u)| = |\langle -t\varepsilon,0\rangle| = |-t\varepsilon| = t$

we obtain

$$|g(v)| = |\langle -t\varepsilon, |z|\varepsilon\rangle| = |\langle -t, |z|\rangle| = \frac{t+|z|}{1+t|z|} \leqq S_t(u,v),$$

which is at once transformed into the assertion $|z| \leqq c_t(u,v)$. To prove the second inequality it suffices to assume that $\|f\|<1$. Then combine 3.7.iv) with the first inequality to obtain

$$\frac{|f(u)|-|f(v)|}{1-|f(u)||f(v)|} \leqq |\langle f(u),f(v)\rangle| \leqq c_t(u,v),$$

which is at once transformed into the assertion.

2) We claim that $S_t(v,u) = S_t(u,v)$ for $u,v \in X$ and $0 \leqq t < 1$. It suffices to prove that $S_t(v,u) \leqq S_t(u,v)$. For $f \in A$ with $\|f\| \leqq 1$ and $|f(v)| \leqq t$ we obtain from 1)

$$|f(u)| \leqq \frac{c_t(u,v)+|f(v)|}{1+c_t(u,v)|f(v)|} \leqq \frac{c_t(u,v)+t}{1+tc_t(u,v)} = S_t(u,v).$$

It follows that $S_t(v,u) \leqq S_t(u,v)$. 3) Likewise we obtain from 1)

$$|f(u)| \leqq \frac{c_t(u,v)+S_t(w,v)}{1+c_t(u,v)S_t(w,v)} \quad \text{for } f \in A \text{ with } \|f\| \leqq 1 \text{ and } |f(w)| \leqq t,$$

$$S_t(w,u) \leqq \frac{c_t(u,v)+S_t(w,v)}{1+c_t(u,v)S_t(w,v)} \quad \text{or} \quad c_t(w,u) \leqq \frac{c_t(u,v)+c_t(w,v)}{1+c_t(u,v)c_t(w,v)},$$

and hence $c_t(w,u) \leqq c_t(u,v)+c_t(w,v)$. In particular since $c_o = S_o$ it follows that S_o is a semi-metric on X as claimed in 3.5.

4) We prove $\geqq$ in 3.3. First deduce from 1) that

$$1-|f(u)| \geqq \frac{1-c_t(u,v)}{1+c_t(u,v)|f(v)|}(1-|f(v)|) \quad \forall\ f \in A \text{ with } \|f\| \leqq 1.$$

Fix now $k \in A$ with $K:= \operatorname{Re} k > 0$. For each $s>0$ then $e^{-sK} \in A$ and hence

$$\frac{1}{s}(1-e^{-sK(u)}) \geqq \frac{1-c_t(u,v)}{1+c_t(u,v)e^{-sK(v)}}\ \frac{1}{s}(1-e^{-sK(v)}).$$

For $s \downarrow 0$ it follows that

$$K(u) \geq \frac{1-c_t(u,v)}{1+c_t(u,v)} K(v) \quad \text{or} \quad \frac{K(u)}{K(v)} \geq \frac{1-c_t(u,v)}{1+c_t(u,v)} = \frac{1-t}{1+t} \frac{1+S_t(u,v)}{1-S_t(u,v)} .$$

5) Next we prove $\leq$ in 3.3. Let $f \in A$ with $\|f\|<1$ and $|f(u)| \leq t$, and assume that $f(v) = |f(v)|$. Then

$$k := \frac{1+f}{1-f} \in A \quad \text{with} \quad K := \operatorname{Re} k = \frac{1-|f|^2}{|1-f|^2} > 0 .$$

It follows that

$$\frac{1+|f(v)|}{1-|f(v)|} = \frac{1+f(v)}{1-f(v)} = k(v) = K(v) \leq H(u,v)K(u) =$$

$$= H(u,v)\frac{1-|f(u)|^2}{|1-f(u)|^2} \leq H(u,v)\frac{1-|f(u)|^2}{(1-|f(u)|)^2} = H(u,v)\frac{1+|f(u)|}{1-|f(u)|} \leq H(u,v)\frac{1+t}{1-t} .$$

The inequality which results remains true for all $f \in A$ with $|f| \leq 1$ and $|f(u)| \leq t$. It follows that

$$\frac{1+S_t(u,v)}{1-S_t(u,v)} \leq H(u,v)\frac{1+t}{1-t} .$$

6) It remains to prove 3.2. We claim that

$$S_o(u,v) = \frac{4G(u,v)}{4+G^2(u,v)} \qquad \forall u,v \in X,$$

which combines at once with 3.3 for $t=0$ to yield the assertion. To prove $\geq$ let $f \in A$ with $\|f\|<1$ and combine $|<f(u),f(v)>| \leq c_o(u,v) = S_o(u,v)$ from 1) with the elementary inequality

$$4|1-\bar{a}b| = 4|1-|a|^2+\bar{a}(a-b)| \leq 4(1-|a|^2) + 2(2|a|)|a-b| \leq$$
$$\leq 4(1-|a|^2) + 4|a|^2 + |a-b|^2 = 4 + |a-b|^2 \quad \forall\ a,b \in \mathbb{C} \text{ with } |a| \leq 1 .$$

It follows that

$$\frac{4|f(u)-f(v)|}{4+|f(u)-f(v)|^2} \leq S_o(u,v) \text{ even for all } f \in A \text{ with } \|f\| \leq 1 ,$$

and hence the assertion. To prove $\leq$ let $f \in A$ with $\|f\|<1$ and $f(u)=0$. Put $a := f(v)$ and define

$$b := \frac{a}{1+\sqrt{1-|a|^2}}, \quad \text{so that} \quad a = \frac{2b}{1+|b|^2} \quad \text{or } \langle b,a\rangle = -b .$$

Then $h := \langle b,f\rangle \in A$ satisfies $h(u) = b$ and $h(v) = -b$. Thus we obtain $2|b| = |h(u)-h(v)| \leq G(u,v)$. It follows that

$$|f(v)|=|a| = \frac{2|b|}{1+|b|^2} \leq \frac{4G(u,v)}{4+G^2(u,v)} \quad \text{even for all } f\in A \text{ with } |f|\leq 1,$$

and hence the assertion. QED.

3.8 RETURN to the UNIT DISK: We consider the algebra $A = \mathrm{CHol}(D) \subset \subset C(DUS) \subset B(DUS)$ which is supnorm isometric to $A(D) \subset C(S)$ via restriction $f\mapsto f|S$. 1) The easiest to compute of the functions G,H,S_t considered above is S_o. From the H.A.Schwarz lemma (see CARATHÉODORY [1950] Vol. I p.137) we obtain at once

$$S_o(0,z) = \mathrm{Sup}\{|f(z)|:f\in A \text{ with } \|f\|\leq 1 \text{ and } f(0)=0\} = |z| \quad \forall z\in DUS.$$

2) For fixed $u\in D$ the functions $h\in A$ with $h(u)=0$ are in one-to-one correspondence to the functions $f\in A$ with $f(0)=0$ via $h(z)=f(\langle u,z\rangle)$ $\forall z\in DUS$. From 3.7.i) then $f(z)=h(\langle u,z\rangle)$ $\forall z\in DUS$ as well. It follows that

$$\begin{aligned} S_o(u,z) &= \mathrm{Sup}\{|h(z)|:h\in A \text{ with } \|h\|\leq 1 \text{ and } h(u)=0\} \\ &= \mathrm{Sup}\{|f(\langle u,z\rangle)|:f\in A \text{ with } \|f\|\leq 1 \text{ and } f(0)=0\} \\ &= S_o(0,\langle u,z\rangle) = |\langle u,z\rangle| = \left|\frac{u-z}{1-\bar{u}z}\right| \quad \forall z\in DUS. \end{aligned}$$

3) For different points $u,v\in S$ the easiest to prove is that $H(u,v) = \infty$. In fact, for $\varepsilon>0$ the function f: $f(z) = 1 - \frac{z}{u} + \varepsilon$ is $\in A$ with $F := \mathrm{Re}\, f > 0$. Thus

$$\frac{F(v)}{F(u)} = \frac{1}{\varepsilon}(1 - \mathrm{Re}\,\frac{v}{u} + \varepsilon) \leq H(u,v) ,$$

and for $\varepsilon\downarrow 0$ it follows that $H(u,v)=\infty$. 4) Thus the Gleason-Harnack parts of DUS for $A=\mathrm{CHol}(D)$ are seen to be D and the one-point parts $\{z\}$ with $z\in S$.

4. Comparison of the two Gleason Part Decompositions

The Gleason part decomposition of Section II.4 was for the bounded-measurable situation: On a measurable space (X,Σ) we assumed a complex subalgebra $A\subset B(X,\Sigma)$ which contained the constants, and the Gleason part decomposition took place on the spectrum $\Sigma(A)$. In order to compare it with the Gleason-Harnack part decomposition of Section 3 we have to put the latter theory into a comparable position. This is done by the Gelfand transformation: The function $f\in A$ is sent into the function $\hat{f}\in B(\Sigma(A))$ defined via evaluation $\hat{f}(\varphi)=\varphi(f)$ $\forall\varphi\in\Sigma(A)$. Then

$$\|\hat{f}\| := \sup_{\varphi\in\Sigma(A)} |\hat{f}(\varphi)| = \sup_{\varphi\in\Sigma(A)} |\varphi(f)| = \|f\|,$$

since $\Sigma(A)$ contains in particular the point evaluations $\varphi_x: f\mapsto f(x) = \int f d\delta_x$ in the points $x\in X$. Thus the Gelfand transformation $f\mapsto\hat{f}$ is an algebra isomorphism $A\to\hat{A} := \{\hat{f}: f\in A\}\subset B(\Sigma(A))$ which is supnorm isometric. The theory of Section 3 then has to be applied to $\hat{A}\subset B(\Sigma(A))$. That means to consider on $\Sigma(A)$ the functions

$$G : G(\varphi,\psi) = \sup\{|\varphi(f)-\psi(f)| : f\in A \text{ with } \|f\|\leq 1\} = \|\varphi-\psi\|,$$

$$H : H(\varphi,\psi) = \sup\left\{\frac{\mathrm{Re}\psi(f)}{\mathrm{Re}\varphi(f)} : f\in A \text{ with } \mathrm{Re}\, f > 0\right\},$$

where for $f\in A$ we note that $\mathrm{Re}\, f > 0$ on X is equivalent to $\mathrm{Re}\,\hat{f} > 0$ on $\Sigma(A)$, and for $0\leq t<1$ the functions

$$S_t : S_t(\varphi,\psi) = \sup\{|\psi(f)| : f\in A \text{ with } \|f\|\leq 1 \text{ and } |\varphi(f)|\leq t\} \quad \forall\varphi,\psi\in\Sigma(A).$$

Now the comparison between the two equivalence relations on $\Sigma(A)$ in question is simple in one direction.

4.1 <u>THEOREM</u>: Assume the bounded-measurable situation. For $\varphi,\psi\in\Sigma(A)$ then $(M(\varphi))^\vee = (M(\psi))^\vee$ implies that $G(\varphi,\psi) = \|\varphi-\psi\| < 2$.

Proof: Fix $\tau\in M(\psi)$. Then there exists $\sigma\in M(\varphi)$ such that $\tau\ll\sigma$ or $\tau = h\sigma$ with $0\leq h\in L^1(\sigma)$. Thus $\varphi(f)-\psi(f) = \int f(1-h)d\sigma$ $\forall f\in A$. It follows that

$$\|\varphi-\psi\| - 2 \leq \int|1-h|d\sigma - 2 = \int(|1-h|-(1+h))d\sigma = -\int\frac{4h}{|1-h|+(1+h)}d\sigma ,$$

which is < 0. QED.

In the bounded-measurable situation the converse statement need not be true. We shall present a counter-example. It requires the subsequent simple lemma.

4.2 LEMMA: Let $\varphi,\psi:A\to\mathbb{C}$ be multiplicative linear functionals on a complex algebra A. If $(1-c)\varphi + c\psi$ is multiplicative on A for some $0<c<1$ then $\varphi=\psi$.

Proof: For $u\in A$ we have

$$(1-c)(\varphi(u))^2 + c(\psi(u))^2 = (1-c)\varphi(u^2) + c\psi(u^2) = ((1-c)\varphi+c\psi)(u^2) =$$
$$= (((1-c)\varphi+c\psi)(u))^2 = ((1-c)\varphi(u) + c\psi(u))^2 =$$
$$= (1-c)^2(\varphi(u))^2 + c^2(\psi(u))^2 + 2(1-c)c\varphi(u)\psi(u),$$

and hence $(\varphi(u)-\psi(u))^2(1-c)c = 0$. QED.

4.3 EXAMPLE: Let $X=\{a,b\}\cup T$ with different points $a,b\in D$ and $T\subset S$ a countable dense subset. And let Σ consist of all subsets of X. Define $A:=CHol(D)|X \subset B(X,\Sigma)=B(X)$. By the maximum modulus principle the restriction $f\mapsto f|X$ is an algebra isomorphism $CHol(D)\to A$ which is supnorm isometric. Hence A has the same nonzero multiplicative linear functionals as CHol(D), which are the point evaluations φ_z at the points $z\in D\cup S$.
1) We assert that

$$M(A,\varphi_z) = \left\{ \begin{matrix} \emptyset & \text{if } z\notin X \\ \{\delta_z\} & \text{if } z\in X \end{matrix} \right\} \quad \forall z\in D\cup S.$$

In particular $\Sigma(A)=\{\varphi_z:z\in X\}$. The inclusion $\supset$ is trivial. To see $\subset$ let $\sigma\in Prob(X,\Sigma)$ and write $\sigma(\{x\})=:c_x\geqq 0$ $\forall x\in X$. For $z\in D\cup S$ then $\sigma\in M(A,\varphi_z)$ means that

$$f(z) = c_a f(a) + c_b f(b) + \sum_{t\in T} c_t f(t) \quad \forall f\in CHol(D).$$

In case $z\in D$ we see from II.3.6 that

$$P(z,.)\lambda = c_a P(a,.)\lambda + c_b P(b,.)\lambda + \sum_{t\in T} c_t\delta_t \text{ , so that } c_t=0 \quad \forall t\in T,$$

and then from 4.2 that either $z=a$ and $\sigma=\delta_a$ or $z=b$ and $\sigma=\delta_b$.

And in case $z\in S$ we see that

$$\delta_z = c_a P(a,.)\lambda + c_b P(b,.)\lambda + \sum_{t\in T} c_t\delta_t ,$$

so that $z\in T$ and $\sigma = \delta_z$ as well. 2) Since the restriction $CHol(D)\to A$ is supnorm isometric it follows from 3.8 that

$$G(\varphi_a,\varphi_b) = \|\varphi_a-\varphi_b\|_{A'} = \|\varphi_a-\varphi_b\|_{CHol(D)'} < 2 .$$

But 1) tells us that $(M(A,\varphi_a))^{\vee} = \mathbb{C}\delta_a$ and $(M(A,\varphi_b))^{\vee} = \mathbb{C}\delta_b$ are singular to each other. Thus in the present example the converse to 4.1 is false indeed. QED.

In contrast to the above, in the compact-continuous situation the two equivalence relations on $\Sigma(A)$ in question can be proved to be identical. The results of Section 3 lead to a pleasant equivalence theorem.

4.4 THEOREM: Assume the compact-continuous situation. For $\varphi,\psi\in\Sigma(A)$ the subsequent assertions are equivalent.

i) $G(\varphi,\psi) = \|\varphi-\psi\|<2$.

ii) $H(\varphi,\psi)<\infty$. And there exist $\sigma\in M(\varphi)$ and $\tau\in M(\varphi)$ such that $\sigma\leqq H(\varphi,\psi)\tau$ and $\tau\leqq H(\varphi,\psi)\sigma$.

iii) $H(\varphi,\psi)<\infty$. And to each $\tau\in M(\psi)$ there exists $\sigma\in M(\varphi)$ such that $\tau\leqq H(\varphi,\psi)\sigma$, and likewise to each $\sigma\in M(\varphi)$ there exists $\tau\in M(\psi)$ such that $\sigma\leqq H(\varphi,\psi)\tau$.

iv) $(M(\varphi))^{\vee} = (M(\psi))^{\vee}$.

Proof: i) $\Rightarrow$ ii) We can assume that $\varphi\neq\psi$. Then $1<H(\varphi,\psi) = H(\psi,\varphi)<\infty$ from 3.2. The definition of the function H shows that

$$\mathrm{Re}\Big(H(\varphi,\psi)\varphi(f)-\psi(f)\Big),\ \mathrm{Re}\Big(H(\varphi,\psi)\psi(f)-\varphi(f)\Big) \geqq 0 \quad \forall\ f\in A \text{ with } \mathrm{Re}\, f \geqq 0 .$$

Thus from A.2.6 we obtain measures $\alpha,\beta \in Pos(X)$ such that

$$H(\varphi,\psi)\varphi(f)-\psi(f) = \int f d\alpha ,$$
$$H(\varphi,\psi)\psi(f)-\varphi(f) = \int f d\beta \qquad \forall f\in A,$$

$$\varphi(f) = \frac{1}{H^2(\varphi,\psi)-1} \int f d(H(\varphi,\psi)\alpha+\beta) ,$$

$$\psi(f) = \frac{1}{H^2(\varphi,\psi)-1} \int f d(\alpha+H(\varphi,\psi)\beta) \qquad \forall f\in A.$$

It follows that

$$\sigma := \frac{H(\varphi,\psi)\alpha+\beta}{H^2(\varphi,\psi)-1} \in M(\varphi) \quad \text{and} \quad \tau := \frac{\alpha+H(\varphi,\psi)\beta}{H^2(\varphi,\psi)-1} \in M(\psi).$$

Furthermore $H(\varphi,\psi)\sigma - \tau = \alpha \geqq 0$ and $H(\varphi,\psi)\tau - \sigma = \beta \geqq 0$ so that the estimations required in ii) are satisfied. ii) $\Rightarrow$ iii) Let $\sigma^o \in M(\varphi)$ and $\tau^o \in M(\psi)$ be representative measures as required in ii). For $\tau \in M(\psi)$ then

$$\sigma := \sigma^o + \frac{1}{H(\varphi,\psi)}(\tau-\tau^o) \quad \text{is} \geqq \frac{1}{H(\varphi,\psi)}\tau \geqq 0,$$

so that $\sigma \in M(\varphi)$ and is as required. iii) $\Rightarrow$ iv) is obvious, and iv) $\Rightarrow$ i) is 4.1. QED.

We retain the compact-continuous situation. Let us conclude the section with a look at the Gleason parts for A in the Gelfand topology of $\Sigma(A)$. The Gelfand topology on $\Sigma(A)$ is defined to be the weakest topology in which for each $f\in A$ the Gelfand transform $\hat{f}\in B(\Sigma(A)): \varphi \mapsto \varphi(f)$ is continuous. Hence it is the weak*topology $\sigma(A',A)$ restricted to $\Sigma(A) \subset$ closed unit ball of A'. The Banach-Alaoglu theorem implies that $\Sigma(A)$ is compact in the Gelfand topology.

<u>4.5 REMARK</u>: Assume the compact-continuous situation. Then each Gleason part for A is a countable union of compact subsets of $\Sigma(A)$.

Proof: Let $P = Gl(A,\varphi) \subset \Sigma(A)$ be the Gleason part of $\varphi \in \Sigma(A)$. Note that for each $c>0$ the set

$$\{\psi\in\Sigma(A): \|\varphi-\psi\| \leqq c\} = \bigcap_{f\in A} \{\psi\in\Sigma(A): |\varphi(f)-\psi(f)| \leqq c\|f\|\}$$

is closed in the Gelfand topology. Thus the assertion follows from

$$P = \{\psi\in\Sigma(A): \|\varphi-\psi\| < 2\} = \bigcup_{n=1}^{\infty} \{\psi\in\Sigma(A): \|\varphi-\psi\| \leqq 2-\tfrac{1}{n}\} . \quad \text{QED.}$$

Notes

The present Notes are of course to be read in connection with the Notes to Chapter II. The existence of Jensen measures was discovered by several authors, the first of which appears to be BISHOP [1963]. The present proof of 1.1 is from KÖNIG [1970b] and is of utmost shortness as to its deduction from the standard theorems of functional analysis. Lemma 1.5 is a version of the modification technique due to LUMER [1968].

The first part of 2.1 (for one-point T) and its consequence 2.2 are due to RAINWATER [1969]. The second part of 2.1 (for one-point T as well) and the subsequent proof of the abstract F.and M.Riesz theorem 2.3 are from KÖNIG [1970b]. It is an ameliorated variant of the Forelli lemma which is a traditional basis for the proof of the abstract F.and M.Riesz theorem. The present form of 2.1 is such that it can also be applied in Section X.5. The systematic use of the Hahn-Banach versions described in the Appendix is due to KÖNIG [1970b]. It replaces the application of the minimax theorem as in GLICKSBERG [1967], RAINWATER [1969] and GAMELIN [1969] Chapter II.

GLEASON [1957] noticed that the relation $G(\varphi,\psi) = \|\varphi-\psi\| < 2$ is an equivalence relation on the spectrum $\Sigma(A)$ of A and took this as the definition of the parts named after him. The essentials of 3.5 and of the fundamental 4.4 are due to BISHOP [1964]. Explicitely 3.5 is in BEAR [1965] and BEAR-WEISS [1967]. The quantitative versions 3.2 and 3.3 of the principal results are due to KÖNIG [1966a][1969c]. See also the Lecture Notes BEAR [1970]. In connection with 4.5 we refer to the topological characterization of Gleason parts due to GARNETT [1967].

Chapter IV

The Abstract Hardy Algebra Situation

The abstract Hardy algebra theory will occupy Chapters IV-IX, with Chapters IV-VI devoted to the universal aspects of the theory and Chapters VII-IX to particular topics under additional assumptions. It makes essential use of Chapter II, while Chapter III is not referred to (except the existence of Jensen measures III.1.1 in the proof of V.6.1).

We fix a finite positive measure space (X,Σ,m), of course with $m(X)>0$. The abstract Hardy algebra situation is defined to consist of a complex subalgebra $H\subset L^{\infty}(m)$ which contains the constants and is closed in the weak* topology $\sigma(L^{\infty}(m),L^{1}(m))$, and of a nonzero multiplicative linear functional $\varphi:H\to\mathbb{C}$ which is weak* continuous, that is which can be represented in the form $\varphi(u)=\int uFdm\ \forall u\in H$ for some $F\in L^{1}(m)$. It will soon become clear that φ then can be represented by nonnegative functions $0\leqq F\in L^{1}(m)$.

There are obvious connections with the abstract function algebra situation in both directions. The direct image construction exhibits the Hardy algebra situation as a localization of the bounded-measurable function algebra situation: it arises when $A\subset B(X,\Sigma)$ is reduced modulo a fixed $m\in Pos(X,\Sigma)$ and when a fixed $\varphi\in\Sigma(A)$ is chosen, of course with m and φ in adequate relation to each other. Under this connection Hardy algebra results can be applied to problems of function algebra theory. The abstract F. and M.Riesz theorem II.3.2 implies that function algebra problems can always be localized, that is reduced to Hardy algebra problems in this way (at least modulo completely singular measures in $A^{\perp}$ which however can often be shown to be$=0$). The vehicle of reduction is the main F. and M.Riesz consequence II.4.5. We shall be more specific in Section X.1.

The inverse image construction simply transforms an $H\subset L^{\infty}(m)$ into the complex subalgebra of those functions in $B(X,\Sigma)$ which mod m are in H. We use this connection to transfer the main theorems of Chapter II into the Hardy algebra situation. The abstract Szegö-Kolmogorov-Krein theorem thus obtained will then form the source of most of the deeper results of the theory. In contrast, the abstract F. and M.Riesz theorem will not be used after the transfer, as it can be expected, but for an isolated particular purpose (to wipe out the remainders of uncomplete localization). An alternative for the inverse image construction would be to introduce the Gelfand structure space K of the commutative B* algebra $L^\infty(m)$ and thus to transform an $H \subset L^\infty(m)$ into a closed complex subalgebra of $C(K) \cong L^\infty(m)$. This connection would permit to deduce the abstract Hardy algebra theory from the well-esteemed compact-continuous situation of Chapter III - but the price would be to work in a complicated artificial space K and thus to loose quite some directness. So we adopt the former approach.

After the fundamentals the present chapter introduces the function classes $H^\# \subset L^\# \subset L(m)$. The algebra $H^\#$ appears to be the appropriate extension of H into the domain of unbounded functions and its systematic use to be responsible for a transparent formulation of the theory. The $L^p(m)$-norm closures H^p of H for $1 \leq p < \infty$ are much less important. The main theorem of the chapter will be the inequality 3.9 for the functional α.

The chapter ends with a first look at the special situation named after Szegö: that φ admits a unique representative function $0 \leq F \in L^1(m)$. This is the situation where the most prominent classical theorems remain true in the same form as for the unit disk algebra $H^\infty(D)$ (with φ= the point evaluation φ_o in the origin or φ_a in any point $a \in D$). The specialization of the theory to the Szegö situation will be always immediate. In Chapter VII then the question will be treated how far the Szegö situation can be apart from the classical $H^\infty(D)$.

1. Basic Notions and Connections with the Function Algebra Situation

Let (H,φ) be a Hardy algebra situation. We introduce

$$K := \{f \in L^1(m) : \int uf dm = \varphi(u) \int f dm \ \forall u \in H\},$$

$$M := \{0 \leq F \in L^1(m) : \varphi(u) = \int uF dm \ \forall u \in H\} = \{F \in K : F \geq 0 \text{ and } \int F dm = 1\},$$

the classes of analytic and of representative functions (=densities), and

$$N := \text{real-linear span}(M-M) = \{c(U-V) : U,V \in M \text{ and } c>0\}.$$

Likewise we introduce

$$MJ := \{0 \leq F \in L^1(m) : \log|\varphi(u)| \leq \int (\log|u|) F dm \quad \forall u \in H\},$$

the class of Jensen functions (=densities), and NJ as above. Exponentiation shows that $MJ \subset M$. At last we define for $1 \leq p < \infty$ the Szegö functional

$$d^p : d^p(F) = \operatorname{Inf}\{\int |u|^p F dm : u \in H \text{ with } \varphi(u) = 1\} \;\forall\; 0 \leq F \in L^1(m).$$

We list some simple properties.

1.1 REMARK: i) $HK \subset K$. ii) $H^{\perp}$:= the annihilator of H in $L^1(m)$ = $=\{f \in K : \int f dm = 0\}$. iii) $H = \{u \in L^{\infty}(m) : \int u f dm = 0 \;\forall f \in K \text{ with } \int f dm = 0\}$.

1.2 REMARK: i) $|\int f dm| \leq d^1(|f|)$ for all $f \in K$. ii) For each $0 \leq F \in L^1(m)$ there exists an $f \in K$ such that $|f| \leq F$ and $\int f dm = d^1(|f|) = d^1(F)$.

Proof of 1.1: i) has an obvious one-line proof based on the multiplicativity of φ. ii) is immediate from the definition, and iii) follows from ii) in view of the bipolar theorem. Proof of 1.2:i) is immediate from the definitions. ii) From $|\varphi(u)| d^1(F) \leq \int |u| F dm \;\forall u \in H$ we conclude that on the linear subspace $\{uF : u \in H\} \subset L^1(m)$ the linear functional $uF \mapsto \varphi(u) d^1(F)$ is well-defined and of norm ≤ 1. Therefore there exists a function $P \in L^{\infty}(m)$ with $|P| \leq 1$ such that $\varphi(u) d^1(F) = \int u F P dm \;\forall u \in H$. In particular $d^1(F) = \int F P dm$. It follows that $f := FP \in K$ fulfills the assertion. QED.

1.3 THE DIRECT IMAGE CONSTRUCTION: Assume that $A \subset B(X,\Sigma)$ is a complex subalgebra which contains the constants and that $\varphi \in \Sigma(A)$. And fix an $m \in \operatorname{Pos}(X,\Sigma)$ such that there are measures in $M(A,\varphi)$ which are $\ll m$.
i) Then

$$A^m := \overline{A \text{ modulo } m}^{\text{weak*}} \subset L^{\infty}(m)$$

is a weak* closed complex subalgebra which contains the constants. And there is a unique weak* continuous linear functional $\varphi^m : A^m \to \mathbb{C}$ with $\varphi^m(u \bmod m) = \varphi(u) \;\forall u \in A$. The functional φ^m is $\neq 0$ and multiplicative. Thus (A^m, φ^m) is a Hardy algebra situation. ii) For (A^m, φ^m) we have

$$K = \{f \in L^1(m) : fm \in an(A,\varphi)\},$$

$$M = \{0 \leq F \in L^1(m) : Fm \in M(A,\varphi)\},$$

$$N \subset \{f \in \operatorname{Re}L^1(m) : fm \in N(A,\varphi)\},$$

with equality when $M(A,\varphi)\subset\{m\}^{\vee}$. Here of course $N(A,\varphi) :=$ real-linear span$(M(A,\varphi)-M(A,\varphi))=\{c(\sigma-\tau):\sigma,\tau\in M(A,\varphi)$ and $c>0\}$. Furthermore

$$\{f\in ReL^1(m):fm\in N(A,\varphi)\}\subset\{f\in K\cap ReL^1(m):\int fdm=0\},$$

which in VI.4.4 will be seen to be the $L^1(m)$-norm closure of N provided that m is chosen so that (A^m,φ^m) is reduced (see below). iii) We define $MJ(A,\varphi):=\{\sigma\in Pos(X,\Sigma):\log|\varphi(u)|\leqq\int\log|u|d\sigma\ \forall u\in A\}$ as in III.1, and $NJ(A,\varphi)$ as above. It will at once be clear that $MJ(A,\varphi)\subset M(A,\varphi)$ as before. We claim that

$$MJ = \{0\leqq F\in L^1(m):Fm\in MJ(A,\varphi)\},$$

$$NJ \subset \{f\in ReL^1(m):fm\in NJ(A,\varphi)\},$$

with equality when $MJ(A,\varphi)\subset\{m\}^{\vee}$. iv) At last we have $d^p(F)=D^p(Fm)$ for all $0\leqq F\in L^1(m)$ and $1\leqq p<\infty$.

Proof: i) and ii) follow from direct verification. In iii) and iv) the approximation of functions $u\in A^m$ by members of A requires more attention. In iii) one has to proceed as in I.3.7. In both cases the subsequent remark will suffice: For $0\leqq F\in L^1(m)$ and $1\leqq p<\infty$ each $u\in A^m$ is in the $L^p(Fm)$-norm closure of A mod Fm. In fact, if this were false then there were an $h\in L^q(Fm)$ with $\int vhFdm=0\ \forall v\in A$ and $\int uhFdm\neq0$. Thus $hF\in L^1(m)$ annihilates A and hence A^m so that we have a contradiction. The further details are immediate and can be omitted. QED.

<u>1.4 THE INVERSE IMAGE CONSTRUCTION:</u> Assume that (H,φ) is a Hardy algebra situation on (X,Σ,m). i) Then

$$\overleftarrow{H} := \{f\in B(X,\Sigma):f \bmod m \in H\}$$

is a complex subalgebra of $B(X,\Sigma)$ which contains the constants. And $\overleftarrow{\varphi}:\overleftarrow{\varphi}(u) = \varphi(u \bmod m)\ \forall u\in\overleftarrow{H}$ defines a functional $\overleftarrow{\varphi}\in\Sigma(\overleftarrow{H})$. ii) We have

$$an(\overleftarrow{H},\overleftarrow{\varphi}) = \{fm:f\in K\},$$

$$M(\overleftarrow{H},\overleftarrow{\varphi}) = \{Fm:F\in M\},$$

$$N(\overleftarrow{H},\overleftarrow{\varphi}) = \{fm:f\in N\}.$$

In particular $M\neq\emptyset$. iii) We have $D^p(\overleftarrow{H},\overleftarrow{\varphi},Fm)=d^p(F)$ for all $0\leqq F\in L^1(m)$ and $1\leqq p<\infty$.

Proof: i) and iii) are obvious. In ii) it suffices to prove the first assertion. Here the inclusion $\supset$ is immediate from the definitions. The same is true for $\subset$ once we have shown that each $\theta\in an(\overleftarrow{H},\overleftarrow{\varphi})$ must be $\ll m$. To prove this let $E\in\Sigma$ with $m(E)=0$. For all $f\in B(X,\Sigma)$ then $f\chi_E \bmod m=0$ and hence $f\chi_E\in\overleftarrow{H}$ with $\overleftarrow{\varphi}(f\chi_E)=0$. Thus $\int f\chi_E d\theta=0$. It follows that $|\theta|(E)=0$. QED.

We retain the Hardy algebra situation (H,φ). In the remainder of the present section we use 1.4 to transfer the main theorems of Chapter II.

1.5 ABSTRACT SZEGÖ-KOLMOGOROV-KREIN THEOREM: Assume that $0\leq P\in L^1(m)$ and $0\leq F\in L(m)$ with $FP\in L^1(m)$ and $1\leq p<\infty$ such that $d^p(P)$ and $d^p(FP)$ are not both $=0$. Then there exist functions $V\in M$ with $[V>0]\subset[P>0]$ (of course modulo m-null sets) such that $\int(\log F)Vdm$ exists in the extended sense. And

$$\operatorname{Inf}\left\{\exp\left(\int(\log F)Vdm\right):\text{all these } V\right\}\leq\frac{d^p(FP)}{d^p(P)}\leq$$

$$\leq\operatorname{Sup}\left\{\exp\left(\int(\log F)Vdm\right):\text{ all these } V\right\}.$$

1.6 THEOREM: Let $0\leq h\in L^1(m)$ and $1\leq p<\infty$. Assume that $F\in M$ is such that $[V>0]\subset[F>0]$ $\forall V\in M$ with $[V>0]\subset[h>0]$. Then

$$\operatorname{Inf}\left\{\exp\left(\int_{[F>0]}(\log\tfrac{h}{F})Vdm\right):V\in M\text{ with }[V>0]\subset[F>0]\text{ and }\exists\right\}$$

$$\leq d^p(h)=d^p(h\chi_{[F>0]})\leq$$

$$\leq\operatorname{Sup}\left\{\exp\left(\int_{[F>0]}(\log\tfrac{h}{F})Vdm\right):V\in M\text{ with }[V>0]\subset[F>0]\text{ and }\exists\right\}.$$

If $d^p(h)>0$ then there exist $F\in M$ with $[F>0]\subset[h>0]$ such that $[V>0]\subset$ $\subset[F>0]$ $\forall V\in M$ with $[V>0]\subset[h>0]$ as required.

1.7 UNIVERSAL JENSEN INEQUALITY: Let $h\in K$. Assume that $F\in M$ is such that $[V>0]\subset[F>0]$ $\forall V\in M$ with $[V>0]\subset[h\neq0]$. Then

$$\log\left|\int hdm\right|\leq\operatorname{Sup}\left\{\int_{[F>0]}(\log\frac{|h|}{F})Vdm:V\in M\text{ with }[V>0]\subset[F>0]\text{ and }\exists\right\}.$$

If $\int hdm\neq0$ then there exist $F\in M$ with $[F>0]\subset[h\neq0]$ such that $[V>0]\subset[F>0]$ $\forall V\in M$ with $[V>0]\subset[h\neq0]$ as required.

1.8 COROLLARY: For $u\in H$ and $F\in M$ we have

$$\log|\varphi(u)| \leqq \operatorname{Sup}\left\{\int(\log|u|)V dm : V\in M \text{ with } [V>0]\subset[F>0]\right\}.$$

A set $E\in\Sigma$ is defined to be full iff there exist functions $V\in M$ with $[V>0]\subset E$. Application of II.2.2. ii) to $\chi_E m$ and Mm then shows: There exist functions $F\in M$ with $[F>0]\subset E$ such that $[V>0]\subset[F>0]\ \forall\, V\in M$ with $[V>0]\subset E$. These $F\in M$ are called dominant over E. The subset $Y(E) := [F>0]\subset E$ is of course independent from the particular dominant $F\in M$.

From the above we know: If $0\leqq h\in L^1(m)$ with $d^p(h)>0$ for some $1\leqq p<\infty$ then $[h>0]$ is full. And if $h\in K$ with $\int h dm \neq 0$ then $[h\neq 0]$ is full.

1.9 F. and M. RIESZ CONSEQUENCE: Assume that $E\in\Sigma$ is full. For $h\in K$ with $[h\neq 0]\subset E$ then $h\chi_{Y(E)}\in K$ and $\int h\chi_{Y(E)} dm = \int h dm$.

Proof: The band $B:=\{\chi_E m\}^\vee$ satisfies $B\cap Mm = B\cap M(\overleftarrow{H},\overleftarrow{\varphi})\neq\emptyset$. Choose an $F\in M$ dominant over E. Then after II.2.2.ii)

$$\theta_{B\cap Mm} = \theta_{Fm} \quad \forall\ \theta\in ca(X,\Sigma), \text{ in particular}$$

$$(fm)_{B\cap Mm} = f\chi_{[F>0]}m = f\chi_{Y(E)}m \quad \forall\ f\in L^1(m).$$

Now $hm\in B\cap an(\overleftarrow{H},\overleftarrow{\varphi})$ and hence $hm-(\int h dm)Fm\in B\cap(\overleftarrow{H})^\perp$. From the abstract F. and M.Riesz theorem II.3.1 thus $h\chi_{Y(E)}m-(\int h dm)Fm\in(\overleftarrow{H})^\perp$. But this is the assertion. QED.

1.10 COROLLARY: Assume that $E\in\Sigma$ is full. Then

$$H\chi_{Y(E)} \subset \overline{H\chi_E}^{\text{weak*}}.$$

In particular $H\chi_{Y(X)}\subset H$, which simply means that $\chi_{Y(X)}\in H$.

Proof: Follows from 1.9 by duality. For $h\in(H\chi_E)^\perp\subset L^1(m)$ we have $\int u\chi_E h dm=0\ \forall u\in H$ or $h\chi_E\in K$ with $\int h\chi_E dm=0$. From 1.9 thus $h\chi_{Y(E)}\in K$ with $\int h\chi_{Y(E)} dm=0$. But this means $H\chi_{Y(E)}\perp h$. The result follows from the bipolar theorem. QED.

The Hardy algebra situation (H,φ) is defined to be reduced iff $Y(X)=X$, that is iff there exist functions $F\in M$ which are >0 on the whole of X. An equivalent condition is that all real-valued functions in H be constant. In fact, for real-valued $u \in H$ and $F \in M$ the equation

$$\int (u-\varphi(u))^2 F dm = \left(\int (u-\varphi(u)) F dm\right)^2 = \left(\int u F dm - \varphi(u)\right)^2 = 0$$

implies that $u=\varphi(u)=\text{const}$ on $[F>0]$, hence on X if (H,φ) is reduced and $F \in M$ is chosen to be dominant on X. And the converse follows from $\chi_{Y(X)} \in H$.

Important parts of the Hardy algebra theory are much easier to formulate in the reduced situation. Also the localization of the function algebra situation via the main F.and M.Riesz consequence II.4.5 can be made to lead to the reduced Hardy algebra situation, even with $1 \in M$. In Chapters V-VIII we shall restrict ourselves to the reduced situation. But in certain parts of the theory reducedness cannot be assumed a priori, in particular in connection with the extension of Hardy algebra situations as in Chapter IX. We deduce from 1.10 that each (H,φ) can be transformed into a reduced situation simply by the wipe-out of the inessential part $X-Y(X)$ of X.

1.11 THEOREM: The restriction

$$H_* := \{u|Y(X) : u \in H\} = \{u|Y(X) : u \in H \text{ with } u=0 \text{ on } X-Y(X)\}$$

is a weak* closed complex subalgebra of $L^\infty(m|Y(X))$ which contains the constants. And $\varphi_*: \varphi_*(u|Y(X))=\varphi(u)$ $\forall u \in H$ defines a nonzero weak* continuous multiplicative linear functional on H_*. We have $M_*=\{F|Y(X): F \in M\}$. Therefore (H_*,φ_*) is a reduced Hardy algebra situation on the measure space $\left(Y(X), \Sigma|Y(X), m|Y(X)\right)$.

Proof: The nontrivial point is the weak* closedness of H_*. To prove this let $v \in L^\infty(m|Y(X))$ be in the weak* closure of H_*. Extend v to $V \in L^\infty(m)$ by $V|(X-Y(X))=0$. For $f \in K$ with $\int f dm=0$ then $\int u\chi_{Y(X)} f dm=0$ $\forall u \in H$ from 1.10 and hence $f|Y(X) \in H_*^\perp$. Thus $0=\int v(f|Y(X))dm=\int V f dm$. It follows that $V \in H$ and hence $v \in H_*$. QED.

1.12 RETURN TO THE UNIT DISK: The direct image construction 1.3 applied to the unit disk algebra $A(D) \subset C(S) \subset B(S,\text{Baire})$, to the point evaluation φ_a in some point $a \in D$, and to Lebesgue measure λ leads to the Hardy algebra situation $(H^\infty(D),\varphi_a)$ because of I.3.3.iii). We have $M_a=\{P(a,\cdot)\}$. In particular the situation is reduced.

2. The Functional α

The functional $\alpha:\operatorname{Re}L(m)\to[-\infty,\infty]$ is defined to be

$$\alpha(f) = \operatorname{Inf}\{-\log|\varphi(u)|: u\in H \text{ with } -\log|u| \geq f\}$$

$$= \operatorname{Inf}\{\operatorname{Sup}(f+\log|u|): u\in H \text{ with } \varphi(u)=1\} \qquad \forall f\in\operatorname{Re}L(m).$$

In fact, from both definitions $\alpha(f)=\infty$ in case that all functions $u\in H$ with $f+\log|u|$ bounded above have $\varphi(u)=0$, and in the opposite case the two definitions can at once be transformed into each other. The functional α is to reflect the essentials of the Hardy algebra situation (H,φ) under consideration. The present section contains some preliminaries. The main theorem will then be obtained in Section 3 as a consequence of the abstract Szegö-Kolmogorow-Krein theorem 1.5.

We list some immediate properties. i) $\operatorname{Inf} f\leq\alpha(f)\leq\operatorname{Sup} f \;\forall f\in\operatorname{Re}L(m)$. ii) α is subadditive, provided that the convention $\infty+(-\infty)=\infty$ is adopted. But we do not claim that $\alpha(tf)=t\alpha(f)\;\forall f\in\operatorname{Re}L(m)$ and $t>0$. iii) α is isotone, that is $f\leq g$ implies that $\alpha(f)\leq\alpha(g)$.

iv) $\alpha(\log|u|)=\log|\varphi(u)|$ for all $u\in H^{\times}:=$ the set of invertible elements of the algebra H. In particular $\alpha(\operatorname{Re} u)=\operatorname{Re}\varphi(u)\;\forall u\in H$ by exponentiation.

2.1 <u>REMARK</u>: i) Let $V\in\operatorname{Re}L^1(m)$. Then $\int fVdm\leq\alpha(f)\;\forall f\in\operatorname{Re}L^\infty(m)\Leftrightarrow V\in MJ$.

ii) Let $f\in\operatorname{Re}L(m)$ with $\alpha(f)<\infty$ and $F\in M$. Then there exist functions $V\in M$ with $[V>0]\subset[F>0]$ such that $\int f^+Vdm<\infty$. And

$$\operatorname{Inf}\left\{\int fVdm : V\in M \text{ with } [V>0]\subset[F>0] \text{ and } \int f^+Vdm<\infty\right\}\leq\alpha(f).$$

Proof: i) In order to prove $\Rightarrow$ note that $\int fVdm\leq\alpha(f)\leq\operatorname{Sup} f\;\forall f\in\operatorname{Re}L^\infty(m)$ implies that $V\geq 0$. The other details are immediate. ii) follows at once from the Jensen inequality 1.8. QED.

Let us remark that the above 2.1.ii) is a rather weak relation compared with the main inequality 3.9 which is in the opposite direction.

It is sometimes more convenient to work with an exponentiated version of the functional α. We introduce the functional

$$a: a(F) = \operatorname{Sup}\left\{|\varphi(u)|: u\in H \text{ with } |u|\leq F\right\} \qquad \forall\ 0\leq F\in L(m),$$

which appears to be a natural definition. The connection between a and α is

$$a(e^{-f}) = e^{-\alpha(f)} \qquad \forall\ f\in \operatorname{Re} L(m).$$

Thus the domain of α is somewhat more restrictive, but this is unimportant for our purposes. Let us reformulate some of the above properties. i) $0\leq \operatorname{Inf} F\leq a(F)\leq \operatorname{Sup} F\leq\infty$ $\forall\ 0\leq F\in L(m)$. ii) $a(F)a(G)\leq a(FG)$ $\forall\ 0\leq F,G\in L(m)$, provided that the convention $0\infty=0$ is adopted. iv) $a(|u|)=|\varphi(u)|$ $\forall\ u\in H^{\times}$.

<u>2.2</u> <u>REMARK</u>: Let $0\leq P\in L^1(m)$ and $0\leq F\in L(m)$ with $PF\in L^1(m)$. Then $d^1(P)a(F)\leq d^1(PF)$, with the convention $0\infty=0$.

Proof: For $u,v\in H$ with $\varphi(u)=1$ and $|v|\leq F$ we have $d^1(P)|\varphi(v)|= d^1(P)|\varphi(uv)|\leq\int|uv|Pdm\leq\int|u|PFdm$ and hence $d^1(P)|\varphi(v)|\leq d^1(PF)$. The assertion follows. QED.

The next aim is to derive from α the sublinear limit functionals $\alpha^0,\alpha^\infty: \operatorname{Re} L^\infty(m)\to\bar{\mathbb{R}}$. We need the subsequent simple lemma.

<u>2.3</u> <u>LEMMA</u>: Assume that the function $h:]0,\infty[\to\bar{\mathbb{R}}$ satisfies

$$h(u+v)\leq h(u)+h(v) \quad \forall\ u,v>0 \text{ and } \sup_{s>0}\frac{h(s)}{s}<\infty.$$

Then

$$\frac{h(t)}{t}\to\sup_{s>0}\frac{h(s)}{s} \text{ for } t\downarrow 0,$$

$$\frac{h(t)}{t}\to\inf_{s>0}\frac{h(s)}{s} \text{ for } t\uparrow\infty.$$

Proof: Let S and I denote the Sup and Inf in question. i) Take a sequence $s(\ell)\downarrow 0$ with

$$\frac{h(s(\ell))}{s(\ell)}\to\liminf_{s\downarrow 0}\frac{h(s)}{s}=: c\leq\limsup_{s\downarrow 0}\frac{h(s)}{s}\leq S.$$

Fix $t>0$. For ℓ sufficiently large then

$$t = m(\ell)s(\ell)+u(\ell) \text{ with } m(\ell)\in\mathbb{N} \text{ and } 0<u(\ell) \leq s(\ell),$$

$$h(t) \leq m(\ell)h(s(\ell))+h(u(\ell))\leq m(\ell)h(s(\ell))+Su(\ell),$$

$$\frac{h(t)}{t} \leq \frac{m(\ell)s(\ell)}{t}\frac{h(s(\ell))}{s(\ell)} + \frac{Su(\ell)}{t}.$$

From $m(\ell)s(\ell)\to t$ for $\ell\to\infty$ and $S<\infty$ we obtain $\frac{h(t)}{t}\leq c$. It follows that $S\leq c$ and hence the assertion for $t\downarrow 0$. ii) Fix $s>0$. For $t>s$ then

$$t = ms + u \text{ with } m\in\mathbb{N} \text{ and } 0<u\leq s,$$

$$h(t) \leq mh(s) + h(u)\leq mh(s) + Su,$$

$$\frac{h(t)}{t} \leq \frac{ms}{t}\frac{h(s)}{s} + \frac{Su}{t} = \frac{t-u}{t}\frac{h(s)}{s} + \frac{Su}{t} = \frac{h(s)}{s} - \frac{u}{t}\frac{h(s)}{s} + \frac{Su}{t}.$$

It follows that

$$\limsup_{t\uparrow\infty} \frac{h(t)}{t} \leq \frac{h(s)}{s} \quad \forall\, s>0 \text{ and hence } \leq I \leq \liminf_{t\uparrow\infty} \frac{h(t)}{t}.$$

Thus we obtain the assertion for $t\uparrow\infty$. QED.

The above lemma permits to form the functionals $\alpha^0,\alpha^\infty:\operatorname{Re} L^\infty(m)\to\mathbb{R}$, defined to be

$$\alpha^0(f) = \lim_{t\downarrow 0} \frac{1}{t}\alpha(tf) = \operatorname{Sup}_{t>0} \frac{1}{t}\alpha(tf),$$

$$\alpha^\infty(f) = \lim_{t\uparrow\infty} \frac{1}{t}\alpha(tf) = \operatorname{Inf}_{t>0} \frac{1}{t}\alpha(tf) \qquad \forall\, f\in\operatorname{Re} L^\infty(m).$$

The subsequent properties are then immediate. i) $\operatorname{Inf} f\leq\alpha^\infty(f)\leq\alpha(f)\leq\alpha^0(f)\leq$ $\leq\operatorname{Sup} f\ \forall\, f\in\operatorname{Re} L^\infty(m)$. ii) α^0 and α^∞ are sublinear. iii) α^0 and α^∞ are isotone. iv) $\alpha^0(\operatorname{Re} u)=\alpha^\infty(\operatorname{Re} u)=\operatorname{Re}\varphi(u)\ \forall\, u\in H$.

<u>2.4</u> <u>REMARK</u>: Let $V\in\operatorname{Re} L^1(m)$. Then $\int fV dm \leq \alpha^\infty(f)\ \forall\, f\in\operatorname{Re} L^\infty(m) \Leftrightarrow V\in MJ$.

A similar result for the functional α^0 can be deduced from the fundamental fact which will now be established.

<u>2.5</u> <u>PROPOSITION</u>: We have

$$\alpha^0(f) = \operatorname{Inf}\left\{\operatorname{Re}\varphi(u):u\in H \text{ with } \operatorname{Re} u\geq f\right\} \qquad \forall\, f \in \operatorname{Re} L^\infty(m).$$

Proof: 1) Let $\sigma(f)$ denote the Inf in question $\forall\, f\in\operatorname{Re} L^\infty(m)$. Then σ:

$\operatorname{Re} L^\infty(m) \to \mathbb{R}$ is a functional with the immediate properties i) $\operatorname{Inf} f \leq \sigma(f) \leq \operatorname{Sup} f \ \forall f \in \operatorname{Re} L^\infty(m)$, ii) sublinear, and iii) isotone. We need two more properties. 2) $\alpha(f) \leq \sigma(f) \ \forall f \in \operatorname{Re} L^\infty(m)$, and hence $\alpha^o(f) \leq \sigma(f) \ \forall f \in \operatorname{Re} L^\infty(m)$. In fact, for $u \in H$ with $\operatorname{Re} u \geq f$ we have $\operatorname{Re}\varphi(u) = \alpha(\operatorname{Re} u) \geq \alpha(f)$, so that $\sigma(f) \geq \alpha(f)$. 3) We claim that

$$\sigma(-e^{-f}) \leq -e^{-\alpha(f)} \qquad \forall f \in \operatorname{Re} L^\infty(m).$$

To see this let $u \in H$ with $|u| \leq e^{-f}$, and let c be a complex number of $|c|=1$ such that $|\varphi(u)| = c\varphi(u) = \varphi(cu) = \operatorname{Re}\varphi(cu)$. Then

$$\operatorname{Re}(cu) \leq |cu| = |u| \leq e^{-f} \quad \text{or} \quad \operatorname{Re}(-cu) \geq -e^{-f},$$

$$e^{\log|\varphi(u)|} = |\varphi(u)| = \operatorname{Re}\varphi(cu) = -\operatorname{Re}\varphi(-cu) \leq -\sigma(-e^{-f}),$$

and from this the assertion follows. 4) For $t>0$ we can estimate

$$\sigma(1-e^{-tf}) \leq \sigma(1) + \sigma(-e^{-tf}) \leq 1-e^{-\alpha(tf)} \leq \alpha(tf) \leq \alpha^o(tf) = t\alpha^o(f),$$

$$\sigma(f) = \sigma\Big(\frac{e^{-tf}-1+tf}{t} + \frac{1-e^{-tf}}{t}\Big) \leq \sigma\Big(\frac{e^{-tf}-1+tf}{t}\Big) + \alpha^o(f).$$

Now after the Taylor remainder term formula

$$\frac{e^{-tf}-1+tf}{t} \leq \frac{t}{2} c^2 e^{tc} \qquad \text{when } |f| \leq c,$$

and hence $\sigma(f) \leq \frac{t}{2} c^2 e^{tc} + \alpha^o(f)$ for all $t>0$. It follows that $\sigma(f) \leq \alpha^o(f)$ and hence $\sigma(f) = \alpha^o(f)$. QED.

<u>2.6 COROLLARY</u>: Let $V \in \operatorname{Re} L^1(m)$. Then $\int fV dm \leq \alpha^o(f) \ \forall f \in \operatorname{Re} L^\infty(m)$ iff $V \in M$.

3. The Function Classes $H^\#$ and $L^\#$

A function $f \in L(m)$ is defined to be of class $L^\#$ iff there exists a sequence of functions $u_n \in H$ with $|u_n| \leq 1$, $u_n \to 1$ pointwise (of course in the $L(m)$ sense), and $u_n f \in L^\infty(m)$ for all $n \geq 1$. Then $L^\infty(m) \subset L^\# \subset L(m)$, and $L^\#$ is an algebra. Furthermore $F \in L^\#$ and $|f| \leq |F|$ imply that $f \in L^\#$.

A function $f \in L(m)$ is defined to be of class $H^\#$ iff there exists a sequence of functions $u_n \in H$ with $|u_n| \leq 1$, $u_n \to 1$ pointwise, and $u_n f \in H$ for all $n \geq 1$. Then $H \subset H^{\#} \subset L^{\#}$, and $H^{\#}$ is an algebra.

3.1 REMARK: $H^{\#} \cap L^{\infty}(m) = H$.

Proof: $\supset$ is obvious. To prove $\subset$ let $f \in H^{\#} \cap L^{\infty}(m)$ and take functions $u_n \in H$ as required in the definition. Then $u_n f \to f$ pointwise and $|u_n f| \leq |f| \leq$ $\leq$const so that $f \in H$. QED.

3.2 REMARK: For $f \in L(m)$ the subsequent properties are equivalent.
i) $f \in H^{\#}$.
ii) There exist functions $f_n \in H$ with $f_n \to f$ and $|f_n| \leq |f|$, and $f \in L^{\#}$.
iii) There exist functions $f_n \in H^{\#}$ with $f_n \to f$ and $|f_n| \leq$ some $F \in L^{\#}$.

Proof: It suffices to prove iii)$\Rightarrow$i). Take functions $u_\ell \in H$ with $|u_\ell| \leq 1$, $u_\ell \to 1$ and $|u_\ell| F \leq c_\ell$ $\forall \ell \geq 1$. Then $u_\ell f_n \in H^{\#} \cap L^{\infty}(m) = H$ and $|u_\ell f_n| \leq c_\ell$, and $u_\ell f_n \to u_\ell f$ for $n \to \infty$. Thus $u_\ell f \in H$ $\forall \ell \geq 1$. It follows that $f \in H^{\#}$. QED.

Let us remark that $H^{\#}$ and $L^{\#}$ do not depend on the functional φ. We prove that $\varphi: H \to \mathbb{C}$ possesses a unique continuation $\varphi: H^{\#} \to \mathbb{C}$ which is continuous in the adequate sense.

3.3 PROPOSITION: There exists a unique continuation $\phi: H^{\#} \to \mathbb{C}$ of φ which is continuous under $L^{\#}$-majorized convergence: If $f_n \in H^{\#}$ with $f_n \to f$ and $|f_n| \leq$ $\leq F \in L^{\#}$ then ($f \in H^{\#}$ from 3.2 and) $\phi(f_n) \to \phi(f)$. ϕ is a multiplicative linear functional. In the sequel it will be named φ as well.

Proof: i) For $f \in H^{\#}$ we define

$$\phi(f) := \frac{\varphi(uf)}{\varphi(u)} \text{ for any } u \in H \text{ with } uf \in H \text{ and } \varphi(u) \neq 0.$$

Such functions $u \in H$ exist after the definition of $H^{\#}$, and it is obvious from the multiplicativity of φ that the quotient in question does not depend on the particular $u \in H$. The functional $\phi: H^{\#} \to \mathbb{C}$ thus defined is multiplicative linear and $\phi|H = \varphi$. Also ϕ is continuous as claimed: If f_n, $f \in H^{\#}$ with $f_n \to f$ and $|f_n|, |f| \leq F \in L^{\#}$ then choose $u \in H$ with $\varphi(u) \neq 0$ such that $uF \in L^{\infty}(m)$, and it follows at once that

$$\phi(f_n) = \frac{\varphi(uf_n)}{\varphi(u)} \to \frac{\varphi(uf)}{\varphi(u)} = \phi(f).$$

ii) The uniqueness assertion is clear from 3.2. QED.

3.4 REMARK: Assume that $F \in L^1(m)$ with $\varphi(u) = \int uFdm \ \forall u \in H$. Then $\varphi(u) = \int uFdm$ for all $u \in H^\#$ with $uF \in L^1(m)$.

The above remark is obvious from 3.2. But it is important to note that for $u \in H^\#$ there need not exist an $F \in M$ such that $uF \in L^1(m)$. In fact, in V.3.5 we shall obtain examples of real-valued functions $u \in H^\#$ such that $\varphi(u)$ is not real (even in the unit disk situation).

An other simple remark is that the versions of the Jensen inequality we met hitherto remain true when H is extended to $H^\#$. The proofs are immediate so that we can omit the details.

3.5 REMARK: Assume that $u \in H^\#$ with $\varphi(u) \neq 0$. i) For $F \in MJ$ we have $\int (\log|u|)^- Fdm < \infty$ and $\log|\varphi(u)| \leq \int (\log|u|)Fdm$. ii) For $F \in M$ there exist functions $V \in M$ with $[V>0] \subset [F>0]$ such that $\int (\log|u|)^- Vdm < \infty$. And

$$\log|\varphi(u)| \leq \operatorname{Sup}\left\{\int (\log|u|)Vdm : V \in M \text{ with} [V>0] \subset [F>0] \text{ and } \int (\log|u|)^- Vdm < \infty\right\}.$$

We turn to the main results. From the above it is obvious that

$$a(F) = \operatorname{Sup}\left\{|\varphi(u)| : u \in H^\# \text{ with } |u| \leq F\right\} \qquad \forall\ 0 \leq F \in L(m).$$

This can be sharpened for the functions $0 \leq F \in L^\#$.

3.6 PROPOSITION: For each $0 \leq F \in L^\#$ there exists an $f \in H^\#$ with $|f| \leq F$ and $\varphi(f) = a(F)$. In particular $a(F) < \infty$.

3.7 PROPOSITION: For each $0 \leq F \in L^\#$ there exist functions $0 \leq P \in L^1(m)$ with $FP \in L^1(m)$ and $d^1(P) > 0$. And

$$a(F) = \operatorname{Inf}\left\{\frac{d^1(FP)}{d^1(P)} : \text{all these } 0 \leq P \in L^1(m)\right\}.$$

Proof of 3.6 and 3.7: i) Let us fix $0 \leq F \in L^\#$. We first prove the existence of functions $0 \leq P \in L^1(m)$ with $FP \in L^1(m)$ and $d^1(P) > 0$. In fact, put $P := |u|G$ where $G \in M$ and $u \in H$ with $|u|F \leq c < \infty$ and $\varphi(u) \neq 0$ as above. Then $FP = F|u|G \leq \leq cG \in L^1(m)$ and $d^1(P) = d^1(|uG|) \geq |\int uGdm| = |\varphi(u)| > 0$ in view of $uG \in K$. ii) The Inf in 3.7 is now well-defined. It will be called I. From 2.2 we have $a(F) \leq I$. We claim that there exists an $f \in H^\#$ with $|f| \leq F$ and $\varphi(f) = I$. Then both 3.6 and 3.7 will be proved. iii) We have

$$Id^1(P) \leq d^1(FP) \leq \int FPdm \qquad \forall\ 0 \leq P \in L^1(m) \text{ with } FP \in L^1(m),$$

$$I|\int hdm| \leq Id^1(|h|) \leq \int |h|Fdm \ \forall\ h \in K \text{ with } hF \in L^1(m).$$

Thus on the linear subspace $\{hF: h\in K \text{ with } hF\in L^1(m)\} \subset L^1(m)$ the linear functional $hF \mapsto I\int hdm$ is well-defined and of norm ≤ 1. Therefore there exists a function $Q\in L^\infty(m)$ with $|Q|\leq 1$ such that

$$I\int hdm = \int hFQdm \qquad \forall\ h\in K \text{ with } hF\in L^1(m).$$

Thus $f:=FQ\in L^\#$ with $|f|\leq F$. We claim that f fulfills the assertion. To see this take funct ons $u_n\in H$ with $|u_n|\leq 1$, $u_n\to 1$ and $|u_n|F \leq c_n<\infty$. For each $h\in K$ then $u_nh\in K$ with $|u_nh|F\leq c_n|h|\in L^1(m)$ and hence

$$\int (u_nf)hdm = \int (u_nh)fdm = I\int u_nhdm = I\varphi(u_n)\int hdm.$$

It follows that $u_nf\in H$ and $\varphi(u_nf)=I\varphi(u_n)$. Therefore $f\in H^\#$ and $\varphi(f)=I$. QED.

We combine 3.7 with 1.5 to obtain the main result of the present chapter.

<u>3.8</u> <u>THEOREM</u>: For each $0\leq F\in L^\#$ there exist functions $V\in M$ such that $\int(\log F)^+Vdm<\infty$. And

$$\text{Inf}\left\{\exp\left(\int(\log F)Vdm\right): V\in M \text{ with } \int(\log F)^+Vdm<\infty\right\} \leq a(F)<\infty.$$

For the subsequent reformulation define C to consist of the functions $f\in \text{Re}\,L(m)$ such that $e^{-f}\in L^\#$. Then $C\subset \text{Re}\,L(m)$ is a cone which in particular contains the functions which are bounded below. From 3.8 we see that for each $f\in C$ there exist functions $V\in M$ such that $\int f^-Vdm<\infty$. We define

$$\theta: \theta(f) = \text{Sup}\left\{\int fVdm: V\in M \text{ with } \int f^-Vdm<\infty\right\} \qquad \forall\ f\in C.$$

Thus $-\infty<\theta(f)\leq\infty$. We list some simple properties. i) $\theta(tf)=t\theta(f)$ $\forall f\in C$ and $t>0$. And $\theta(f+g)\leq\theta(f)+\theta(g)$ $\forall f,g\in C$ at least under the additional assumption that $\int f^-Vdm<\infty$ $\forall V\in M$. ii) θ is isotone. iii) If (H,φ) is reduced then $\theta(|f|)=0 \Rightarrow f=0$ for all $f\in \text{Re}\,L(m)$.

<u>3.9</u> <u>REFORMULATION</u>: $-\infty < \alpha(f) \leq \theta(f)$ for all $f\in C$.

3.10 CONSEQUENCE: $\alpha^{O}(f)=\theta(f)$ for all $f\in \mathrm{Re}\, L^{\infty}(m)$.

Proof: $\leq$ follows from 3.9 and $\geq$ from 2.6. QED.

Another consequence of 3.7 is the subsequent useful Fatou type theorem.

3.11 PROPOSITION: Let $0\leq F, F_n \leq G\in L^{\#}$. Then

$$\limsup_{n\to\infty} F_n \leq F \text{ implies that } \limsup_{n\to\infty} a(F_n) \leq a(F).$$

3.12 REFORMULATION: Assume that $-g\leq f, f_n \in \mathrm{Re}\, L(m)$ with $e^{g}\in L^{\#}$. Then

$$f \leq \liminf_{n\to\infty} f_n \text{ implies that } \alpha(f) \leq \liminf_{n\to\infty} \alpha(f_n).$$

Proof of 3.11: In view of 3.7 it suffices to prove

$$\limsup_{n\to\infty} a(F_n) \leq \frac{d^1(FP)}{d^1(P)} \quad \forall\, 0\leq P\in L^1(m) \text{ with } FP\in L^1(m) \text{ and } d^1(P)>0.$$

Let us fix such a function $0\leq P\in L^1(m)$. And take functions $u_\ell \in H$ with $|u_\ell|\leq 1$, $u_\ell \to 1$ and $|u_\ell|G\leq c_\ell$. We can of course assume that $\varphi(u_\ell)>0$. Also let $v\in H$ with $\varphi(v)=1$. Then

$$|\varphi(u_\ell)\varphi(u)|d^1(P) = |\varphi(u_\ell uv)|d^1(P) \leq \int |u_\ell uv| P dm$$

$$\leq \int |u_\ell v| PF_n dm \quad \forall\, u\in H \text{ with } |u|\leq F_n,$$

$$a(F_n) \leq \frac{1}{d^1(P)\varphi(u_\ell)} \int |u_\ell v| PF_n dm.$$

Now put $G_n := \mathrm{Sup}\{F_s : s\geq n\}$ so that $F_n\leq G_n \leq G$ and $|u_\ell|G_n\leq |u_\ell|G\leq c_\ell$. Furthermore $G_n \downarrow \limsup_{n\to\infty} F_n \leq F$. We thus obtain

$$a(F_n) \leq \frac{1}{d^1(P)\varphi(u_\ell)} \int |v|P|u_\ell|G_n dm,$$

$$\limsup_{n\to\infty} a(F_n) \leq \frac{1}{d^1(P)\varphi(u_\ell)} \int |v|P|u_\ell|F dm \leq \frac{\int |v|PFdm}{d^1(P)\varphi(u_\ell)},$$

$$\limsup_{n\to\infty} a(F_n) \leq \frac{\int |v|PFdm}{d^1(P)} \quad \forall\, v\in H \text{ with } \varphi(v)=1.$$

Now pass to the infimum over these $v\in H$. QED.

The next result is a related technical characterization of $L^{\#}$.

It is much simpler and in fact a direct consequence of the definitions involved.

3.13 REMARK: For $f \in \operatorname{Re} L(m)$ consider the subsequent properties.

i) $e^f \in L^\#$.

ii) There exist real constants t_n such that $\alpha((f-t_n)^+)\to 0$.

iii) There exist functions $f_n \in \operatorname{Re} L(m)$ with $\exp f_n \in L^\#$ such that $\alpha((f-f_n)^+)\to 0$.

Then i) $\Rightarrow$ ii) $\Rightarrow$ iii), and iii) $\Rightarrow$ i) when (H,φ) is reduced.

Proof: i) $\Rightarrow$ ii) Take functions $u_n \in H$ with $|u_n| \leqq 1$, $u_n \to 1$ and $|u_n| e^f \leqq c_n$ and put $t_n := \log c_n$. Then $|u_n| \exp((f-t_n)^+) \leqq 1$ and hence $0 \leqq \alpha((f-t_n)^+) \leqq$ $\leqq -\log|\varphi(u_n)|$. From $\varphi(u_n)\to 1$ we obtain $\alpha((f-t_n)^+)\to 0$. ii) $\Rightarrow$ iii) is obvious. iii) $\Rightarrow$ i) We can find functions $u_n \in H$ with $|u_n| \leqq 1$, $|u_n| \exp f_n \leqq c_n$ and $\varphi(u_n)\to 1$. Also there are functions $v_n \in H$ with $|v_n| \exp((f-f_n)^+) \leqq 1$, that is $|v_n| \leqq 1$ and $|v_n| \exp f \leqq \exp f_n$, and with $-\log|\varphi(v_n)| < \alpha((f-f_n)^+) + \frac{1}{n}$ and hence $|\varphi(v_n)|\to 1$. We can assume that $\varphi(v_n)\to 1$. The products $w_n := u_n v_n \in H$ then fulfill $|w_n| \leqq 1$, $\varphi(w_n)\to 1$ and $|w_n| \exp f \leqq c_n$. It remains to prove that $w_n \to 1$. But for $F \in M$ we have $\int |w_n - 1|^2 F dm \leqq 2 - 2\operatorname{Re}\int w_n F dm = 2(1-\operatorname{Re}\varphi(w_n))\to 0$, so that for a suitable subsequence the desired convergence occurs on $[F>0]$, and hence on X if (H,φ) is reduced and $F \in M$ is chosen to be dominant on X. QED.

The remainder of the section is devoted to an alternative idea of proof which at least leads to the special case 3.10 of the main result 3.9. It is independent from Chapter II but uses instead a beautiful particular version of the Krein-Smulian theorem (see DUNFORD-SCHWARTZ [1958] Vol.I p.429) valid in $L^\infty(m)$-spaces.

3.14 KREIN-SMULIAN CONSEQUENCE: For a convex subset $S \subset L^\infty(m)$ we have: S is weak* closed $\Leftrightarrow$ if $f_n \in S$ with $|f_n| \leqq c < \infty$ and $f_n \to f$ pointwise then $f \in S$.

Proof of 3.14: $\Rightarrow$ is obvious. In order to prove $\Leftarrow$ is suffices after Krein-Smulian to prove that $S(R) := \{f \in S : |f| \leqq R\}$ is weak* closed for each $R>0$. Let $f \in L^\infty(m)$ be in the weak* closure of $S(R)$. Then it is a fortiori in the $\sigma(L^\infty(m), L^2(m)) = \sigma(L^2(m), L^2(m))|L^\infty(m)$ closure of $S(R)$ and hence in the $L^2(m)$-norm closure of $S(R)$ since $S(R)$ is convex. Thus there exist $f_n \in S(R)$

with $f_n \to f$ pointwise. The assumption implies that $f \in S$. QED.

Proof of 3.10 from 3.14 (and from 2.6 and 3.12): i) Let us consider

$$S := \left\{ f \in \operatorname{Re} L^\infty(m) : \alpha^O(f) \leq 0 \right\} = \left\{ f \in \operatorname{Re} L^\infty(m) : \alpha(tf) \leq 0 \ \forall t > 0 \right\},$$

which is a cone in $\operatorname{Re} L^\infty(m)$. We deduce from 3.14 that S is weak* closed. In fact, if $f_n \in S$ with $|f_n| \leq c$ and $f_n \to f$ then 3.12 implies that $\alpha(tf) \leq \liminf_{n\to\infty} \alpha(tf_n) \leq 0 \ \forall t>0$ and hence $f \in S$. ii) We determine the dual cone $D := \{h \in \operatorname{Re} L^1(m) : \int hf\, dm \leq 0 \ \forall f \in S\}$. From 2.6 we know that $M \subset D$ and hence $\{cV : V \in M \text{ and } c \geq 0\} \subset D$. In order to prove the converse let $h \in D$. Then $h \geq 0$ since S contains all functions $f \leq 0$ in $\operatorname{Re} L^\infty(m)$. Furthermore S contains $\operatorname{Re} cu$ for all $u \in H$ with $\varphi(u)=0$ and complex c, so that $\operatorname{Re} c \int uh\,dm \leq 0$ for all complex c and hence $\int uh\,dm = 0$ for all $u \in H$ with $\varphi(u)=0$. This means that $h \in K$. It follows that $D = \{cV : V \in M \text{ and } c \geq 0\}$. iii) Now we obtain from the bipolar theorem A.1.7

$$S = \bar{S}^{\text{weak}*} = \left\{ v \in \operatorname{Re} L^\infty(m) : \int hv\,dm \leq \sup_{f \in S} \int hf\,dm \text{ for all } h \in \operatorname{Re} L^1(m) \right\}$$

$$= \left\{ v \in \operatorname{Re} L^\infty(m) : \int hv\,dm \leq 0 \text{ for all } h \in \operatorname{Re} L^1(m) \text{ with } \int hf\,dm \leq 0 \ \forall\ f \in S \right\}$$

$$= \left\{ v \in \operatorname{Re} L^\infty(m) : \int hv\,dm \leq 0 \ \forall\ h \in D \right\}$$

$$= \left\{ v \in \operatorname{Re} L^\infty(m) : \int hv\,dm \leq 0 \ \forall\ h \in M \right\} = \left\{ v \in \operatorname{Re} L^\infty(m) : \theta(v) \leq 0 \right\},$$

where $\theta(v) := \operatorname{Sup}\{\int hv\,dm : h \in M\} \ \forall v \in \operatorname{Re} L^\infty(m)$ as above. Thus for $v \in \operatorname{Re} L^\infty(m)$ we have $\theta(v-\theta(v)) = \theta(v) - \theta(v) = 0$ and hence $\alpha^O(v) - \theta(v) = \alpha^O(v - \theta(v)) \leq 0$ or $\alpha^O(v) \leq \theta(v)$. And $\alpha^O(v) \geq \theta(v)$ is clear from 2.6. Thus $\alpha^O(v) = \theta(v) \ \forall\ v \in \operatorname{Re} L^\infty(m)$. QED.

<u>3.15</u> <u>RETURN to the UNIT DISK</u>: Consider for fixed $a \in D$ the Hardy algebra situation $(H^\infty(D), \varphi_a)$ on $(S, \text{Baire}, \lambda)$. From 1.12 we know that $M_a = \{P(a,\cdot)\}$. We see that $H^\#$ is $= H^\#(D)$ as defined in Section I.4, and it will result from 4.1 below that $L^\#$ is $= L^O(P(A,\cdot)\lambda) = L^O(\lambda)$, both independent from $a \in D$ as it must be. Now what is the extended functional φ_a on $H^\#(D)$? Recall from I.4.1 the bijective correspondence $H^\#(D) \leftrightarrow \operatorname{Hol}^\#(D)$ which in the direction $\leftarrow$ is via radial limits. The restriction $H^\infty(D) \leftrightarrow \operatorname{Hol}^\infty(D)$ is in the direction $\to$ described by $F \mapsto f = \langle F\lambda \rangle$, that is to $F \in H^\infty(D)$ there corresponds the function $f: f(a) = \varphi_a(F)\ \forall\ a \in D$. It is almost obvious that this remains true on $H^\#(D)$: for $F \in H^\#(D)$ we have $\varphi_a(F) = f(a)\ \forall\ a \in D$ where $f \in \operatorname{Hol}^\#(D)$ corresponds to F. In fact, there are functions $u, v \in H^\infty(D)$ with $uF = v$ and $\varphi_a(u) \neq 0$. Then

$\langle u\lambda\rangle f=\langle v\lambda\rangle$, so that $\varphi_a(u)f(a)=\varphi_a(v)=\varphi_a(u)\varphi_a(F)$ and hence $f(a)=\varphi_a(F)$. So we have incorporated $H^{\#}(D)$ and $Hol^{\#}(D)$ into the abstract Hardy algebra theory.

4. The Szegö Situation

4.1 THEOREM: For $F\in M$ and $1\leq p<\infty$ the subsequent properties are equivalent.

i) $M=\{F\}$.

ii) $L^{\#}\subset L^{O}(Fm)$ and $\alpha(f)=\int fFdm\ \forall\ f\in C$.

iii) $\alpha(f)\leqq\int fFdm\ \forall\ f\in Re\,L^{\infty}(m)$.

iv) $d^{p}(h)=\exp\left(\int_{[F>0]}(\log\frac{h}{F})Fdm\right)$ for all $0\leqq h\in L^{1}(m)$.

v) $d^{p}(h)\leq\exp\left(\int_{[F>0]}(\log\frac{h}{F})Fdm\right)$ for all $0\leq h\in L^{1}(m)$.

vi) $\log|\int hdm|\leqq\int_{[F>0]}(\log\frac{|h|}{F})Fdm$ for all $h\in K$.

If (H,φ) is reduced then the equivalence extends to

ii*) $L^{\#}=L^{O}(Fm)$, or $C=\{f\in Re\,L(m):\int f^{-}Fdm<\infty\}$. And $\alpha(f)=\int fFdm\ \forall f\in C$.

In the subsequent chapters we shall exhibit several other equivalent conditions.

Proof: We shall prove i) ⇒ iv) ⇒ v) ⇒ i) for each $1\leq p<\infty$, then i) ⇒ vi) ⇒ i) and i) ⇒ ii) ⇒ iii) ⇒ v) for p=1. i) ⇒ iv) is immediate from 1.6, and iv) ⇒ v) is trivial. v) ⇒ i) For $V\in M$ we have

$$1=d^{p}(V)\leqq\exp\left(\int_{[F>0]}(\log\frac{V}{F})Fdm\right)\leqq\int_{[F>0]}\frac{V}{F}Fdm=\int_{[F>0]}Vdm\leqq\int Vdm=1.$$

Thus V=O on [F=O], and $\frac{V}{F}$=const on [F>O] from II.5.1. It follows that V=F. i) ⇒ vi) is immediate from 1.7, and vi) ⇒ i) follows as v) ⇒ i) above.

i) ⇒ ii) For $f\in C$ we obtain from 3.8 and 3.9 that $\int f^{-}Fdm<\infty$ or $e^{-f}\in L^{O}(Fm)$ and $\alpha(f)\leqq\int fFdm$. If $\alpha(f)=\infty$ we are done. If $\alpha(f)<\infty$ then 2.1.ii) implies that $\int f^{+}Fdm<\infty$ and $\int fFdm\leqq\alpha(f)$. ii) ⇒ iii) is trivial. iii) ⇒ v) for p=1: First we conclude from 3.12 that $\alpha(f)\leq\int fFdm$ for all $f\in Re\,L(m)$ which are bounded from below. Fix now $0\leqq h\in L^{1}(m)$. We can of course assume that

$d^1(h)>0$. For $\delta>0$ and $R>0$ the function

$$f:=\begin{cases} \log\frac{h+\delta F}{F} & \text{on } [F>0] \\ R & \text{on } [F=0] \end{cases}$$

is bounded from below so that $\alpha(f)\leqq\int fFdm$. It follows that

$$a(e^{-f}) = e^{-\alpha(f)} \geqq \exp(-\int fFdm) = \exp\Big(-\int_{[F>0]}(\log\frac{h+\delta F}{F}Fdm\Big).$$

Thus from 2.2 we have

$$\exp\Big(-\int_{[F>0]}(\log\frac{h+\delta F}{F})Fdm\Big) \leqq a(e^{-f}) \leqq \frac{d^1(he^{-f})}{d^1(h)} \leqq \frac{1}{d^1(h)}\int|u|he^{-f}dm$$

$$\leqq \frac{1}{d^1(h)}\Big(\int|u|Fdm + e^{-R}\int|u|hdm\Big) \quad \forall\ u\in H \text{ with } \varphi(u)=1,$$

since $he^{-f}=h\frac{F}{h+\delta F}\leqq F$ on $[F>0]$. Now let $R\uparrow\infty$ and then take the infimum over the $u\in H$ with $\varphi(u)=1$. In view of $d^1(F)=1$ we obtain

$$d^1(h) \leqq \exp\Big(\int_{[F>0]}(\log\frac{h+\delta F}{F})Fdm\Big) \quad \forall\ \delta>0.$$

Then for $\delta\downarrow0$ the assertion follows from Beppo Levi.

It remains to prove i) $\Rightarrow L^o(Fm)\subset L^\#$ if (H,φ) is reduced. Assume that $f\in \mathrm{Re}\,L(m)$ with $e^f\in L^o(Fm)$, or $\int f^+Fdm<\infty$. Let $f_n:=\min(f,n)$ so that $0\leqq(f-f_n)^+\downarrow0$ and $\exp\big(-(f-f_n)^+\big)\in L^\infty(m)\subset L^\#$. Thus from 3.9 and i) we have $\alpha\big((f-f_n)^+\big)\leqq$ $\leqq\int(f-f_n)^+Fdm$, and therefore $\int f^+Fdm<\infty$ implies $\alpha\big((f-f_n)^+\big)\to0$. Since $\exp f_n$ $\in L^\infty(m)\subset L^\#$ we see from 3.13 that $e^f\in L^\#$. QED.

In this connection we formulate an immediate consequence of 3.8.

4.2 REMARK: We have $L^\#\subset \bigcup_{V\in M} L^o(Vm)$.

It is natural to ask for the inequalities which are opposite to iii) and v) in 4.1. The answer is as follows.

4.3 THEOREM: For $F\in M$ and $1\leqq p<\infty$ the subsequent properties are equivalent.

i) $F\in MJ$.

ii) For $f\in \mathrm{Re}\,L(m)$ with $\alpha(f)<\infty$ we have $\int f^+Fdm<\infty$ and $\int fFdm\leqq\alpha(f)$.

iii) $\int fFdm \leq \alpha(f)\ \forall\ f \in \operatorname{Re} L^{\infty}(m)$.

iv) $\exp\left(\int_{[F>0]} (\log\frac{h}{F})Fdm\right) \leq d^p(h)\ \forall\ 0 \leq h \in L^1(m)$.

Proof: i) ⇒ ii) ⇒ iii) ⇒ i) is as immediate as 2.1. i) ⇒ iv) For $u \in H$ with $\varphi(u)=1$ we have

$$\int |u|^p h dm \geq \int_{[F>0]} |u|^p \frac{h}{F} Fdm \geq \exp\left(\int_{[F>0]} (\log(|u|^p \frac{h}{F}))\, Fdm\right)$$

$$= \exp\left(p\int(\log|u|)Fdm\right)\exp\left(\int_{[F>0]} (\log\frac{h}{F})Fdm\right) \geq \exp\left(\int_{[F>0]} (\log\frac{h}{F})Fdm\right),$$

from which the assertion is immediate. iv) ⇒ i) For $u \in H$ with $\varphi(u) \neq 0$ and $\delta > 0$ we put $0 \leq h := F(|u|+\delta)^{-p} \in L^1(m)$. Then

$$|\varphi(u)|\exp\left(-\int(\log(|u|+\delta))Fdm\right) = |\varphi(u)|\exp\left(\frac{1}{p}\int_{[F>0]} (\log\frac{h}{F})Fdm\right)$$

$$\leq |\varphi(u)|\left(d^p(h)\right)^{\frac{1}{p}} \leq \left(\int |u|^p h dm\right)^{\frac{1}{p}} = \left(\int\left(\frac{|u|}{|u|+\delta}\right)^p Fdm\right)^{\frac{1}{p}} \leq 1,$$

$$|\varphi(u)| \leq \exp\left(\int(\log(|u|+\delta))Fdm\right)\ \forall\ \delta > 0,$$

from which the assertion follows for $\delta \downarrow 0$. QED.

The Szegö situation is defined to be the case that $M=\{F\}$, in which case $F \in M$ will be called Szegö as well. The above results show that then $F \in MJ$. But there are worlds between $F \in MJ$ and the Szegö situation $M=\{F\}$. We present a simple example.

4.4 EXAMPLE: In the unit disk algebra $H^{\infty}(D) \subset L^{\infty}(\lambda)$ the functions $f \in H^{\infty}(D)$ with $\int f\bar{Z}d\lambda = \langle f\lambda\rangle'(0) = 0$ form a weak* closed subalgebra $T \subset H^{\infty}(D)$ of codimension one. One verifies that for the Hardy situation (T,φ_o) the set M consists of the functions $F_c := 1 - \operatorname{Re}(cZ) \geq 0$ for the complex numbers c of modulus $|c| \leq 1$, and that on the other hand $MJ = \{F_o\}$.

We conclude the section with a first look at a special situation which is much less restrictive than the Szegö situation but still permits important additional conclusions. We shall come back to it in Chapter VIII.

4.5 PROPOSITION: The subsequent properties of a Hardy algebra situation (H,φ) are equivalent.

i) M is compact in $\sigma(\operatorname{Re} L^1(m), \operatorname{Re} L^{\infty}(m))$.

ii) Each linear functional $\psi \in (\operatorname{Re} L^{\infty}(m))^*$ with $\psi \leq \alpha^o = \theta|\operatorname{Re} L^{\infty}(m)$ is of the form $\psi(f)=\int fVdm\ \forall f\in \mathrm{Re}\,L^{\infty}(m)$ for some $V\in M$.

In this situation we have $\alpha^{o}(f)=\theta(f)=\mathrm{Max}\{\int fVdm: V\in M\}\ \forall\ f\in \mathrm{Re}\,L^{\infty}(m)$. Furthermore $MJ\neq\emptyset$ and $\alpha^{\infty}(f)=\mathrm{Max}\ \{\int fVdm: V\in MJ\}\ \forall\ f\in \mathrm{Re}\,L^{\infty}(m)$.

Proof: We consider the dual space $(\mathrm{Re}\,L^{\infty}(m))^*$ with the weak* topology $\sigma((\mathrm{Re}\,L^{\infty}(m))^*, \mathrm{Re}\,L^{\infty}(m))=:\tau$. It is well-known that $S(\theta):=\{\psi\in(\mathrm{Re}\,L^{\infty}(m))^*:$ $\psi\leqq\theta\}$ is τ-compact. Under the usual embedding $\mathrm{Re}\,L^{1}(m)\subset(\mathrm{Re}\,L^{\infty}(m))^*$ we have $M\subset S(\theta)$ from 2.6, and the bipolar theorem tells us that

$$\bar{M}^{\tau} = \left\{\psi\in(\mathrm{Re}\,L^{\infty}(m))^*:\ \psi(f)\leqq \operatorname*{Sup}_{V\in M}\int fVdm=\theta(f)\ \forall\ f\in \mathrm{Re}\,L^{\infty}(m)\right\} = S(\theta).$$

Therefore $M=S(\theta)$ iff M is τ-closed, that is τ-compact, and that is compact in $\tau|\mathrm{Re}\,L^{1}(m)=\sigma(\mathrm{Re}\,L^{1}(m), \mathrm{Re}\,L^{\infty}(m))$. Thus i) and ii) are in fact equivalent. The remainder is then immediate from the Hahn-Banach version that each sublinear functional is the pointwise maximum of the linear functionals which are below it, combined with 2.4. QED.

$M\subset \mathrm{Re}\,L^{1}(m)$ is always $\sigma(\mathrm{Re}\,L^{1}(m), \mathrm{Re}\,L^{\infty}(m))$ closed and bounded. Therefore we are in the situation described in 4.5 whenever $\dim N<\infty$.

Notes

In 1965 the abstract Hardy algebra theory achieved complete clearness about the Szegö situation. The last steps were HOFFMAN [1962b], LUMER [1964], and HOFFMAN-ROSSI [1965] and KÖNIG [1965]. The main result of KÖNIG [1965] was the equivalence of properties i) and iv)-vi) in 4.1 and those in VI.4.9. The first step beyond was AHERN-SARASON [1967a] in a still rather special situation: $\dim N<\infty$ and an additional uniqueness assumption. The paper introduced important further ideas which will be dealt with in Sections VI.6 and IX.1. The universal abstract Hardy algebra situation was then elaborated in GAMELIN-LUMER [1968], LUMER [1968], and GAMELIN [1969] as well as in KÖNIG [1967b][1967c][1969a][1969b][1970a] in parallel developments which of course penetrated each other. In the Gamelin-Lumer approach a basic tool is the Krein-Smulian consequence 3.14 and the resultant access to the core of the Hardy algebra theory as discovered in HOFFMANN-ROSSI [1967] and described above in form of an alternative proof of 3.10. In the present work we adopt the other approach.

The function class $H^{\#}$ was introduced in KÖNIG [1966b][1967a] in the Szegö situation, while KÖNIG [1967c] has the present definition and the main ideas. At the same time LUMER [1966] introduced his universal Hardy class which is fundamental in the Gamelin-Lumer approach. The two function classes coincide in important special situations, but the definitions are quite different. In contrast to the usual spaces H^p for $1 \leq p < \infty$ the class $H^{\#}$ is an algebra and appears to be superior for several purposes, in particular for substitution theorems, for outer functions (:=invertible elements of $H^{\#}$!) and factorization, and for the abstract conjugation. All these topics will be dealt with in the subsequent chapters.

Chapter V

Elements of Abstract Hardy Algebra Theory

In the present chapter we fix a Hardy algebra situation (H,φ) on (X,Σ,m). We assume (H,φ) to be reduced from the start. The aim is to develop a basic portion of the abstract Hardy algebra theory. The selection is motivated in part by certain famous classical theorems.

We start with the problem to characterize the moduli of the invertible elements of the algebra $H^{\#}$, which replace the outer functions of the earlier theory. In the Szegö situation the result specializes at once to the respective classical theorem. The easiness of its proof demonstrates the power of our main theorem IV.3.8=3.9. Then substitution theorems are established: the simpler one is on the substitution of functions $f\in H^{\#}$ into entire functions, while the more delicate one is on the substitution of functions $f\in H$ with $|f|\leq 1$ into functions of class $Hol^{\#}(D)$. Both theorems have applications to the value distribution of functions in H and $H^{\#}$. Next we introduce the important subclass H^{+} of $H^{\#}$. The functions in H^{+} possess certain weak-L^1 properties which in the unit disk situation extend the respective classical theorems. At last we define an appropriate notion of spectrum for the functions in $H^{\#}$ to derive results on their value distribution. In particular, in the Szegö situation (except in the trivial case $H=\mathbb{C}$) the existence of nonconstant inner functions (=functions in H of modulus one) is proved.

1. The Moduli of the invertible Elements of $H^{\#}$

<u>1.1</u> <u>REMARK</u>: For $u\in H^{\#}$ we have $|\varphi(u)|\leq a(|u|)<\infty$. For $u\in(H^{\#})^{\times}$ we have $0<|\varphi(u)|=a(|u|)<\infty$.

Proof: The first assertion is obvious from the definition of the functional a with the remark which preceeds IV.3.6 and from IV.3.6. Let now $u\in(H^{\#})^{\times}$ and $v\in H^{\#}$ with $uv=1$. From $\varphi(u)\varphi(v)=\varphi(uv)=1$ it follows that $\varphi(u),\varphi(v)\neq 0$. Then multiplication of $0<|\varphi(u)|\leq a(|u|)<\infty$ and $0<|\varphi(v)|\leq a(|v|)<\infty$ leads to $1=|\varphi(u)|\,|\varphi(v)|\leq a(|u|)a(|v|)\leq a(|uv|)=a(1)=1$. Thus $|\varphi(u)|=a(|u|)$. QED.

1.2 LEMMA: Assume that $0 \leq F, G \in L(m)$ with $FG \leq 1$ and $a(F)a(G)=1$ (recall our convention $0\infty=0$ after which $0<a(F),a(G)<\infty$). Then $F,G \in L^{\#}$ and $FG=1$.

Proof: Take functions $u_n, v_n \in H$ with $|u_n| \leq F$, $|v_n| \leq G$ and $0<\varphi(u_n) \to a(F)$, $0<\varphi(v_n) \to a(G)$. Then $u_n v_n \in H$ with $|u_n v_n| \leq 1$ and $\varphi(u_n v_n) \to 1$. Thus in view of $\int |u_n v_n - 1|^2 V dm \leq 2(1-\varphi(u_n v_n)) \to 0$ $\forall V \in M$ and of the reducedness of (H,φ) we have $u_n v_n \to 1$ pointwise after transition to an appropriate subsequence. Therefore $|u_n v_n| \leq FG$ shows that $FG=1$. Furthermore from $|u_n v_n| F \leq |u_n| FG = |u_n|$ and $|u_n v_n| G \leq |v_n| FG = |v_n|$ we see that $F,G \in L^{\#}$. QED.

1.3 THEOREM: Assume that $0<F \in L(m)$. Then there exist functions $f \in (H^{\#})^{\times}$ with $|f|=F$ iff $a(F)a(\frac{1}{F})=1$. In this case $f \in (H^{\#})^{\times}$ with $|f|=F$ is unique up to a constant factor of modulus one. Thus there exists a unique $F^{\times} \in (H^{\#})^{\times}$ with $|F^{\times}|=F$ and $\varphi(F^{\times})=a(F)$. Furthermore $u \in H^{\#}$ with $|u| \leq F$ and $\varphi(u)=a(F)$ implies that $u=F^{\times}$.

Proof: i) Let $f \in (H^{\#})^{\times}$ with $|f|=F$. From 1.1 then $1=|\varphi(f)||\varphi(\frac{1}{f})|= =a(F)a(\frac{1}{F})$. ii) Assume now that $a(F)a(\frac{1}{F})=1$. From 1.2 then $F,\frac{1}{F} \in L^{\#}$. Thus after IV.3.6 there exist functions $u,v \in H^{\#}$ with $|u| \leq F$ and $\varphi(u)=a(F)$, and $|v| \leq \frac{1}{F}$ and $\varphi(v)=a(\frac{1}{F})$. Take any pair of such functions $u,v \in H^{\#}$. Then $uv \in H^{\#}$ with $|uv| \leq 1$ and $\varphi(uv)=a(F)a(\frac{1}{F})=1$, so that $uv=1$ since (H,φ) is reduced. Thus $u \in (H^{\#})^{\times}$ and $|u|=F$. iii) We know from ii) that there exists a unique $u \in H^{\#}$ with $|u| \leq F$ and $\varphi(u)=a(F)$. Then 1.1 implies that each $f \in (H^{\#})^{\times}$ with $|f|=F$ must be $=cu$ with complex c of modulus $|c|=1$. Now all assertions are clear. QED.

1.4 COROLLARY: For $f \in H^{\#}$ with $m([f=0])=0$ we have

$$\log|\varphi(f)| \leq -\alpha(-\log|f|) \leq \alpha(\log|f|),$$

$$-\infty < \log|\varphi(f)| = \alpha(\log|f|) \Leftrightarrow f \in (H^{\#})^{\times}.$$

Proof: The first assertion means that

$$|\varphi(f)| \leq a(|f|) \leq \frac{1}{a(|\frac{1}{f}|)},$$

which is immediate from the definitions. Thus $-\infty<\log|\varphi(f)|=\alpha(\log|f|)$ is equivalent to

$$0<|\varphi(f)| = a(|f|) = \frac{1}{a(|\frac{1}{f}|)},$$

which in particular implies that $0<a(|f|)$, $a(|\frac{1}{f}|)<\infty$ and hence is equivalent to $0<|\varphi(f)|=a(|f|)$ and $a(|f|)a(|\frac{1}{f}|)=1$. But after 1.3 this is indeed equivalent to $f\in(H^{\#})^{\times}$. QED.

We combine 1.3 with IV.3.8 to obtain the subsequent result.

1.5 THEOREM: Assume that $0<F\in L(m)$ satisfies

i) $F,\frac{1}{F}\in L^{\#}$ and

ii) the integral $\int(\log F)Vdm$ has the same value $c\in[-\infty,\infty]$ for all those $V\in M$ for which it exists in the extended sense.

Then $c\in\mathbb{R}$, and $a(F)=e^{c}$ and $a(\frac{1}{F})=e^{-c}$. Thus there exist functions $f\in(H^{\#})^{\times}$ with $|f|=F$.

Proof: From IV.3.8 applied to F and $\frac{1}{F}$ we obtain $e^{c}\leq a(F)<\infty$ and $e^{-c}\leq$ $\leq a(\frac{1}{F})<\infty$. Hence the assertions. QED.

For $0<F\in L(m)$ condition i) is also necessary for the existence of functions $f\in(H^{\#})^{\times}$ with $|f|=F$, while this cannot be said about condition ii) Assume now the Szegö situation $M=\{V\}$. Then ii) is void, so that i) is a necessary and sufficient condition. But now i) means that $F,\frac{1}{F}\in L^{o}(Vm)$ and hence that $\log F\in L^{1}(Vm)$.

1.6 SPECIALIZATION: Assume the Szegö situation $M=\{V\}$. Let $0<F\in L(m)$. Then there exist functions $f\in(H^{\#})^{\times}$ with $|f|=F$ iff $\log F\in L^{1}(Vm)$.

1.7 COROLLARY: Assume the Szegö situation $M=\{V\}$. For $f\in H^{\#}$ we have $-\infty<\log|\varphi(f)|=\int(\log|f|)Vdm \Leftrightarrow f\in(H^{\#})^{\times}$.

Proof: $\Leftarrow$ is clear from IV.3.5.i) since $V\in MJ$. $\Rightarrow$) We obtain $\log|f|\in L^{1}(Vm)$ and $0<|\varphi(f)|=\exp(\int(\log|f|)Vdm)=\exp(-\alpha(-\log|f|))=a(|f|)$ from IV.4.1, so that 1.6 and 1.3 imply that $f\in(H^{\#})^{\times}$. QED.

1.8 FACTORIZATION THEOREM: Assume the Szegö situation $M=\{V\}$. Each function $u\in H^{\#}$ with $\log|u|\in L^{1}(Vm)$ possesses a unique factorization $u=Pf$, where $P\in H$ with $|P|=1$ (an inner function) and $f\in(H^{\#})^{\times}$ with $\varphi(f)>0$. In fact $f=|u|^{\times}$. Furthermore $\log|u|\in L^{1}(Vm)$ is fulfilled for $u\in H^{\#}$ whenever $\varphi(u)\neq 0$.

Proof: We habe to prove the last assertion. But $(\log|u|)^{+}\in L^{1}(Vm)$ is equivalent to $|u|\in L^{o}(Vm)=L^{\#}$, while $(\log|u|)^{-}\in L^{1}(Vm)$ results from IV.3.5.i). QED.

1.9 RETURN to the UNIT DISK: In the special situation $(H^{\infty}(D),\varphi_{o})$ we see from I.4.6 that $\log|u|\in L^{1}(\lambda)$ is fulfilled for all nonzero $u\in H^{\#}(D)$.

An immediate consequence of 1.8 is the subsequent characterization of $H^{\#}$.

1.10 PROPOSITION: Assume the Szegö situation $M=\{V\}$. Then

$$H^{\#} = \left\{\frac{u}{v}: u,v\in H \text{ with } v\in (H^{\#})^{\times}\right\}.$$

Proof: $\supset$ is obvious. To prove $\subset$ let $h\in H^{\#}$. Take a function $u\in H$ with $uh\in H$ and $\varphi(u)\neq 0$. Then $|u|^{\times}h\in H$ as well. Therefore $h=\frac{|u|^{\times}h}{|u|^{\times}}$ is a representation of h as claimed. QED.

2. Substitution into entire Functions

2.1 SUBSTITUTION THEOREM: Let $f: ¢\to ¢$ be an entire function and

$$F: F(t) = \operatorname{Max}\left\{|f(z)|: z\in ¢ \text{ with } |z|\leq t\right\} \ \forall t\geq 0.$$

For $u\in H^{\#}$ with $F(|u|)\in L^{\#}$ then $f(u)\in H^{\#}$ and $\varphi(f(u))=f(\varphi(u))$.

Proof: Take functions $u_n\in H$ with $|u_n|\leq 1$, $u_n\to 1$ and $u_n u\in H$. Since on each compact subset of the complex plane f is the uniform limit of polynomials it is clear that $f(u_n u)\in H$ and $\varphi(f(u_n u))=f(\varphi(u_n u))$. Now $|f(u_n u)|\leq$ $\leq F(|u|)\in L^{\#}$, and $f(u_n u)\to f(u)$ pointwise and $f(\varphi(u_n u))\to f(\varphi(u))$. Thus $f(u)\in H^{\#}$ from IV.3.2 and $\varphi(f(u))=f(\varphi(u))$ from IV.3.3. QED.

2.2 SPECIAL CASE: Assume that $u\in H^{\#}$ with $e^{|u|}\in L^{\#}$. Then $e^{u}\in H^{\#}$ and $\varphi(e^{u})=$ $=e^{\varphi(u)}$.

The above substitution theorem is simple but efficient. It can of course be extended to entire functions of several complex variables.

As a first example we turn to the question whether there can be nonconstant real-valued functions in $H^{\#}$. Recall from Section IV.1 that the reducedness of (H,φ) means that all real-valued functions in H are constant.

2.3 PROPOSITION: Assume that $u\in H^{\#}$ is real-valued and $e^{|u|}\in L^{\#}$. Then $u=\text{const}$.

Proof: For $f:f(z)=\sin z \ \forall z\in\mathbb{C}$ we have $F(t)\leqq e^{t} \ \forall t\geqq 0$. For $t>0$ therefore $F(|tu|)\leqq e^{t|u|}\in L^{\#}$ so that $\sin(tu)\in H^{\#}$ from 2.1. But $\sin(tu)\in H^{\#}\cap L^{\infty}(m)=H$ and hence $\sin(tu)=\text{const}=c(t)$ since u is real-valued. Now $\frac{1}{t}\sin(tu)\to u$ for $t\downarrow 0$ so that $u=\text{const}$. QED.

It is of interest to note the subsequent variant which of course implies the above assertion.

2.4 PROPOSITION: Assume that $u\in H^{\#}$ is $\geqq 0$ and $e^{\sqrt{u}}\in L^{\#}$. Then $u=\text{const}$.

Proof: For $f:f(z)=\sum_{\ell=0}^{\infty}(-1)^{\ell}\frac{z^{\ell}}{(2\ell)!} \ \forall z\in\mathbb{C}$ we have $F(t)\leqq e^{\sqrt{t}} \ \forall t\geqq 0$ and of course $f(z^2)=\cos z \ \forall z\in\mathbb{C}$. For $t>0$ therefore $F(|tu|)=F(tu)\leqq e^{\sqrt{tu}}\in L^{\#}$ so that $f(tu)\in H^{\#}$ from 2.1. But $f(tu)=\cos\sqrt{tu}\in H^{\#}\cap L^{\infty}(m)=H$ and hence $\cos\sqrt{tu}=$ $=\text{const}=c(t)$ since $u\geqq 0$. Now $\frac{1}{t}(1-\cos\sqrt{tu})\to\frac{u}{2}$ for $t\downarrow 0$ so that $u=\text{const}$. QED.

In the Szegö situation $M=\{V\}$ the above 2.3 says that the real-valued functions $u\in H^{\#}\cap L^{1}(Vm)$ are constant. This assertion is sharp in the sense that there always (except in the trivial case $H=\mathbb{C}$) exist nonconstant real-valued functions $u\in H^{\#}$ such that $|u|^{\tau}\in L^{1}(Vm)$ for all $0<\tau<1$. We shall find such examples in 5.7. It follows that 2.4 is sharp in the respective sense as well.

3. Substitution into Functions of Class $Hol^{\#}(D)$

We want to substitute the functions $u\in H$ with $|u|\leqq 1$ into functions $f\in Hol(D)$, and to obtain functions $f(u)\in H^{\#}$ and the formula $\varphi(f(u))=f(\varphi(u))$. This will not be possible without restrictions: as in the simple substitution theorem 2.1 where the restriction $F(|u|)\in L^{\#}$ had to be imposed. Let us seek to explore what can be expected.

If $|u|\leqq$ some $R<1$ then it is clear that we obtain $f(u)\in H$ and $\varphi(f(u))=$ $f(\varphi(u))$ for all $f\in Hol(D)$. If $|u|<1$ then the definition of $f(u)\in L(m)$ is still obvious, but $f(u)\in H^\#$ ceases to be true for all $f\in Hol(D)$. Let us present an example: The function $g: g(z)=\exp(-\frac{1+z}{1-z})$ $\forall z\in D$ is in $Hol^\infty(D)$ since $|g(z)|<1$ $\forall z\in D$. Its boundary function $G=\exp(-\frac{1+Z}{1-Z})\in H^\infty(D): G(e^{it})=$ $=\exp(-i\cot an\ \frac{t}{2})$ $\forall t\in\mathbb{R}$ fulfills $|G|=1$. Let $f=\frac{1}{g}: f(z)=\exp(\frac{1+z}{1-z})$ $\forall z\in D$ with the radial limit function $F=\frac{1}{G}=\exp(\frac{1+Z}{1-Z})$. Then $F\notin H^\#(D)$ from the uniqueness assertion in 1.3. Thus $f\notin Hol^\#(D)$, which also follows from I.4.3. Take now $(H,\varphi)=(H^\infty(D),\varphi_o)$ on (S,Baire,λ) and $u=\frac{1}{2}(1+Z)$. Then $|u|<1$ but $f(u)=f(\frac{1}{2}(1+Z))=\exp(\frac{3+Z}{1-Z})=\exp(1+2\frac{1+Z}{1-Z})=eF^2\in L^\infty(\lambda)$ is not in $H^\#(D)$.

Therefore even for functions $u\in H$ with $|u|<1$ some restriction on the functions $f\in Hol(D)$ is needed. But it is important to admit functions $u\in H$ with $|u|\leqq 1$ as well, in particular those with $|u|=1$ on the whole of X. This requires $f\in Hol(D)$ to have a nice boundary behaviour. Take for example $(H,\varphi)=(H^\infty(D),\varphi_o)$ and $u=Z$: for $f\in Hol(D)$ to be admitted the function $f(u)=f(Z)\in L(\lambda)$ should be at least the pointwise radial limit function of f on S. Since we require this $f(u)=f(Z)$ to be $\in H^\#(D)$ it is natural to decide that we restrict ourselves to the functions $f\in Hol^\#(D)$.

Now the boundary behaviour of the functions $f\in Hol^\#(D)$ is nice but not perfect, a fact which requires some restriction on the part of the functions $u\in H$ with $|u|\leqq 1$. For $f\in Hol^\#(D)$ we know from I.4.1 that

$$S-N:=\left\{s\in S:\ \text{the radial limit}\ f(s):=\lim_{R\uparrow 1} f(Rs)\ \text{exists}\right\}$$

defines a Baire set $N\subset S$ with $\lambda(N)=0$, and the function $s\mapsto f(s)$ on $S-N$ is a representative of the boundary function $F\in H^\#(D)\subset L(\lambda)$. Therefore in order to allow the definition of $f(u)\in L(m)$ for all $f\in Hol^\#(D)$, the function $u\in H$ with $|u|\leqq 1$ should be such that for each Baire set $N\subset S$ with $\lambda(N)=0$ the inverse set $[u\in N]:=\{x\in X: u(x)\in N\}$ (determined of course modulo m-null sets) can be made void, that is have $m([u\in N])=0$. In particular, the constant functions $u=\text{const}=c$ with $|c|=1$ cannot be admitted.

There is another reason for restriction on the part of the functions $u\in H$ with $|u|\leqq 1$: We want to have $f(u)\in H^\#$, but also $\varphi(f(u))=f(\varphi(u))$ for all $f\in Hol^\#(D)$. Thus the same reason as above requires that $|\varphi(u)|$ be <1. In particular, once more the constant functions $u=\text{const}=c$ with $|c|=1$ cannot be admitted.

Let us now turn to positive results. We first show that we do not have to exclude other functions $u\in H$ with $|u|\leqq 1$ than the above constants.

3.1 REMARK: Let $u \in H$ with $|u| \leq 1$. Then $|\varphi(u)|=1 \Leftrightarrow u=\text{const}=c$ with $|c|=1$.

Proof: We have to prove $\Rightarrow$. Let $\varphi(u)=:c$ with $|c|=1$. Then $0=1-\bar{c}c=1-\bar{c}\varphi(u)=$ $=\varphi(1-\bar{c}u)=\int(1-\bar{c}u)Fdm$ and $\mathrm{Re}\,\bar{c}u \leq |\bar{c}u|=|u| \leq 1$ imply that $\mathrm{Re}(1-\bar{c}u)=0$ or $\mathrm{Re}\,\bar{c}u=1$, provided that $F \in M$ is chosen to be dominant over X. From this and from $|\bar{c}u| \leq 1$ we obtain $\mathrm{Im}\,\bar{c}u=0$. Thus $\bar{c}u=1$ or $u=c$. QED.

3.2 PROPOSITION: Let $u \in H$ with $|u| \leq 1$ and $|\varphi(u)| < 1$. Then

$$(Vm)([u \in B]) \leq \frac{1+|\varphi(u)|}{1-|\varphi(u)|}\lambda(B) \qquad \forall \text{ Baire sets } B \subset S \text{ and } \forall\, V \in M.$$

In particular $m([u \in B])=0$ for all Baire sets $B \subset S$ with $\lambda(B)=0$.

Proof: i) The set function $B \mapsto m([u \in B])$ is seen to be a measure on the Baire subsets of S. Since Baire measures are outer regular we can assume that B is an open subset of S and hence that it is an open interval $B=$ $=\{e^{it}: a<t<b\}$ with $\varepsilon<a<b<\varepsilon+2\pi$ for some real ε. ii) The function

$$f(e^{it}) = \left\{\begin{array}{ll} 1 & \text{for } a<t<b \\ \frac{1}{2} & \text{for } t=a \text{ and } t=b \\ 0 & \text{for } \varepsilon \leq t<a \text{ and } b<t \leq \varepsilon+2\pi \end{array}\right\}$$

defines an $F \in \mathrm{Re}\,L^{\infty}(\lambda)$ with $0 \leq F \leq 1$ and hence $f:=\langle F\lambda \rangle \in \mathrm{Re}\,\mathrm{Harm}^{\infty}(D)$ with $0 \leq f \leq 1$. From I.2.1 we have $f(Rs) \to f(s)$ for $R \uparrow 1$ in all points $s \in S$. We further introduce

$$h: h(z) = \int_S \frac{s+z}{s-z} f(s)d\lambda(s) \qquad \forall\, z \in D,$$

so that $h \in \mathrm{Hol}(D)$ and $f=\mathrm{Re}\,h$. For $0<R<1$ we can form $h(Ru) \in H$ and obtain

$$\int h(Ru)Vdm = \varphi\big(h(Ru)\big) = h\big(R\varphi(u)\big),$$
$$\int f(Ru)Vdm = f\big(R\varphi(u)\big) \qquad \forall\, V \in M.$$

For $R \uparrow 1$ it follows that $\int f(u)Vdm=f(\varphi(u))$ from dominated convergence. Now we have the estimations

$$f\big(\varphi(u)\big) = \int_S P\big(\varphi(u),s\big)f(s)d\lambda(s) \leq \frac{1+|\varphi(u)|}{1-|\varphi(u)|}\lambda(B),$$

$$\int f(u)Vdm \geq \int_{[u \in B]} f(u)Vdm = \int_{[u \in B]} Vdm = (Vm)([u \in B]).$$

The combination of all these implies the assertion. QED.

The next result is the basic preparation for the second half of the desired substitution theorem. The idea of proof is the same as above.

3.3 PROPOSITION: Let $u\in H$ with $|u|\leqq 1$ and $|\varphi(u)|<1$. Let $F\in L^1(\lambda)$ and $f=<F\lambda>\in Harm^1(D)$. Then $f(u)\in L(m)$ is well-defined after 3.2. We claim that

$$\int|f(u)|Vdm<\infty \text{ and } \int f(u)Vdm=f(\varphi(u)) \quad \forall\ V\in M.$$

Proof: We can assume that $F\geqq 0$ so that $f\geqq 0$ as well. Put $F_n:=\min(F,n)$ and $f_n:=<F_n\lambda>\in Harm^\infty(D)$ so that $0\leqq F_n\leqq n$ and $F_n\uparrow F$ and hence $0\leqq f_n\leqq n$ and $f_n\uparrow f$. We can find a Baire set $N\subset S$ with $\lambda(N)=0$ such that for all $s\in S-N$
i) the radial limits

$$\lim_{R\uparrow 1} f(Rs)=:f(s) \text{ and } \lim_{R\uparrow 1} f_n(Rs)=:f_n(s) \quad \forall\ n\geqq 1 \text{ exist,}$$

so that the functions $s\mapsto f(s)$ and $s\mapsto f_n(s)$ on $S-N$ are representatives of F and F_n, and ii) it is $f_n(s)=\min(f(s),n)$ $\forall n\geqq 1$. Thus $f_n(s)\uparrow f(s)$ $\forall s\in S-N$ for $n\to\infty$. Now choose a representative function $x\mapsto u(x)$ of $u\in H$ with $|u(x)|\leqq 1$ and $u(x)\notin N$ $\forall$ $x\in X$ which can be done after 3.2. Then for $0<R<1$ we obtain as in the proof of 3.2

$$\int f(Ru)Vdm=f(R\varphi(u)) \text{ and } \int f_n(Ru)Vdm=f_n(R\varphi(u)) \ \forall\ V\in M.$$

The first equation is of no direct use. From the other ones we obtain for $R\uparrow 1$ in view of dominated convergence $\int f_n(u)Vdm=f_n(\varphi(u))$ $\forall$ $V\in M$. For $n\to\infty$ thus $\int f(u)Vdm=f(\varphi(u))$ from Beppo Levi. QED.

3.4 SUBSTITUTION THEOREM: Assume that $u\in H$ with $|u|\leqq 1$ and $|\varphi(u)|<1$. Let $f\in Hol^\#(D)$. Then $f(u)\in L(m)$ is well-defined after 3.2. We claim that $f(u)\in H^\#$ and $\varphi(f(u))=f(\varphi(u))$. If $f\in(Hol^\#(D))^\times$ then $f(u)\in(H^\#)^\times$, and furthermore

$$\int|\log|f(u)||\ Vdm<\infty \text{ and } \int(\log|f(u)|)\ Vdm=\log|f(\varphi(u))|\ \forall V\in M.$$

Proof: i) Take functions $f_n\in Hol^\infty(D)$ with $|f_n|\leqq 1$, $f_n\to 1$ and $f_nf=:g_n\in$ $\in Hol^\infty(D)$. Then choose a Baire set $N\subset S$ with $\lambda(N)=0$ such that the radial limits

$$\lim_{R\uparrow 1} f_n(Rs)=:f_n(s) \text{ and } \lim_{R\uparrow 1} f(Rs)=:f(s)$$

$$\text{and hence } \lim_{R\uparrow 1} g_n(Rs) =: g_n(s) = f_n(s)f(s)$$

exist for all $s\in S-N$. Furthermore in view of

$$\int_{S-N} |f_n(s)-1|^2 d\lambda(s) \leq 2\bigl(1-\mathrm{Re}\, f_n(0)\bigr)\to 0$$

we can achieve after transition to an appropriate subsequence and to a larger N that $f_n(s)\to 1$ for $n\to\infty$ $\forall s\in S-N$. Let $x\mapsto u(x)$ be a representative function of $u\in H$ with $|u(x)|\leq 1$ and $u(x)\notin N$ $\forall$ $x\in X$. Then for $0<R<1$ it is clear that $f_n(Ru)$, $g_n(Ru)\in H$ with

$$\varphi\bigl(f_n(Ru)\bigr) = f_n\bigl(R\varphi(u)\bigr) \text{ and } \varphi\bigl(g_n(Ru)\bigr) = g_n\bigl(R\varphi(u)\bigr).$$

For $R\uparrow 1$ we obtain in view of dominated convergence $f_n(u)$, $g_n(u)\in H$ and

$$\varphi\bigl(f_n(u)\bigr) = f_n\bigl(\varphi(u)\bigr) \text{ and } \varphi\bigl(g_n(u)\bigr) = g_n\bigl(\varphi(u)\bigr).$$

Furthermore $|f_n(u)|\leq 1$ and $f_n(u)\to 1$. Since $f_n(u)f(u)=g_n(u)$ it follows that $f(u)\in H^\#$. And

$$f_n\bigl(\varphi(u)\bigr)\varphi\bigl(f(u)\bigr) = \varphi\bigl(f_n(u)\bigr)\varphi\bigl(f(u)\bigr) = \varphi\bigl(f_n(u)f(u)\bigr) =$$

$$= \varphi\bigl(g_n(u)\bigr) = g_n\bigl(\varphi(u)\bigr) = f_n\bigl(\varphi(u)\bigr)\, f\bigl(\varphi(u)\bigr)$$

shows that $\varphi(f(u))=f(\varphi(u))$.

ii) If $f\in(\mathrm{Hol}^\#(D))^\times$ then $f(u)\in(H^\#)^\times$ is obvious. Now from I.4.7 we have $f=ce^h$, where $c\in\mathbb{C}$ with $|c|=1$ and

$$h{:}h(z) = \int_S \frac{s+z}{s-z} G(s)d\lambda(s) \ \forall \ z\in D \quad \text{with } G\in \mathrm{Re}\, L^1(\lambda).$$

Then $\log|f|=\mathrm{Re}\, h = \langle G\lambda\rangle =: g\in\mathrm{Harm}^1(D)$. It is clear that $\log|f(u)|=g(u)$ for the substitutions $f(u)\in H^\#$ after i) and $g(u)\in L(m)$ after 3.3. Thus the last assertion follows from 3.3. QED.

In connection with the above theorem we recall from Section I.4 that all functions $f\in\mathrm{Hol}^+(D)$ are in $\mathrm{Hol}^\#(D)$ and hence (except the constant 0) in $(\mathrm{Hol}^\#(D))^\times$.

3.5 EXAMPLE: We consider functions $u \in H$ with $|u|=1$, the so-called inner functions. In 6.9 the existence of nonconstant inner functions will be established in all Szegö situations (except in the trivial case $H=\mathbb{C}$). Of course such functions exist in the unit disk situation (for example $u=Z$). i) Let $u \in H$ be a nonconstant inner function, so that $|\varphi(u)|<1$ from 3.1. Then for $c \in D$ the function

$$v := \frac{u-c}{1-\bar{c}u} = (u-c)\sum_{\ell=0}^{\infty}(\bar{c}u)^{\ell} \in H \text{ with } \varphi(v) = \frac{\varphi(u)-c}{1-\bar{c}\varphi(u)}$$

is a nonconstant inner function as well. If $c=\varphi(u)$ then $\varphi(v)=0$, and in case $\varphi(u)=0$ we have $\varphi(v)=-c$. Thus from each nonconstant inner function $u \in H$ we can produce for each $c \in D$ a nonconstant inner function $v \in H$ with $\varphi(v)=c$. ii) Let $u \in H$ be a nonconstant inner function. Note that the function

$$f: f(z) = \frac{1-z^2}{1+z^2} \quad \forall\, z \in D$$

is in $\mathrm{Hol}^{+}(D) \subset \mathrm{Hol}^{\#}(D)$. Thus it follows from 3.4 that

$$f(u) = \frac{1-u^2}{1+u^2} = \frac{1}{i}\,\frac{\mathrm{Im}\,u}{\mathrm{Re}\,u} \in H^{\#} \text{ with } \varphi\big(f(u)\big) = \frac{1-(\varphi(u))^2}{1+(\varphi(u))^2}.$$

So we obtain a nonconstant real-valued function $\frac{\mathrm{Im}\,u}{\mathrm{Re}\,u} \in H^{\#}$. Since $\varphi(u)$ attains all numbers in D the value $\varphi(\frac{\mathrm{Im}\,u}{\mathrm{Re}\,u})$ attains all numbers in the open upper halfplane. Furthermore $(\frac{\mathrm{Im}\,u}{\mathrm{Re}\,u})^2 \in H^{\#}$ is a positive function in $H^{\#}$ and $\varphi\big((\frac{\mathrm{Im}\,u}{\mathrm{Re}\,u})^2\big)$ attains all complex numbers except the real numbers $\geqq 0$.

An important consequence is the subsequent value distribution theorem.

3.6 THEOREM: Assume that $E \subset I \subset S$ where E is a Baire set with $\lambda(E)>0$ and $I=\{e^{it}: a<t<b\}$ an open interval. Let $u \in H$ with $|u| \leqq 1$ and $|\varphi(u)|<1$ fulfill

$$m([u \in E]) = 0,$$
$$m([u \in U \cap D]) = 0 \text{ for some open } U \subset \mathbb{C} \text{ with } I \subset U.$$

Then $m([u \in I])=0$.

Proof: i) Let us fix $V \in M$. Then

$$G: G(z) = -\int_{[|u|<1]} \frac{u+z}{u-z}\, V dm \quad \forall\, z \in U$$

is a holomorphic function on U. Therefore Re G is a real-analytic function on I, that is $t \mapsto \operatorname{Re} G(e^{it})$ is real-analytic on $a<t<b$. Note that

$$\operatorname{Re} G(z) = \int_{[|u|<1]} \frac{1-|u|^2}{|z-u|^2} Vdm \quad \forall\ z \in I.$$

ii) Let $F \in L^1(\lambda)$ vanish outside of I and $f = \langle F\lambda \rangle \in \operatorname{Harm}^1(D)$. From 3.3 we have $\int |f(u)| Vdm < \infty$ and $\int f(u) Vdm = f(\varphi(u))$. Now

$$\int f(u) Vdm = \int_{[|u|=1]} F(u) Vdm + \int_{[|u|<1]} \left(\int_I \frac{1-|u|^2}{|s-u|^2} F(s) d\lambda(s) \right) Vdm$$

$$= \int_{[|u|=1]} F(u) Vdm + \int_I \left(\operatorname{Re} G(s) \right) F(s) d\lambda(s),$$

$$f(\varphi(u)) = \int_I \frac{1-|\varphi(u)|^2}{|s-\varphi(u)|^2} F(s) d\lambda(s),$$

where the Fubini theorem has been applied. If in particular F vanishes outside of E then

$$\int_{[|u|=1]} F(u) Vdm = 0 \text{ because of } m([u \in E]) = 0.$$

It follows that

$$\operatorname{Re} G(s) = \frac{1-|\varphi(u)|^2}{|s-\varphi(u)|^2} \text{ for } \lambda\text{-almost all } s \in E \text{ and hence for all } s \in I,$$

where the real-analyticity of Re G on I has been applied. But this implies that

$$\int_{[|u|=1]} F(u) Vdm = 0 \ \forall\ F \in L^1(\lambda) \text{ which vanish outside of I.}$$

For $F = \chi_I$ we obtain $(Vm)([u \in I]) = 0$ for all $V \in M$. Therefore $m([u \in I]) = 0$ since (H, φ) is reduced. QED.

3.7 EXAMPLE: Let $u \in H$ be a nonconstant inner function. For each Baire set $E \subset S$ with $\lambda(E) > 0$ then $m([u \in E]) > 0$. This follows at once from 3.6 applied to I=S and $U = \mathbb{C}$. Thus u attains all values of modulus one in quite a sharp sense.

4. The Function Class H^+

4.1 PROPOSITION: For $h \in L(m)$ with $\operatorname{Re} h \geq 0$ the subsequent properties are equivalent.

i) $e^{-th} \in H$ for all $t \geq 0$.

ii) $\frac{1}{h+s} \in H$ for all $s \in \Delta :=$ the open halfplane $\operatorname{Re} s > 0$.

iii) $\frac{1}{h+s} \in H$ for some $s \in \Delta$.

In this case 1) $h \in H^{\#}$ with $\operatorname{Re}\varphi(h) \geq 0$.

2) $\varphi(e^{-th}) = e^{-t\varphi(h)}$ $\forall\, t \geq 0$ and $\varphi(\frac{1}{h+s}) = \frac{1}{\varphi(h)+s} = \forall s \in \Delta$.

3) $\int (\operatorname{Re} h) V dm \leq \operatorname{Re}\varphi(h)$ for all $V \in M$.

4) If $\operatorname{Re}\varphi(h) = 0$ then h is a purely imaginary constant.

Proof: i)⇒ii) For each $s \in \Delta$ the pointwise integral

$$\frac{1}{h+s} = \int_0^\infty e^{-t(h+s)} dt \text{ is in } H,$$

since for $0 < R < \infty$ the integral $\int_0^R e^{-t(h+s)} dt$ is the limit of Riemannian sums

$$\sum_{\ell=1}^{r} e^{-t(\ell)(h+s)} \big(t(\ell) - t(\ell-1)\big) \in H \text{ with } 0 = t(0) < t(1) < \ldots < t(r) = R$$

under pointwise and bounded convergence and hence is $\in H$, and $\int_0^R e^{-t(h+s)} dt \to$ $\to \int_0^\infty e^{-t(h+s)} dt$ even in $L^\infty(m)$-norm for $R \uparrow \infty$. Furthermore $\varphi(e^{-th}) = e^{-tc}$ $\forall\, t \geq 0$ for some complex c with $\operatorname{Re} c \geq 0$, since $t \mapsto \varphi(e^{-th})$ in $t \geq 0$ is a continuous complex-valued function which satisfies the functional equation of the exponential function and is of modulus ≤ 1. The above approximations then show that $\varphi(\frac{1}{h+s}) = \frac{1}{c+s}$ for all $s \in \Delta$. ii)⇒i) For $t \geq 0$ we have

$$e^{-th} = \frac{1}{e^{th}} = \lim_{n\to\infty} \frac{1}{(1+\frac{t}{n}h)^n},$$

where the convergence is pointwise and bounded. ii)⇒iii) is trivial. iii)⇒ii) The set $\{s \in \Delta : \frac{1}{h+s} \in H\}$ is closed in Δ so that we have to prove that it is open. Let $\sigma \in \Delta$ with $\frac{1}{h+\sigma} \in H$. For $|s-\sigma| < \operatorname{Re}\sigma$ then $\left|\frac{s-\sigma}{h+\sigma}\right| \leq \frac{|s-\sigma|}{\operatorname{Re}\sigma} < 1$ and hence

$$\frac{1}{h+s} = \frac{1}{(h+\sigma)+(s-\sigma)} = \frac{1}{h+\sigma}\,\frac{1}{1+\frac{s-\sigma}{h+\sigma}} = \frac{1}{h+\sigma} \sum_{\ell=0}^{\infty} \left(\frac{s-\sigma}{h+\sigma}\right)^{\ell} \in H.$$

Thus i) ii) iii) have been proved to be equivalent. Let us now assume these properties and retain the complex c with $\operatorname{Re} c \geq 0$ as in i)⇒ii) above. 1) For $\varepsilon > 0$ we have

$$\frac{1}{1+\varepsilon h} \in H \text{ with } \left|\frac{1}{1+\varepsilon h}\right| \leq 1,\ \frac{1}{1+\varepsilon h} \to 1 \text{ for } \varepsilon \downarrow 0, \text{ and } \frac{h}{1+\varepsilon h} = \frac{1}{\varepsilon}\left(1 - \frac{1}{1+\varepsilon h}\right) \in H.$$

Therefore $h\in H^{\#}$. Then $1=\varphi((h+s)\frac{1}{h+s})=(\varphi(h)+s)\frac{1}{c+s}$ for $s\in\Delta$ implies that $\varphi(h)=c$. 2) is then clear. 3) For $V\in M$ and $t>0$ we have

$$e^{-t\mathrm{Re}c}=|e^{-tc}|=|\varphi(e^{-th})| = |\int e^{-th}Vdm| \leqq \int e^{-t\mathrm{Re}\,h}\,Vdm,$$

$$e^{-\mathrm{Re}c}\leqq (\int (e^{-\mathrm{Re}h})^{t}Vdm)^{\frac{1}{t}},$$

$$e^{-\mathrm{Re}c}\leqq \exp(\int (-\mathrm{Re}h)Vdm) \quad \text{for } t\downarrow 0 \text{ after II.5.1.}$$

Hence $\int (\mathrm{Re}h)Vdm \leqq \mathrm{Re}\,c$. 4) In case $\mathrm{Re}\,c=0$ we have $\mathrm{Re}h=0$ from 3). For $t>0$ then $|e^{-th}|=1$ and $|\varphi(e^{-th})|=|e^{-tc}|=1$ so that $e^{-th}=\mathrm{const}=c(t)$ from 3.1. It follows that $h=\mathrm{const}=c$. QED.

The function class H^{+} is defined to consist of the functions $h\in L(m)$ with $\mathrm{Re}h\geqq 0$ which possess the equivalent properties i) ii) iii) in 4.1. It is in close relation to the class of functions $u\in H$ with $|u|\leqq 1$ considered in the last section.

<u>4.2 PROPOSITION</u>: We have

$$H^{+} = \left\{\frac{1-u}{1+u}: u\in H \text{ with } |u|\leq 1 \text{ and } u\neq \text{ the constant } -1\right\}.$$

Proof: $\supset$) The function $f:f(z)=\frac{1-z}{1+z}\ \forall\ z\in D$ has $\mathrm{Re}\,f>0$ and is therefore in $\mathrm{Hol}^{\#}(D)$. And $\frac{1}{f+s}$ for $s\in\Delta$ is even in $\mathrm{Hol}^{\infty}(D)$. Thus for $u\in H$ with $|u|\leqq 1$ and $|\varphi(u)|<1$ we have $\frac{1}{f(u)+s}\in H$ for $s\in\Delta$ from 3.4 and hence $f(u)=\frac{1-u}{1+u}\in H^{+}$. The case $|\varphi(u)|=1$ where u is constant $\neq -1$ is trivial. $\subset$) For $h\in H^{+}$ we have $u:=\frac{1-h}{1+h}=\frac{2}{1+h}-1\in H$ with $|u|\leqq 1$ and $u\neq -1$, and one verifies that $h=\frac{1-u}{1+u}$. QED.

<u>4.3 RETURN to the UNIT DISK</u>: Consider the Hardy algebra situations $(H^{\infty}(D),\varphi_a)$ with $a\in D$ on $(S,\mathrm{Baire},\lambda)$. From IV.3.15 we know that $H^{\#}=H^{\#}(D)$ and the connection with $\mathrm{Hol}^{\#}(D)$. Now I.4.8 shows that $H^{+}=H^{+}(D)$. Thus we have incorporated $H^{+}(D)$ and $\mathrm{Hol}^{+}(D)$ into the abstract Hardy algebra theory.

<u>4.4 REMARK</u>: If $h\in H^{\#}$ with $\mathrm{Re}\,h\geqq 0$ satisfies $e^{|h|}\in L^{\#}$ then $h\in H^{+}$. This is immediate from 2.1. In particular $h\in H$ with $\mathrm{Re}\,h\geqq 0$ implies that $h\in H^{+}$. But for $h\in H^{\#}$ with $\mathrm{Re}\,h\geqq 0$ the restriction $e^{|h|}\in L^{\#}$ cannot be dropped: we have seen this in Section I.4 in the unit disk situation. And 3.5 shows that even $h\in H^{\#}$ with $h\geqq 0$ does not enforce that $\mathrm{Re}\varphi(h)\geqq 0$ so that $h\in H^{+}$ cannot

be concluded.

4.5 REMARK: Assume that $h_n \in H^+$ and $h_n \to h \in L(m)$ pointwise. Then $h \in H^+$ and $\varphi(h_n) \to \varphi(h)$.

4.6 REMARK: For $h \in L(m)$ the subsequent properties are equivalent.

i) $h \in H^+$.

ii) There exist functions $h_n \in H^+$ such that $h_n \to h$.

iii) There exist functions $h_n \in H$ with $\operatorname{Re} h_n \geqq 0$ such that $|h_n| \leqq |h|$ and $h_n \to h$.

Proof of 4.5: Obvious from the definition. Proof of 4.6:i)⇒iii) We can take the functions $h_n = \frac{nh}{n+h} = n - \frac{n^2}{n+h} \in H$. QED.

The next aim is to transfer the substitution theorem 3.4 into a theorem about H^+. To do this it is natural in view of 4.2 to transform the unit disk D onto the halfplane Δ via the fractional-linear map h: $h(s) = \frac{1-s}{1+s}$ $\forall s \in \mathbb{C}$ as in Section I.2. Let us write up the transition in all briefness.

The function h is equal to its inverse. It maps $\Delta \to D$ and $i\mathbb{R} \to S - \{-1\}$ with

$$h(ix) = e^{it} \Leftrightarrow x = -\tan\frac{t}{2} \text{ or } t = -2\arctan x \quad \forall x \in \mathbb{R} \text{ and } |t| < \pi.$$

1) The substitution formula

$$\frac{1}{2\pi}\int_{-\pi}^{\pi} f(e^{it})\,dt = \frac{1}{\pi}\int_{-\infty}^{\infty} f\big(h(ix)\big)\frac{dx}{1+x^2} \quad \forall \text{ bounded measurable } f: S \to \mathbb{C}$$

shows that the measure $\lambda: d\lambda(e^{it}) = \frac{1}{2\pi}dt$ on S is transformed into the measure $\Lambda: d\Lambda(ix) = \frac{1}{\pi}\frac{dx}{1+x^2}$ on $i\mathbb{R}$. Thus if ℓ denotes Lebesgue measure on $\mathbb{R}$, then for a Baire set $N \subset \mathbb{R}$ we have $\ell(N) = 0 \Leftrightarrow \Lambda(iN) = \lambda(h(iN)) = 0$.
2) Let $U \in H^+$ and $u \in H$ with $|u| \leqq 1$ and $u \neq -1$ such that $U = \frac{1-u}{1+u} = h(u)$ after 4.2. Then $\varphi(U) = \frac{1-\varphi(u)}{1+\varphi(u)} = h(\varphi(u))$ so that $\operatorname{Re}\varphi(U) > 0 \Leftrightarrow |\varphi(u)| < 1$. Let us assume this. For a Baire set $N \subset \mathbb{R}$ we have $[U \in iN] = [u = h(U) \in h(iN)]$ so that $\ell(N) = 0$ implies that $m([U \in iN]) = 0$ after 1) and 3.2. 3) Assume that $F \in \operatorname{Hol}(\Delta)$ and $f \in \operatorname{Hol}(D)$ such that $F = f \circ h: F(s) = f(z)$ $\forall$ corresponding $s \in \Delta$ and $z \in D$. Then $F \in \operatorname{Hol}^{\#}(\Delta) \Leftrightarrow f \in \operatorname{Hol}^{\#}(D)$ after Section I.4. 4) In corresponding boundary points $ix \in i\mathbb{R}$ and $h(ix) = e^{it} = :\tau \in S$ the radial limits

$$\lim_{\sigma\downarrow 0} F(\sigma+ix) =: F(ix) \text{ and } \lim_{R\uparrow 1} f(R\tau) =: f(\tau)$$

are not in obvious connection with each other. But from the conformal character of the map h it is clear that the existence of the angular limits

$$\lim_{\substack{s\to ix \\ s\in\Omega(ix,\alpha)}} F(s)=:F(ix) \text{ on } \Omega(ix,\alpha):=\{s\in\Delta : \mathrm{Re}(s-ix)\geq|s-ix|\cos\alpha\}\ \forall\, 0<\alpha<\tfrac{\pi}{2}$$

$$\text{and } \lim_{\substack{z\to\tau \\ z\in\omega(\tau,\alpha)}} f(z)=:f(\tau) \text{ on } \omega(\tau,\alpha) \text{ (as in I.4.2) } \forall\, 0<\alpha<\tfrac{\pi}{2}$$

is equivalent to each other, and then of course $F(ix)=f(\tau)=f(h(ix))$. We know from I.4.2 that in case of $\mathrm{Hol}^{\#}$ functions these limits exist for λ-almost all $\tau\in S$ and hence for ℓ-almost all $x\in\bar{\mathbb{R}}$. 5) Let us combine 2)-4): If $F\in\mathrm{Hol}^{\#}(\Delta)$ and $\mathrm{Re}\varphi(U)>0$ then $F(U)\in L(m)$ is well-defined in the same sense as $f(u)$ was in the last section and $F(U)=(f\circ h)(U)=f(h(U))=$ $=f(u)$. Thus $F(U)\in H^{\#}$ after 3.4. Furthermore

$$\varphi(F(U))=\varphi(f(u))=f(\varphi(u))=f(h(\varphi(U))=(f\circ h)(\varphi(U))=F(\varphi(U)).$$

We can therefore formulate the substitution theorem as follows.

<u>4.7</u> <u>SUBSTITUTION THEOREM</u>: Assume that $u\in H^{+}$ with $\mathrm{Re}\varphi(u)>0$. Let $f\in\mathrm{Hol}^{\#}(\Delta)$. Then $f(u)\in L(m)$ is well-defined and $f(u)\in H^{\#}$ with $\varphi(f(u))=f(\varphi(u))$. If $f\in(\mathrm{Hol}^{\#}(\Delta))^{\times}$ then $f(u)\in(H^{\#})^{\times}$, and furthermore

$$\int|\log|f(u)||Vdm<\infty \text{ and } \int(\log|f(u)|)Vdm = \log|f(\varphi(u))|\ \forall V\in M.$$

<u>4.8</u> <u>COROLLARY</u>: If in particular $\mathrm{Re}\, f\geq 0$ then $f(u)\in H^{+}$. The same holds true for the substitution theorem 3.4.

Proof: Apply the theorem to the functions $\frac{1}{f+s}$ with $s\in\Delta$. QED.

Let us apply the result to the main branch of the power function $f_\tau : f_\tau(s)=s^\tau$ $\forall s\in\Delta$ for $\tau\in\bar{\mathbb{R}}$. We have $f_\tau\in\mathrm{Hol}^{+}(\Delta)$ for $|\tau|\leq 1$ and hence $f_\tau\in$ $\mathrm{Hol}^{\#}(\Delta)$ for all $\tau\in\bar{\mathbb{R}}$. In particular the special case $\tau=-1$ demonstrates the power of the substitution theorem. We shall come back to the case $|\tau|<1$ in the next section.

<u>4.9</u> <u>PROPOSITION</u>: Let $u\in H^{+}$ be $\neq 0$. For all $\tau\in\bar{\mathbb{R}}$ then $u^\tau\in H^{\#}$ and $\varphi(u^\tau)=$ $=(\varphi(u))^\tau$. In particular $u^\tau\in H^{+}$ if $|\tau|\leq 1$.

4.10 SPECIAL CASE: Let $u \in H^+$ be $\neq 0$. Then $\frac{1}{u} \in H^+$.

In addition we see from 4.7 that $\int |\log|u||Vdm<\infty$ and $\int(\log|u|)Vdm= =\log|\varphi(u)|\ \forall V\in M$. We combine this with the application of 4.7 to the main branch $f: f(s)=\log s\ \forall s\in\Delta$ which is in $Hol^\#(\Delta)$ since $\frac{\pi}{2} \pm if$ has real part ≥ 0, and with IV.3.4 to obtain the subsequent result.

4.11 PROPOSITION: Let $u\in H^+$ be $\neq 0$. Then $\log u\in H^\#$ and $\varphi(\log u)=\log\varphi(u)$. Furthermore

$$\int|\log u|\ Vdm<\infty \text{ and } \int(\log u)\ Vdm=\log\varphi(u)\ \ \forall\ V\in M.$$

From 4.10 we obtain another remarkable closedness property of the class H^+.

4.12 REMARK: Let $u, v\in H^+$ with $\mathrm{Re}(uv)\geq 0$. Then $uv\in H^+$.

Proof: We can assume that $v\neq 0$. For $t>0$ then $u+\frac{t}{v}\in H^+$ from 4.10 and $\neq 0$ so that

$$\frac{1}{uv+t} = \frac{1}{v}\,\frac{1}{u+\frac{t}{v}} \in H^\# \text{ and hence } \in H^\#\cap L^\infty(m)=H,$$

once more from 4.10. Thus $uv\in H^+$. QED.

To conclude the section we transfer the value distribution theorem 3.6 into a theorem about H^+. It is clear that we can use the transition from Δ to D described above.

4.13 THEOREM: Assume that $E\subset I\subset\mathbb{R}$ where E is a Baire set with $\ell(E)>0$ and I an open interval. Let $u\in H^+$ with $\mathrm{Re}\varphi(u)>0$ fulfill

$$m([u\in iE]) = 0,$$

$$m([u\in U\cap\Delta]) = 0 \text{ for some open } U\subset\mathbb{C} \text{ with } iI\subset U.$$

Then $m([u\in iI])=0$.

The next section will be devoted to another important problem on the function class H^+. For $h\in H^+$ we know from 4.1 that $\int(\mathrm{Re}h)Vdm\leq\mathrm{Re}\varphi(h)<\infty$ for all $V\in M$. But it is not always true that $\int|h|Vdm<\infty$. Let us present an example in the Szegö situation $M=\{F\}$: If $u\in H$ is a nonconstant inner function then $h=\frac{1-u}{1+u}\in H^+$ with Reh=0. If $\int|h|Fdm$ were $<\infty$ then 2.3 would enforce that h=const and hence that u=const. So it must be $\int|h|Fdm=\infty$. However, the statement that $h\in L^1(Vm)$ for all $V\in M$ and $h\in H^+$ is not far remote from truth: in the next section it will be shown that the functions in H^+ all possess certain weak-L^1 properties.

5. Weak-L^1 Properties of the Functions in H^+

The section starts with a short excursion on distribution functions which is unrelated to the abstract Hardy algebra situation. Let us fix a function $0\leqq f\in L(m)$. The distribution function of f is defined to be

$$W = Wf:W(t) = m([f\geqq t]) \quad \text{for all real } t\geqq 0.$$

Thus W is monotone decreasing on $[0,\infty[$ with $W(0)=m(X)$ and $W(\infty)=0$. And W is continuous from the left in $]0,\infty[$.

5.1 REMARK: For $0<p<\infty$ we have

$$t^pW(t) \leqq \int_{[f\geqq t]} f^p dm \quad \forall\ t\geqq 0.$$

In particular $\int f^p dm<\infty$ implies that $t^pW(t)\to 0$ for $t\uparrow\infty$.

The proof is immediate. In view of the above remark the function f is called of weak-$L^p(m)$ type iff $t\mapsto t^pW(t)$ is bounded for $t\uparrow\infty$. Our aim is to prove the subsequent estimations which are in the opposite direction.

5.2 PROPOSITION: For $0<p<\infty$ we have

$$\frac{p-\tau}{p}\int f^\tau dm \leqq \big(m(X)\big)^{\frac{p-\tau}{p}} \Big(\frac{p}{\tau}\sup_{t\geqq 0} t^pW(t)\Big)^{\frac{\tau}{p}} \text{ for } 0<\tau<p,$$

$$\limsup_{\tau\uparrow p} \frac{p-\tau}{p}\int f^\tau dm \leqq \limsup_{t\uparrow\infty} t^pW(t).$$

In particular if f is of weak-$L^p(m)$ type then $\int f^\tau dm<\infty$ for $0<\tau<p$.

5.3 LEMMA: Let $F:[0,\infty[\to\mathbb{R}$ be continuous and $\geqq 0$. Then

$$\int F(f)dm = -\int_0^{\to\infty} F(t)dW(t).$$

Proof of 5.3: Let $R>0$ and $\tau:0=t_0<t_1<\ldots<t_r=R$ be a decomposition of $[0,R]$. Then

$$I(R):=\int_{[0\leqq f<R]} F(f)\,dm+\int_0^R F(t)\,dW(t)=\sum_{\ell=1}^{r}\left(\int_{[t_{\ell-1}\leqq f<t_\ell]} F(f)\,dm+\int_{t_{\ell-1}}^{t_\ell} F(t)\,dW(t)\right).$$

Here the first integral under the sum is $\leqq \mathrm{Max}(F|[t_{\ell-1},t_\ell])(W(t_{\ell-1})-W(t_\ell))$ and $\geqq \mathrm{Min}(F|[t_{\ell-1},t_\ell])(W(t_{\ell-1})-W(t_\ell))$ and hence $=F(\tau_\ell)(W(t_{\ell-1})-W(t_\ell))$ with $t_{\ell-1}\leqq\tau_\ell\leqq t_\ell$. Thus

$$I(R)=\sum_{\ell=1}^{r}\int_{t_{\ell-1}}^{t_\ell}(F(t)-F(\tau_\ell))\,dW(t)\ \text{ and hence }\ |I(R)|\leqq\omega(\delta(\tau))\,(W(0)-W(R)),$$

where $\delta(\tau)=\max(t_1-t_0,\ldots,t_r-t_{r-1})$ and ω is the modulus of continuity of the function F on $[0,R]$. For $\delta(\tau)\to 0$ we obtain $I(R)=0$. Then the assertion follows for $R\uparrow\infty$. QED.

Proof of 5.2: Define $D: D(s)=\mathrm{Sup}\{t^pW(t):t\geqq s\}$ for $s\geqq 0$. Then D is $\leqq\infty$ and monotone decreasing in $[0,\infty[$ with $D(s)\to\limsup_{t\uparrow\infty} t^pW(t)$ for $s\uparrow\infty$. Thus we can assume that $D(s)<\infty$ in $s\geqq 0$. Now for $0<\tau<p$ and $0<s<R$ we have

$$-\int_s^R t^\tau dW(t)=-R^\tau W(R)+s^\tau W(s)+\tau\int_s^R W(t)t^{\tau-1}dt\leqq\frac{D(s)}{s^{p-\tau}}+\tau D(s)\int_s^R t^{\tau-p-1}dt.$$

It follows for $s>0$ that

$$-\int_s^{\to\infty} t^\tau dW(t)\leqq\frac{D(s)}{s^{p-\tau}}+\tau D(s)\int_s^\infty t^{\tau-p-1}dt=\frac{p}{p-\tau}\,\frac{D(s)}{s^{p-\tau}},$$

$$\int f^\tau dm=-\int_0^{\to\infty} t^\tau dW(t)\leqq-\int_0^s t^\tau dW(t)+\frac{p}{p-\tau}\,\frac{D(s)}{s^{p-\tau}}\quad\text{from 5.3,}$$

$$\frac{p-\tau}{p}\int f^\tau dm\leqq\frac{p-\tau}{p}m(X)s^\tau+\frac{D(s)}{s^{p-\tau}}.$$

From the last relation we deduce at first

$$\limsup_{\tau\uparrow p}\frac{p-\tau}{p}\int f^\tau dm\leqq D(s)\ \ \forall s>0\ \text{ and hence }\leqq\limsup_{t\uparrow\infty} t^pW(t).$$

On the other hand we can assume that $f\neq 0$ and hence $D(0)>0$ and then put $s^p=\frac{p}{\tau}\frac{D(0)}{m(X)}$. It follows that

$$\frac{p-\tau}{p}\int f^\tau dm\leqq\left(\frac{p-\tau}{p}m(X)+\frac{D(0)}{s^p}\right)s^\tau=(m(X))^{\frac{p-\tau}{p}}\left(\frac{p}{\tau}D(0)\right)^{\frac{\tau}{p}}.\ \text{QED.}$$

We return to the reduced Hardy algebra situation (H,φ). A rather simple estimation shows that the functions in H^+ are all of weak-$L^1(Vm)$ type $\forall V\in M$. However, the subsequent precise limit relation is a much deeper result. It depends on the tauberian theorem I.2.5.

5.4 PROPOSITION: Let $h\in H^+$ with $\varphi(h)=a+ib$. For each $V\in M$ then

$$t(Vm)([|h-ib|\geq t]) \leq 2a \quad \forall\ t\geq 0.$$

Thus h is of weak-$L^1(Vm)$ type.

5.5 THEOREM: Let $h=P+iQ\in H^+$ with $\varphi(h)=a+ib$. For each $V\in M$ then

$$\frac{\pi}{2}t(Vm)([|h|\geq t]) \rightarrow a - \int PVdm \text{ for } t\uparrow\infty.$$

Proofs: i) For $t>0$ we have

$$\frac{h}{h+t} = 1-\frac{t}{h+t} \in H \text{ and } \varphi\left(\frac{h}{h+t}\right) = \frac{\varphi(h)}{\varphi(h)+t}.$$

For each $V\in M$ therefore

$$\int\frac{h}{h+t}Vdm = \frac{\varphi(h)}{\varphi(h)+t}, \quad \text{with real part}$$

$$\int\frac{|h|^2+tP}{|h|^2+2tP+t^2}Vdm = \frac{a^2+b^2+ta}{a^2+b^2+2ta+t^2} \quad \forall\ t>0.$$

ii) We prove 5.4 and can assume that $b=0$. Then from i) we obtain

$$\frac{a}{t} \geq \frac{a}{a+t} = \int\frac{|h|^2+tP}{|h|^2+2tP+t^2}Vdm \geq \int_{[|h|\geq t]}\frac{|h|^2+tP}{|h|^2+2tp+t^2}Vdm$$

$$\geq \frac{1}{2}\int_{[|h|\geq t]}Vdm=\frac{1}{2}(Vm)([|h|\geq t]) \quad \forall\ t>0,$$

which is the assertion. iii) For $t>0$ the difference

$$t\left(\frac{|h|^2}{|h|^2+t^2} - \frac{|h|^2+tP}{|h|^2+2tP+t^2}\right) = Pt^2\frac{|h|^2-t^2}{(|h|^2+t^2)(|h|^2+2tP+t^2)}$$

is of modulus $\leq P$ and tends $\rightarrow -P$ pointwise for $t\uparrow\infty$. Since

$$t\int\frac{|h|^2+tP}{|h|^2+2tP+t^2}Vdm \rightarrow a \quad \text{for } t\uparrow\infty$$

after i) and $\int PVdm<\infty$ we see that

$$t\int \frac{|h|^2}{|h|^2+t^2}Vdm \to a - \int PVdm \quad \text{for } t\uparrow\infty.$$

Now let W: $W(x)=(Vm)([|h|\geq x])$ $\forall x\geq 0$ be the distribution function of $|h|$ with respect to the measure Vm and $\phi:\phi(x)=W(\frac{1}{x})$ $\forall x>0$ and $\phi(0)=0$. Then ϕ is monotone increasing and bounded in $[0,\infty[$ and continuous in the origin. From 5.3 we obtain for $t>0$

$$t\int \frac{|h|^2}{|h|^2+t^2}Vdm = -\int_0^{\to\infty} \frac{tx^2}{x^2+t^2}dW(x) = -\int_{0\leftarrow}^{\to\infty} \frac{tx^2}{x^2+t^2}dW(x) =$$

$$= \int_{0\leftarrow}^{\to\infty} \frac{t(\frac{1}{x})^2}{(\frac{1}{x})^2+t^2}d\phi(x) = \int_{0\leftarrow}^{\to\infty} \frac{\frac{1}{t}}{(\frac{1}{t})^2+x^2}d\phi(x) = \int_{0}^{\to\infty} \frac{\frac{1}{t}}{(\frac{1}{t})^2+x^2}d\phi(x).$$

We know that this tends $\to a - \int PVdm$ for $\frac{1}{t}\downarrow 0$. Thus the Loomis theorem I.2.5 says that $\frac{\pi}{2}\frac{1}{x}\phi(x)\to a-\int PVdm$ for $x\downarrow 0$ which means that $\frac{\pi}{2}xW(x)\to a-\int PVdm$ for $x\uparrow\infty$. QED.

Let us add another version of the almost-$L^1(Vm)$ character of the functions in H^+. As before we present a rather simple estimation and a deeper theorem of tauberian nature.

<u>5.6 PROPOSITION</u>: Let $h\in H^+$ be $\neq 0$. For each $V\in M$ and each real τ with $|\tau|<1$ then

$$\cos\frac{\tau\pi}{2}\int|h|^\tau Vdm \leq \mathrm{Re}\left(\varphi(h)\right)^\tau,$$

of course with $(\varphi(h))^\tau$ from the main branch of the power function.

Proof: In view of 4.10 we can restrict ourselves to the case $0<\tau<1$. The result could be deduced from 5.4 and 5.2 with a worse constant. We prefer to base the proof on the substitution theorem 4.7. i) Let $s\in\mathbb{C}$ with $\mathrm{Re}\, s\geq 0$. Thus $s=|s|e^{it}$ with $|t|\leq\frac{\pi}{2}$ and hence $s^\tau=|s|^\tau e^{i\tau t}$ for the main branch. It follows that $\mathrm{Re}\, s^\tau=|s|^\tau\cos\tau t\geq|s|^\tau\cos\frac{\tau\pi}{2}$. ii) For $\varepsilon>0$ we have $\frac{h}{1+\varepsilon h}\in H$ with real part ≥ 0 and hence $\frac{h}{1+\varepsilon h}\in H^+$. Thus from 4.9 we obtain that

$$\left(\frac{h}{1+\varepsilon h}\right)^\tau\in H \text{ and } \varphi\left(\left(\frac{h}{1+\varepsilon h}\right)^\tau\right) = \left(\frac{\varphi(h)}{1+\varepsilon\varphi(h)}\right)^\tau.$$

For each $V\in M$ we thus have

$$\cos\frac{\tau\pi}{2}\int\left|\frac{h}{1+\varepsilon h}\right|^\tau Vdm \leq \int\mathrm{Re}\left(\frac{h}{1+\varepsilon h}\right)^\tau Vdm = \mathrm{Re}\left(\frac{\varphi(h)}{1+\varepsilon\varphi(h)}\right)^\tau.$$

Now let $\varepsilon\downarrow 0$ and apply the Fatou theorem to obtain the result. QED.

5.7 EXAMPLE: Let $u\in H$ be a nonconstant inner function. Then $h=\frac{1-u}{1+u}\in H^+$ is nonconstant with $\operatorname{Re} h=0$. In the Szegö situation $M=\{V\}$ we know from 2.3 that $\int|h|Vdm$ must be $=\infty$. Now 5.6 shows that $\int|h|^\tau Vdm<\infty$ for all $0<\tau<1$. Thus we have an example as announced at the end of Section 2.

5.8 THEOREM: Let $h=P+iQ\in H^+$ with $\varphi(h)=a+ib$. For each $V\in M$ then

$$\cos\frac{\tau\pi}{2}\int|h|^\tau Vdm \to a-\int PVdm \text{ for } \tau\uparrow 1.$$

5.9 LEMMA: Let $s=u+iv\in\mathbb{C}$ with $u\geq 0$ and $0<\tau\leq 1$. Then for the main branch of the power function $|s^\tau-(iv)^\tau|\leq u^\tau$.

Proof of 5.9: We can assume that $u>0$ and $v\neq 0$. Then we have

$$s^\tau-(iv)^\tau = \tau\int_0^u (t+iv)^{\tau-1}dt,$$

$$|s^\tau-(iv)^\tau| \leq \tau\int_0^u |t+iv|^{\tau-1}dt \leq \tau\int_0^u t^{\tau-1}dt=u^\tau. \text{ QED.}$$

Proof of 5.8: i) From 5.2 and 5.5 we obtain

$$\limsup_{\tau\uparrow 1} \cos\frac{\tau\pi}{2}\int|h|^\tau Vdm = \limsup_{\tau\uparrow 1}\frac{\pi}{2}(1-\tau)\int|h|^\tau Vdm$$

$$\leq \limsup_{t\uparrow\infty}\frac{\pi}{2} t\,(Vm)([|h|\geq t]) = a-\int PVdm.$$

ii) From 4.9 and from 5.6 combined with IV.3.4 we see for $0<\tau<1$ that $(\varphi(h))^\tau=\varphi(h^\tau)=\int h^\tau Vdm$ and hence

$$\operatorname{Re}(\varphi(h))^\tau = \int(\operatorname{Re} h^\tau)Vdm \leq \int(\operatorname{Re}(iQ)^\tau+P^\tau)Vdm \text{ after 5.9}$$

$$= \cos\frac{\tau\pi}{2}\int|Q|^\tau Vdm+\int P^\tau Vdm \leq \cos\frac{\tau\pi}{2}\int|h|^\tau Vdm + \left(\int PVdm\right)^\tau.$$

It follows that

$$a = \operatorname{Re}\varphi(h) \leq \liminf_{\tau\uparrow 1} \cos\frac{\tau\pi}{2}\int|h|^\tau Vdm + \int PVdm.$$

From i) and ii) we obtain the assertion. QED.

The connection imposed upon the functions $0\leq P\in L(m)$ and $Q\in\operatorname{Re} L(m)$ in form of the relation $h:=P+iQ\in H^+$ is a rather weak one: in particular it can happen that $P=0$ and $Q\neq 0$ whenever the existence of nonconstant inner functions $\in H$ has been proved. In the unit disk situation the connection has been made explicit at the end of Section I.4. In Section VI.2 it will be obvious from the respective definitions that the connection is much weaker than the requirement that $0 \leqq P \in L(m)$ be conjugable and $Q \in \operatorname{Re} L(m)$ be its conjugate function (up to an additive real constant). For the latter pairs of functions P and Q in the unit disk situation the above results 5.4 and 5.5 as well as 5.6 and 5.8 are well-known classical theorems due to Kolmogorov. Of course these theorems can be expected to extend to conjugate functions in the abstract Hardy algebra situation. But it is a surprise to see them valid under assumptions which are much weaker even in the unit disk situation.

6. Value Carrier and Lumer Spectrum

The present section is to obtain further results on the value distribution of the functions in H and $H^{\#}$. The first part is devoted to preparations which are unrelated to the abstract Hardy algebra situation.

For a function $f \in L(m)$ we define the value carrier

$$\omega(f) := \{z \in \mathbb{C} : m([|f-z|<\delta]) > 0 \ \forall \delta > 0\},$$

$$\mathbb{C} - \omega(f) := \{z \in \mathbb{C} : |f-z| \geqq \delta \text{ for some } \delta > 0\}.$$

We list some simple properties. i) $\omega(f)$ is closed. ii) There exists a representative function $x \mapsto f(x)$ such that $f(x) \in \omega(f)$ for all $x \in X$. iii) $\omega(f) \neq \emptyset$. iv) If $\omega(f) = \{c\}$ then $f = \text{const} = c$.

6.1 LEMMA: Let $G \subset \mathbb{C}$ be a domain and $a \in \mathbb{C}$ in the exterior of G. Let $h \in L^{\circ}(m)$. Assume that

$$\int \log \left|\frac{h-z}{a-z}\right| dm \leqq 0 \quad \forall z \in \partial G.$$

Then the same holds true $\forall z \in G$. And if $\int \log \left|\frac{h-t}{a-t}\right| dm = 0$ in some point $t \in G$ then $\omega(h) \cap G = \emptyset$.

Proof: i) Let us fix $t \in G$ such that $\int \log \left|\frac{h-t}{a-t}\right| dm > -\infty$ and hence $\log \left|\frac{h-t}{a-t}\right| \in L^1(m)$. Choose a representative function $x \mapsto h(x)$ with $h(x) \neq t$ for all $x \in X$. ii) Let K and B denote closure and boundary of G relative to the Riemann sphere. Let $A \subset C(K)$ consist of those functions which are holomorphic on G. In view of the maximum modulus principle the restriction $f \mapsto f|B$ is a supnorm isomorphism $A \to A|B$. Let $\theta \in \mathrm{Prob}(B)$ be a Jensen measure for $f|B \mapsto f(t)$ on $A|B$ after III.1.1. For each $u \in \mathbb{C}$ the function $z \mapsto \frac{u-z}{a-z} = 1 - \frac{u-a}{z-a}$ is in A since $a \notin K$. Thus we have

$$\log\left|\frac{u-t}{a-t}\right| \leqq \int \log\left|\frac{u-z}{a-z}\right| d\theta(z) \quad \forall\ u \in \mathbb{C}.$$

We claim that for all $u \in G$ the strict $<$ relation holds true. To see this note that $u \mapsto \int \log\left|\frac{u-z}{a-z}\right| d\theta(z)$ is a continuous real-valued function on G which satisfies the mean value equation and hence is harmonic on G. Therefore

$$u \mapsto \int \log\left|\frac{u-z}{a-z}\right| d\theta(z) - \log\left|\frac{u-t}{a-t}\right|$$

is a harmonic function $\geqq 0$ on $\{u \in G : u \neq t\}$ which tends $\to \infty$ for $u \to t$. In view of the maximum principle it must therefore be > 0 in all points $u \neq t$ in G as we have claimed. iii) From the above we have

$$(*) \qquad \log\left|\frac{h(x)-t}{a-t}\right| \leqq \int \log\left|\frac{h(x)-z}{a-z}\right| d\theta(z) \quad \forall\ x \in X,$$

where integration extends over B or over ∂G. We want to integrate $(*)$ with respect to m and use the Fubini theorem. To see that this is correct we write

$$\int \left(\log\left|\frac{h(x)-z}{a-z}\right|\right)^- d\theta(z) \leq -\log\left|\frac{h(x)-t}{a-t}\right| + \int \left(\log\left|\frac{h(x)-z}{a-z}\right|\right)^+ d\theta(z),$$

consider the estimation

$$\left(\log\left|\frac{h(x)-z}{a-z}\right|\right)^+ = \left(\log\left|1 - \frac{h(x)-a}{z-a}\right|\right)^+ \leqq \left(\log(1+c|h(x)-a|)\right)^+ \ \forall\ z \in B,$$

with $c>0$ a suitable constant, and take into account the assumption $\log\left|\frac{h-t}{a-t}\right| \in L^1(m)$. It follows that

$$\int \log\left|\frac{h-t}{a-t}\right| dm \leqq \int \left(\int \log\left|\frac{h-z}{a-z}\right| dm\right) d\theta(z) \leqq 0.$$

If in particular the integral on the left is $=0$ then $(*)$ must have been an equality for m-almost all $x \in X$. Then ii) implies that $\{x \in X : h(x) \in G\}$ must be an m-null set. Therefore $\omega(h) \cap G = \emptyset$ since G is open. QED.

We further need a simple fact on connectedness in metric spaces.

6.2 LEMMA: Let E be a metric space and $T\subset E$. Assume that for each pair of disjoint (!) open subsets $U,V\subset E$ we have: $T\subset U\cup V \Rightarrow T\subset U$ or $T\subset V$. Then T is connected.

Proof: Assume that T is not connected. Then there exist nonvoid $P,Q\subset T$ with $T=P\cup Q$ and $\bar{P}\cap\bar{Q}\cap T=\emptyset$. Thus $P\cap\bar{Q}=\bar{P}\cap Q=\emptyset$. Let us put

$$U:=\bigcup_{u\in P} V\left(u,\tfrac{1}{2}\delta(u,\bar{Q})\right) \text{ and } V:=\bigcup_{v\in Q} V\left(v,\tfrac{1}{2}\delta(v,\bar{P})\right),$$

where δ is the metric of E and $V(a,\alpha)$ is the open ball of center $a\in E$ and radius $\alpha>0$. Then $U,V\subset E$ are open with $P\subset U$ and $Q\subset V$, so that $T\subset U\cup V$. It remains to show that $U\cap V=\emptyset$. Indeed for $x\in U\cap V$ we have $\delta(u,x)<\frac{1}{2}\delta(u,\bar{Q})$ for some $u\in P$ and $\delta(v,x)<\frac{1}{2}\delta(v,\bar{P})$ for some $v\in Q$. It follows that $\delta(u,v)\leqq\delta(u,x)+$ $+\delta(v,x)<\frac{1}{2}\delta(u,\bar{Q})+\frac{1}{2}\delta(v,\bar{P})\leqq\frac{1}{2}\delta(u,v)+\frac{1}{2}\delta(v,u)=\delta(u,v)$ which proves that $U\cap V=\emptyset$. We thus arrive at a contradiction. QED.

6.3 PROPOSITION: Let $h\in L^{O}(m)$ and $a\in\mathbb{C}$. Then the set

$$\Delta^{+}:=\{a\}\cup\{z\in\mathbb{C}: z\neq a \text{ and } \int\log\left|\frac{h-z}{a-z}\right|dm>0\}$$

is connected. And the set

$$\Delta:=\{a\}\cup\{z\in\mathbb{C}: z\neq a \text{ and } \int\log\left|\frac{h-z}{a-z}\right|dm\neq 0\}$$

satisfies $\omega(h)\subset\bar{\Delta}$.

Proof: i) Let $U,V\subset\mathbb{C}$ be disjoint open sets with $\Delta^{+}\subset U\cup V$. Assume that $a\in U$. After 6.2 it suffices to prove $\Delta^{+}\cap V=\emptyset$. To see this let G be a component of V. Then G is a domain in $\mathbb{C}$ and a in the exterior of G. Now ∂G is disjoint to V and to U and hence $\partial G\cap\Delta^{+}=\emptyset$. Then 6.1 implies that $G\cap\Delta^{+}=\emptyset$ as well. It follows that $V\cap\Delta^{+}=\emptyset$. ii) Assume that $t\notin\bar{\Delta}$. Then there exists $\delta>0$ such that the closed disk $\{z:|z-t|\leqq\delta\}$ is disjoint with Δ. Thus application of 6.1 to $G:=\{z:|z-t|<\delta\}$ implies that $\omega(h)\cap G=\emptyset$ and in particular $t\notin\omega(h)$. QED.

After these preparations we return to the reduced Hardy algebra situation (H,φ). For a function $h\in H^{\#}$ we define the Lumer spectrum $\Delta(h)$ to consist of the complex numbers $z\in\mathbb{C}$ such that $h-z\notin(H^{\#})^{\times}$. Then in particular $\varphi(h)\in\Delta(h)$ so that $\Delta(h)\neq\emptyset$. We want to prove several results on the relations between $\omega(h)$ and $\Delta(h)$. Note that 6.4 and 6.5 do not depend on the above preparations.

<u>6.4</u> <u>PROPOSITION</u>: For each $h\in H^{\#}$ we have $\partial\Delta(h)\subset\omega(h)$.

Proof: i) We know that $\mathbb{C}-\omega(h)=\{z\in\mathbb{C}:|h-z|\geq \text{ some } \delta>0\}$ is open. Let us put $\mathbb{C}-\omega(h)=P\cup Q$ with

$$P:=\{z\notin\omega(h): \frac{1}{h-z}\in H\} \text{ and } Q:=\{z\notin\omega(h):\frac{1}{h-z}\notin H\}.$$

It is obvious that Q is open. To see that P is open let $s\in P$ with $|h-s|\geq\delta>0$. For $z\in\mathbb{C}$ with $|z-s|\leq\frac{\delta}{2}$ then $|h-z|\geq\frac{\delta}{2}$ so that $z\notin\omega(h)$. And in view of $|\frac{z-s}{h-s}|\leq\frac{1}{2}$ we obtain

$$\frac{1}{h-z}=\frac{1}{(h-s)-(z-s)}=\frac{1}{h-s}\,\frac{1}{1-\frac{z-s}{h-s}}=\frac{1}{h-s}\sum_{\ell=0}^{\infty}\left(\frac{z-s}{h-s}\right)^{\ell}\in H,$$

so that $z\in P$ and hence in fact P is open as well. ii) Now for $s\in\mathbb{C}-\omega(h)$ we have $s\in\Delta(h)\Leftrightarrow\frac{1}{h-s}\notin H^{\#}\Leftrightarrow\frac{1}{h-s}\notin H\Leftrightarrow s\in Q$. Therefore $\Delta(h)\subset Q\cup\omega(h)=\mathbb{C}-P$ which is closed and hence $\overline{\Delta(h)}\subset Q\cup\omega(h)$. Furthermore $Q\subset\Delta(h)$ and hence $Q\subset(\Delta(h))^{\circ}$. It follows that $\partial\Delta(h)$ is contained in $\omega(h)$. QED.

<u>6.5</u> <u>PROPOSITION</u>: Assume that $h\in H^{\#}$ is nonconstant and $e^{|h|}\in L^{\#}$. Then $\Delta(h)\subset(\operatorname{conv}\omega(h))^{\circ}$, where conv denotes the convex hull.

Proof: Assume that $a\in\mathbb{C}$ is not in $(\operatorname{conv}\omega(h))^{\circ}$. Then there is a sequence of points $a_n\notin\operatorname{conv}\omega(h)$ which tends $\to a$. Take complex c_n with $|c_n|=1$ such that $\operatorname{Re} c_n(z-a_n)\geq 0$ $\forall z\in\operatorname{conv}\omega(h)$. We can assume that $c_n\to c$. Then $|c|=1$ and $\operatorname{Re} c(z-a)\geq 0$ $\forall z\in\operatorname{conv}\omega(h)$. Thus $\operatorname{Re} c(h-a)\geq 0$. Now we obtain $c(h-a)\in H^{+}$ from 4.4 and hence $\frac{1}{c(h-a)}\in H^{+}$ from 4.10. Thus $h-a\in(H^{\#})^{\times}$ or $a\notin\Delta(h)$. QED.

The next result is in the opposite direction.

<u>6.6</u> <u>PROPOSITION</u>: Assume that there exists an $F\in MJ$ which is dominant on X. Then for each $h\in H^{\#}$ we have $\omega(h)\subset\overline{\Delta(h)}$.

Proof: We want to apply 6.3 to h and $a:=\varphi(h)$, but to the measure Fm instead of m. This is correct since in view of $[F>0]=X$ the value carrier $\omega(h)$ is the same relative to Fm as relative to m. We can assume that $\Delta(h)\neq\mathbb{C}$. Let $z\notin\Delta(h)$. Then $h-z\in(H^{\#})^{\times}$ so that IV.3.5.i) implies that

$$\int|\log|h-z||\,Fdm<\infty \text{ and } \log|\varphi(h)-z|=\int(\log|h-z|)Fdm.$$

Thus we have $h \in L^{O}(Fm)$. And $z \notin \Delta$ where Δ is as in 6.3. It follows that $\Delta \subset \Delta(h)$. Thus $\omega(h) \subset \bar{\Delta} \subset \overline{\Delta(h)}$ from 6.3. QED.

6.7 PROPOSITION: Assume the Szegö situation $M=\{F\}$. Then $\Delta(h)$ is connected for each $h \in H^{\#}$.

Proof: In view of 6.3 applied to $h \in H^{\#} \subset L^{\#} = L^{O}(Fm)$ and $a := \varphi(h)$ and to the measure Fm instead of m we have to prove that

$$\Delta(h) = \{\varphi(h)\} \cup \{z \neq \varphi(h) : \log|\varphi(h)-z| < \int (\log|h-z|) F dm\}.$$

But this is immediate from 1.7. QED.

The power of the present method becomes visible with the subsequent important theorem.

6.8 THEOREM: Assume the Szegö situation $M=\{F\}$. Let $h \in H^{\#}$ be such that $h\bar{u} \in H^{\#}$ for all inner functions $u \in H$. Then h=const.

6.9 COROLLARY: Assume the Szegö situation $M=\{F\}$ and $H \neq \mathbb{C}$. Then there exist nonconstant inner functions in H.

Proof of 6.8: For each complex $z \neq \varphi(h)$ we have $\varphi(h-z) \neq 0$ and hence from 1.8 the factorization $h-z = u_z h_z$ with $u_z \in H$ an inner function and $h_z \in (H^{\#})^{\times}$. It follows that $h\bar{u}_z - z\bar{u}_z = h_z$ and hence $z\bar{u}_z \in H$ from the assumption. For $z \neq 0$ therefore $\bar{u}_z = \frac{1}{u_z} \in H$ so that u_z=const after 1.3. It follows that $h-z \in (H^{\#})^{\times}$ $\forall z \neq 0, \varphi(h)$ and hence $\Delta(h) \subset \{0, \varphi(h)\}$. Now from 6.7 and 6.6 we obtain $\omega(h) = \{\varphi(h)\}$ and hence h=const (it is clear that 6.6 alone would suffice as well). QED.

6.10 EXAMPLE: Let us determine $\omega(h)$ and $\Delta(h)$ for a nonconstant inner function $h \in H$. We have $\omega(h) \subset S$ and hence $\Delta(h) \subset D$ from 6.5 and $\partial\Delta(h) \subset S$ from 6.4. It follows that $\Delta(h) = D$. Hence $S = \partial\Delta(h) \subset \omega(h)$ from 6.4 so that $\omega(h) = S$. But note that the former 3.7 is a somewhat sharper statement.

Notes

Outer functions and factorization were central themes in the literature within the Szegö situation frame. The introduction of the class $H^{\#}$

freed the theory from the technical burden of permanent labour with H^p inclusions. 1.3 in the present form is in KÖNIG [1967c]. The characterization 1.10 of $H^\#$ in the Szegö situation is in KÖNIG [1967a](with definition and result reversed).

The substitution theorem 2.1 for the Szegö situation is in KÖNIG [1967a]. It was the reason for the introduction of the class $H^\#$. Anterior to it was LUMER [1965] with the idea of substitution theorems and particular cases. The full theorem 2.1 and the consequence 2.3 are in KÖNIG [1967c].

The class H^+ was introduced and most of the main results of Sections 3-5 were announced in KÖNIG [1970a]. The proofs appear in the present volume for the first time. The main exception is the value distribution theorem 3.6=4.13 which is the principal result in YABUTA [1973b]. The subsequent development of his research is in YABUTA [1974a][1974b][1975] [1976]. Several properties of the class H^+ have been elaborated in cooperation with Yabuta, see YABUTA [1973a][1973b]. The important special case 4.10 appeared before in various contexts and abstractions, for example in HOFFMAN [1962a] p.76 and in KÖNIG [1967a].

The classical versionsof the main results of Section 5 are in KOLMOGOROV [1925]. We refer to their presentation in KATZNELSON [1968].

The Lumer spectrum was introduced (under the name of inner spectrum) and the essential of 6.5 was announced in LUMER [1965]. Then 6.6 appeared in KÖNIG [1967a][1967c] and 6.7 in KÖNIG [1966b]. Theorem 6.8 (restricted to L^1 functions which is unessential) is one of the main results in KÖNIG [1965]. For the connectedness lemma 6.2 we have been supplied with independent proofs from our topological friends T.tom Dieck and R. Fritsch.

Chapter VI

The Abstract Conjugation

In the unit disk situation the classical conjugation is the operation which associates with each $p \in \mathrm{ReHarm}(D)$ the unique function $q \in \mathrm{ReHarm}(D)$ such that $p+iq \in \mathrm{Hol}(D)$ and $q(0)=0$. In order to extend the conjugation to the abstract Hardy algebra situation we have to redefine it as an operation which takes place on the unit circle S: that is which associates with each P from a certain subclass E of $\mathrm{ReL}(\lambda)$ a unique function $Q\in\mathrm{ReL}(\lambda)$. The immediate idea to define it via $P+iQ\in H^{\#}(D)$ plus some normalization of Q is bound to fail since there are lots of nonconstant real-valued functions in $H^{\#}(D)$. And there is no obvious idea how to restrict E and $H^{\#}(D)$ in order to escape from these nonconstant functions and still preserve a not too narrow definition. So let us seek to transplant the initial definition from D to S.

For each $P\in\mathrm{ReL}^1(\lambda)$ we can form $p:=\langle P\lambda\rangle\in\mathrm{ReHarm}^1(D)$. Then the classical conjugate function $q\in\mathrm{ReHarm}(D)$ is obtained from

$$(p+iq)(z) = \int_S \frac{s+z}{s-z}\, P(s)d\lambda(s) \quad \forall\ z\in D.$$

We have $p+iq\in\mathrm{Hol}^+(D)-\mathrm{Hol}^+(D)\subset\mathrm{Hol}^{\#}(D)$ and hence a boundary function $P+iQ\in H^{\#}(D)$. Of course we define $Q\in\mathrm{ReL}(\lambda)$ to be the conjugate function of $P\in\mathrm{ReL}^1(\lambda)$. But how can the above transit through D be avoided? To see this recall from I.4.4 that $e^{p+iq}\in\mathrm{Hol}^{\#}(D)$ and also $e^{t(p+iq)}\in\mathrm{Hol}^{\#}(D)$ for all $t\in\mathbb{R}$. Therefore $e^{t(P+iQ)}\in H^{\#}(D)$ for all $t\in\mathbb{R}$. But the latter condition determines $Q\in\mathrm{ReL}(\lambda)$ up to an additive real constant: In fact, if $G\in\mathrm{ReL}(\lambda)$ were such that $e^{t(P+iG)}\in H^{\#}(D)$ for all $t\in\mathbb{R}$, then the functions $e^{it(G-Q)}$ were $\in H^{\#}(D)$ and hence invertible elements of $H^{\#}(D)$ of modulus one, so that $e^{it(G-Q)}=\mathrm{const}=c(t)\ \forall\ t\in\mathbb{R}$ and hence $G-Q=\mathrm{const}$. Thus for $P\in\mathrm{ReL}^1(\lambda)$ the conjugate function can be defined to be the unique $Q\in\mathrm{ReL}(\lambda)$ such that $e^{t(P+iQ)}\in H^{\#}(D)$ for all $t\in\mathbb{R}$ and such that the corresponding functions $\in\mathrm{Hol}^{\#}(D)$ all have values >0 in the origin, that is such that $\varphi_0\left(e^{t(P+iQ)}\right)>0$ for all $t\in\mathbb{R}$. It is clear that this definition can be extended to the abstract Hardy algebra situation. Furthermore we see that each function $P\in\mathrm{ReL}(\lambda)$ which is conjugable in the new sense must be in $\mathrm{ReL}^1(\lambda)$: In fact, from $e^{t(P+iQ)}\in H^{\#}(D)\forall t\in\mathbb{R}$ it follows that all these functions are in $(H^{\#}(D))^{\times}$ so that after I.4.7 the corresponding functions $h_t \in (Hol^{\#}(D))^{\times}$ have the representations

$$h_t(z) = \exp\left(\int_S \frac{s+z}{s-z} P_t(s)d\lambda(s)\right) \quad \forall z \in D \quad \text{with } P_t \in ReL^1(\lambda).$$

It follows that $\log|h_t| = \langle P_t \lambda \rangle$ and hence $tP = \log|e^{t(P+iQ)}| = P_t \; \forall t \in \mathbb{R}$, so that $P \in ReL^1(\lambda)$. It would be even faster to invoke IV.4.1 to conclude that $e^{tP} = |e^{t(P+iQ)}| \in L^{\#} = L^{\circ}(\lambda) \; \forall t \in \mathbb{R}$ and hence that $P \in ReL^1(\lambda)$. In the unit disk situation we thus have $E = ReL^1(\lambda)$.

After the above transcription we know what we have to do. As before we fix a reduced Hardy algebra situation (H,φ) on (X,Σ,m). The chapter starts with the definition of the abstract conjugation combined with several characterizations of its domain $E \subset ReL(m)$. It turns out that E is the largest linear subspace of ReL(m) on which the functional α: $ReL(m) \rightarrow [-\infty,\infty]$ is finite-valued and linear. The main achievement is then the characterization of E with the means of M. It requires the full power of the main theorems of Chapter IV. The principal result is that a function $P \in ReL(m)$ which is bounded, or at least not too far remote from boundedness in some sense or other, is in E iff the integral $\int PVdm$ has the same value for all those $V \in M$ for which it exists in the extended sense. Hereafter we prove for $E^{\infty} := E \cap ReL^{\infty}(m)$ a simple but powerful approximation theorem: that $ReH \subset E^{\infty}$ is dense in E^{∞} in a sense which is much sharper then weak$*$ density. The result has important immediate implications. This is conceivable since in the Szegö situation $M=\{F\}$ we have $E = ReL^1(Fm)$ and hence $E^{\infty} = ReL^{\infty}(m)$, so that here the approximation theorem sharpens the fundamental result that ReH is weak$*$ dense in $ReL^{\infty}(m)$ (the weak$*$Dirichlet property).

The remainder of the chapter then has the aim to extend the classical Marcel Riesz and Kolmogorov estimations for conjugate functions to the abstract Hardy algebra situation. The Kolmogorov estimation will of course be deduced from V.5.6. As to the Marcel Riesz estimation we shall derive certain sharp particular results. In order to obtain more comprehensive versions we have to impose rather sharp restrictions upon (H,φ). The relevant additional conditions and their mutual dependences - which will be of interest in other contexts as well - will be dealt with in a separate section.

1. A Representation Theorem

The subsequent representation theorem will be needed for the main theorem of the next section. We present a simple direct proof.

1.1 THEOREM: Let $h_t \in L^\infty(m)$ with $|h_t|=1$ for all $t\in\mathbb{R}$. Assume that $h_s h_t = h_{s+t}$ $\forall s,t\in\mathbb{R}$ and $h_t\to 1$ in $L^1(m)$-norm for $t\to 0$. Then there exists a unique $Q\in \mathrm{Re}\,L(m)$ such that $h_t=e^{itQ}$ for all $t\in\mathbb{R}$.

Proof: The uniqueness assertion is obvious. The basic idea of the existence proof is careful differentiation of the vector-valued function $t\mapsto h_t$. We have to face technical difficulties due to the fact that the size of the function $Q\in\mathrm{Re}\,L(m)$ to be constructed is not restricted at all.

i) The function $\mathbb{R}\to L^1(m):t\mapsto h_t$ is continuous in $L^1(m)$-norm. Therefore we can form the elementary integral $H_t:=\int_0^t h_u du\in L^1(m)$ $\forall t\in\mathbb{R}$. From the fundamental theorem of calculus we have $\frac{1}{s}(H_{s+t}-H_t)\to h_t$ in $L^1(m)$-norm for $s\to 0$. In particular $\frac{1}{s}H_s\to 1$ in $L^1(m)$-norm for $s\to 0$. Thus from the Egorov theorem we obtain a sequence of subsets $E(r)\in\Sigma$ with $E(r)\uparrow X$ and a sequence of numbers $\varepsilon(n)\downarrow 0$ such that on each fixed $E(r)$ we have $\frac{1}{\varepsilon(n)}H_{\varepsilon(n)}\to 1$ for $n\to\infty$ in $L^\infty(m)$-norm.

ii) For fixed $s\in\mathbb{R}$ the linear operator $L^1(m)\to L^1(m):f\mapsto h_s f$ is continuous. It follows that

$$h_s H_t = \int_0^t h_s h_u du = \int_0^t h_{s+u} du = \int_s^{t+s} h_u du = H_{t+s}-H_s,$$

and hence that $(h_s-1)H_t=H_{t+s}-H_s-H_t=(h_t-1)H_s$ for all $s,t\in\mathbb{R}$.

iii) We claim that there exists a unique function $Q\in L(m)$ such that $h_t-1=iQH_t$ for all $t\in\mathbb{R}$. In view of i) the uniqueness assertion is obvious. To prove the existence define on each $E(r)$ the function $Q_r:=\frac{1}{i}(h_{\varepsilon(n)}-1)/H_{\varepsilon(n)}$ for n sufficiently large which in view of ii) is independent from n. It is then immediate that there exists a well-defined function $Q\in L(m)$ such that $Q|E(r)=Q_r$ $\forall r\geq 1$. For $t\in\mathbb{R}$ it follows on $E(r)$ that $(h_t-1)H_{\varepsilon(n)}=(h_{\varepsilon(n)}-1)H_t=iQH_{\varepsilon(n)}H_t$ for n sufficiently large and hence that $h_t-1=iQH_t$. Thus $h_t-1=iQH_t$ on the whole of X. iv) The function $Q\in L(m)$ obtained in

iii) is real-valued. In fact, we have $\bar{h}_t = h_{-t}$ and hence $\bar{H}_t = -H_{-t}$, from which it follows that $h_t - 1 = i\bar{Q}H_t$ for all $t \in \mathbb{R}$. This implies that $\bar{Q} = Q$ in view of the uniqueness assertion in iii).

v) We claim that on each fixed $E(r)$ we have $\frac{1}{t}(h_t - 1) \to iQ$ for $t \to 0$ in $L^1(m)$-norm. In fact, from i) ii) we see that for n fixed

$$\frac{1}{t}(h_t - 1)H_{\varepsilon(n)} = \frac{1}{t}(H_{\varepsilon(n)+t} - H_{\varepsilon(n)}) - \frac{1}{t}H_t \to h_{\varepsilon(n)} - 1 = iQH_{\varepsilon(n)} \quad \text{for } t \to 0$$

in $L^1(m)$-norm, and on $E(r)$ the function $|H_{\varepsilon(n)}|$ is $\geq$ some $\delta > 0$ for n sufficiently large.

vi) We now prove that $h_t = e^{itQ}$ for all $t \in \mathbb{R}$. For this purpose observe that the functions $h_t^* := h_t e^{-itQ}$ $\forall t \in \mathbb{R}$ fulfill the assumptions of the theorem so that i)-v) can be applied to them as well. Thus we obtain a unique $Q^* \in L(m)$ such that $h_t^* - 1 = iQ^* H_t^*$ $\forall t \in \mathbb{R}$, and on each fixed $E^*(r)$ we have $\frac{1}{t}(h_t^* - 1) \to iQ^*$ for $t \to 0$ in $L^1(m)$-norm. But on each $E(r)$ we have

$$\frac{1}{t}(h_t^* - 1) = \frac{1}{t}(h_t - 1)e^{-itQ} + \frac{1}{t}(e^{-itQ} - 1) \to 0 \quad \text{for } t \to 0$$

in $L^1(m)$-norm. It follows that $Q^* = 0$. Thus $h_t^* = 1$ or $h_t = e^{itQ}$ for all $t \in \mathbb{R}$. QED.

2. Definition of the abstract Conjugation

We return to the reduced Hardy algebra situation (H, φ).

2.1 THEOREM: For $P \in \mathrm{ReL}(m)$ the subsequent properties are equivalent.

i) $a(e^{tP})a(e^{-tP}) = 1$ $\forall t \in \mathbb{R}$.

ii) $\alpha(P) \in \mathbb{R}$ and $\alpha(tP) = t\alpha(P)$ $\forall t \in \mathbb{R}$.

iii) There exists a function $Q \in \mathrm{ReL}(m)$ such that $e^{t(P+iQ)} \in H^\#$ $\forall t \in \mathbb{R}$.

In this case the function $Q \in \mathrm{ReL}(m)$ is unique up to an additive real constant. Hence there exists a unique $Q \in \mathrm{ReL}(m)$ such that in addition $\varphi(e^{t(P+iQ)}) = e^{t\alpha(P)}$ for all $t \in \mathbb{R}$.

The function class E is defined to consist of the functions $P \in \mathrm{ReL}(m)$

which possess the equivalent properties i)ii)iii) in 2.1. The functions $P \in E$ are called conjugable. For $P \in E$ the unique function $Q \in ReL(m)$ such that $e^{t(P+iQ)} \in H^{\#}$ and $\varphi(e^{t(P+iQ)}) = e^{t\alpha(P)}$ $\forall t \in \mathbb{R}$ is called the conjugate function of P and written $Q =: P^*$.

2.2 LEMMA: Consider a sequence of functions $u_n \in H^{\#}$ with $|u_n| \leq$ some $G \in L^{\#}$ such that $\limsup_{n\to\infty} |u_n| \leq 1$ and $\varphi(u_n) \to 1$. Then there exists a subsequence $u_{n(\ell)}$ which tends $\to 1$ for $\ell \to \infty$.

Proof of 2.2: Take functions $v_\ell \in H$ with $|v_\ell| \leq 1$, $v_\ell \to 1$ and $|v_\ell| G \leq c_\ell$ $\forall \ell \geq 1$. We can assume that $\varphi(v_\ell) > 1 - \frac{1}{2^\ell}$. For $V \in M$ then

$$\int |v_\ell(u_n - 1)|^2 V dm = \int |(v_\ell u_n - 1) + (1 - v_\ell)|^2 V dm$$

$$\leq 2\int |v_\ell u_n - 1|^2 V dm + 2\int |1 - v_\ell|^2 V dm$$

$$\leq 2\int |v_\ell u_n|^2 V dm + 2 - 4\varphi(v_\ell) \mathrm{Re}\varphi(u_n) + 4 - 4\varphi(v_\ell).$$

Introduce now $G_n := \mathrm{Sup}\{|u_s| : s \geq n\}$ and observe that $|u_n| \leq G_n \leq G$ and $G_n \downarrow \limsup_{n\to\infty} |u_n| \leq 1$ for $n \to \infty$. Then

$$\limsup_{n\to\infty} \int |v_\ell(u_n - 1)|^2 V dm \leq 8 - 8\varphi(v_\ell) < \frac{8}{2^\ell} \qquad \forall \ell \geq 1.$$

Thus there exists a sequence $1 \leq n(1) < \ldots < n(\ell) < n(\ell+1) < \ldots$ such that

$$\int |v_\ell(u_{n(\ell)} - 1)|^2 V dm < \frac{8}{2^\ell} \qquad \forall \ell \geq 1.$$

It follows that $v_\ell(u_{n(\ell)} - 1) \to 0$ and hence that $u_{n(\ell)} \to 1$ for $\ell \to \infty$. QED.

Proof of 2.1:i) $\Rightarrow$ ii) From V.1.2 we have $0 < a(e^{tP}) < \infty$ $\forall t \in \mathbb{R}$ and hence $\alpha(tP) \in \mathbb{R}$ and $\alpha(tP) + \alpha(-tP) = 0$ $\forall t \in \mathbb{R}$. Furthermore $e^{tP} \in L^{\#}$ $\forall t \in \mathbb{R}$. Now

$$\alpha\big((s+t)P\big) \leq \alpha(sP) + \alpha(tP) \text{ and } \alpha\big(-(s+t)P\big) \leq \alpha(-sP) + \alpha(-tP) \;\forall s,t \in \mathbb{R},$$

so that addition leads to $\alpha\big((s+t)P\big) = \alpha(sP) + \alpha(tP)$ $\forall s,t \in \mathbb{R}$. It remains to show that the function $t \mapsto \alpha(tP)$ is continuous on $\mathbb{R}$. But for $t_n \to t$ with $|t_n|, |t| \leq c$ we have $-c|P| \leq \pm t_n P, \pm tP$ with $e^{c|P|} \in L^{\#}$ and conclude from IV.3.12 that

$$\alpha(tP) \leq \liminf_{n\to\infty} \alpha(t_nP) \leq \limsup_{n\to\infty} \alpha(t_nP) = -\liminf_{n\to\infty} \alpha(-t_nP) \leq -\alpha(-tP) = \alpha(tP).$$

Hence $\alpha(t_nP)\to\alpha(tP)$ so that indeed $t\mapsto\alpha(tP)$ is continuous on $\mathbb{R}$. ii) ⇒ i) is obvious.

iii) ⇒ i) is immediate from V.1.1. i)&ii)⇒iii) For each $t\in\mathbb{R}$ there exists after V.1.3 a unique $f_t\in(H^\#)^\times$ with $|f_t|=e^{tP}$ and $\varphi(f_t)=a(e^{tP})=$ $=e^{-\alpha(-tP)}=e^{t\alpha(P)}$. It follows that $f_sf_t=f_{s+t}$ $\forall s,t\in\mathbb{R}$. The assertion will follow if 1.1 can be applied to the functions $h_t:=f_te^{-tP}$ $\forall t\in\mathbb{R}$. Thus we have to prove that $h_t\to1$ in $L^1(m)$-norm for $t\to0$. Assume that this is false. Then $\int|h_{t(n)}-1|dm\to\varepsilon>0$ for a suitable sequence $t(n)\to0$ with $0<|t(n)|\leq1$. It follows that $|f_{t(n)}|\leq e^{|P|}\in L^\#$ so that from 2.2 we obtain a subsequence $f_{t(n(\ell))}\to1$ for $\ell\to\infty$. Thus $h_{t(n(\ell))}\to1$ pointwise and hence in $L^1(m)$-norm so that we arrive at a contradiction. So we have obtained a function $Q\in ReL(m)$ such that $f_t=e^{t(P+iQ)}\in(H^\#)^\times$ and $\varphi(f_t)=\varphi(e^{t(P+iQ)})=e^{t\alpha(P)}$ for all $t\in\mathbb{R}$.

The uniqueness assertion is seen as in the introduction: If $G\in ReL(m)$ is such that $e^{t(P+iG)}\in H^\#$ $\forall t\in\mathbb{R}$ then the functions $e^{it(G-Q)}$ are $\in H^\#$ and hence invertible elements of $H^\#$ of modulus one, so that $e^{it(G-Q)}=$const$=$ $=c(t)$ $\forall$ $t\in\mathbb{R}$ and hence $G-Q=$const. QED.

We proceed to collect a number of basic consequences.

<u>2.3</u> <u>PROPOSITION</u>: i) For $P\in E$ we have $e^{\pm P}\in L^\#$ and hence $P\in L^\#$.

ii) $E\subset ReL(m)$ is a linear subspace.

iii) The conjugation $E\to ReL(m):P\mapsto P^*$ is a linear operator.

iv) The restriction $\alpha|E:E\to\mathbb{R}$ is a positive linear functional.

v) Assume that $T\subset ReL(m)$ is a linear subspace such that the restriction $\alpha|T$ is finite-valued and linear. Then $T\subset E$. Thus E is the largest linear subspace of $ReL(m)$ on which α is finite-valued and linear.

Proof: It remains to show that $\alpha|E$ is additive. But this is an immediate consequence of $\alpha(P+Q)\leq\alpha(P)+\alpha(Q)$ $\forall P,Q\in E$ and $\alpha(P)+\alpha(-P)=0$ $\forall$ $P\in E$. QED.

2.4 PROPOSITION: Let $P_n \in E(n=1,2,\dots)$ and $P \in ReL(m)$. Assume that at least one of the subsequent conditions be fulfilled.

i) $P_n \to P$ pointwise and $|P_n| \leq$ some $G \in ReL(m)$ with $e^G \in L^\#$.

ii) $\alpha(|P_n - P|) \to 0$.

Then $P \in E$ and $\alpha(P_n) \to \alpha(P)$.

Proof: i) For $t \in \mathbb{R}$ we have $-|t|G \leq \pm tP_n, \pm tP$. At first IV.3.9 implies that $-\infty < \alpha(\pm tP)$. Then from IV.3.12 we obtain

$$-\infty < \alpha(tP) \leq \liminf_{n\to\infty} \alpha(tP_n) \leq \limsup_{n\to\infty} \alpha(tP_n) = -\liminf_{n\to\infty} \alpha(-tP_n) \leq -\alpha(-tP) < \infty.$$

It follows that $\alpha(tP) \in \mathbb{R}$ and $\alpha(tP) + \alpha(-tP) = 0 \ \forall t \in \mathbb{R}$ so that condition i) in 2.1 is fulfilled. Thus $P \in E$. Furthermore $\alpha(P_n) \to \alpha(P)$ from the above for $t=1$. ii) Let $t \in \mathbb{R}$ and $|t| \leq r \in \mathbb{N}$. Then

$$tP \leq tP_n + |t||P-P_n| \leq tP_n + r|P-P_n|,$$

$$\alpha(tP) \leq \alpha(tP_n) + \alpha(r|P-P_n|) \leq \alpha(tP_n) + r\alpha(|P-P_n|).$$

Likewise $\alpha(tP_n) \leq \alpha(tP) + r\alpha(|P-P_n|)$. Thus $\alpha(tP) \in \mathbb{R}$ and $\alpha(tP_n) \to \alpha(tP)$ for $n\to\infty \ \forall t \in \mathbb{R}$. The assertion follows. QED.

2.5 REMARK: Let $h = P + iQ \in H^\#$ such that $e^{|h|} \in L^\#$. Then $P \in E$ and $Q = P^* + \mathrm{Im}\varphi(h)$. This is immediate from the substitution theorem V.2.1. In particular $\mathrm{Re}H \subset E$ and $Q = P^* + \mathrm{Im}\varphi(h)$ for $h = P + iQ \in H$.

There is a certain converse to 2.5 which is somewhat more delicate. For $P \in E$ we have $P \in L^\#$ from 2.4.i), but we cannot be sure that $P^* \in L^\#$ as well. However, we can prove the subsequent assertions.

2.6 PROPOSITION: Assume that $P \in E$ and $P^* \in L^\#$. Then $P + iP^* \in H^\#$ and $\varphi(P+iP^*) = \alpha(P)$.

This is an immediate consequence of the subsequent lemma combined with IV.3.2 and IV.3.3.

2.7 LEMMA: i) For $z \in \mathbb{C}$ with $\mathrm{Re}\, z \leq 0$ we have

$$\frac{1}{t}|e^{tz} - 1| \leq |z| \quad \forall t > 0.$$

ii) For $z\in\mathbb{C}$ we have

$$\frac{1}{t}|e^{tz}-1| \leq \frac{1}{\varepsilon}|e^{(\tau+\varepsilon)z}| + |z| \quad \forall\, 0<t\leq\tau \text{ and } \varepsilon>0.$$

Proof: i) Apply the mean value theorem to the function $s\mapsto e^{sz}$ in $0\leq s\leq t$. ii) We can assume that $z=x+iy$ with $x\geq 0$. Then

$$|e^{tz}-1| = |(e^{tx}-1)e^{ity}+(e^{ity}-1)| \leq (e^{tx}-1) + |e^{ity}-1|.$$

Now $e^{tx}-1\leq txe^{tx}\leq txe^{\tau x}$ and $x\leq\frac{1}{\varepsilon}(1+\varepsilon x)\leq\frac{1}{\varepsilon}e^{\varepsilon x}$ so that $e^{tx}-1\leq\frac{t}{\varepsilon}e^{(\tau+\varepsilon)x}$. And $|e^{ity}-1|\leq 2|\sin\frac{ty}{2}|\leq t|y|$. The assertion follows. QED.

2.8 PROPOSITION: Assume that $P\in E$ is bounded below or bounded above. Then $P+iP^*\in H^{\#}$ and $\varphi(P+iP^*)=\alpha(P)$.

Proof: We can assume that $P\geq 0$. Then V.4.1 can be applied. The assertion follows. QED.

3. Characterization of E with the means of M

In one direction we obtain an instant final result.

3.1 THEOREM: Let $P\in ReL(m)$ with $e^{\pm P}\in L^{\#}$. Assume that the integral $\int PVdm$ has the same value $c\in[-\infty,\infty]$ for all those $V\in M$ for which it exists in the extended sense. Then $P\in E$ and $\alpha(P)=c\in\mathbb{R}$.

Proof: We apply IV.3.9 to tP with real $t\neq 0$. It follows that $-\infty<\alpha(tP)\leq$ $\leq tc$. Thus $c\in\mathbb{R}$ and hence $\alpha(tP)\in\mathbb{R}$ $\forall t\in\mathbb{R}$. Then from $\alpha(tP)\leq tc$ and $\alpha(-tP)\leq -tc$ we obtain $\alpha(tP)=tc$ $\forall t\in\mathbb{R}$. QED.

In the opposite direction it would be most pleasant to deduce from $P\in E$ that

$$\int|P|Vdm<\infty \quad\text{and}\quad \int PVdm = \alpha(P) \quad \text{for all } V\in M,$$

or at least for those $V\in M$ for which $\int PVdm$ exists in the extended sense. But we cannot prove this unless we impose an additional boundedness condition upon $P\in E$ which appears to be sharper than the implicated condition $e^{\pm P}\in L^{\#}$.

Let us start with the immediate implications of the Jensen inequalities as noted in IV.3.5.

3.2 JENSEN CONSEQUENCES: Assume that $P\in E$. i) For $V\in MJ$ we have $\int|P|Vdm<\infty$ and $\alpha(P)=\int PVdm$. ii) For $F\in M$ there exist functions $V\in M$ with $[V>0]\subset[F>0]$ such that $\int PVdm$ exists in the extended sense. And

$$\operatorname{Inf}\{\int PVdm: \text{all these } V\in M\}\leqq\alpha(P)\leqq\operatorname{Sup}\{\int PVdm : \text{all these } V\in M\}.$$

Prior to the main point we note a simple but important assertion which is in obvious relation to V.4.1 on the function class H^+.

3.3 PROPOSITION: o) Let $P\in E$ and $V\in M$ be such that $\int e^{-\delta P}Vdm<\infty$ for some $\delta>0$. Then $\int|P|Vdm<\infty$ and $\int PVdm\leqq\alpha(P)$.

i) If $P\in E$ is bounded below then $\alpha(P)=\operatorname{Sup}\{\int PVdm:V\in M\}$.

ii) If $P\in E^\infty:=E\cap ReL^\infty(m)$ then $\alpha(P)=\int PVdm$ for all $V\in M$.

3.4 COROLLARY: Let $P\in ReL^\infty(m)$. Then $P\in E\Leftrightarrow\int PVdm$ has the same value for all $V\in M\Leftrightarrow\alpha(P)=\int PVdm$ for all $V\in M$.

Proof of 3.3: o) The assumption implies that $\int P^-Vdm<\infty$. From IV.3.4 we have for $0<t\leqq\delta$

$$e^{-t\alpha(P)}=\varphi\left(e^{-t(P+iP^*)}\right)=\int e^{-t(P+iP^*)}Vdm,$$

$$e^{-t\alpha(P)}\leqq\int e^{-tP}Vdm,\quad e^{-\alpha(P)}\leqq\left(\int e^{-tP}Vdm\right)^{\frac{1}{t}},$$

so that II.5.1 implies that $-\alpha(P)\leqq\int(-P)Vdm$. The assertion follows. i) follows from o) and 3.2.ii). ii) is then obvious. QED.

In the next lemma we introduce the sharpened boundedness condition quoted above and prove a certain extension of the basic theorem IV.3.10.

3.5 LEMMA: Define

$$\alpha^+(f):=\liminf_{t\downarrow 0}\frac{1}{t}\alpha\left((tf-1)^+\right)\qquad\forall f\in ReL(m).$$

Thus $0\leqq\alpha^+(f)\leqq\infty$ and $\alpha^+(cf)=c\alpha^+(f)$ $\forall c>0$. i) Then

$$\operatorname{Inf}\{\alpha\left((f-h)^+\right):h\in ReL^\infty(m)\}\leqq\alpha^+(f)\leqq\operatorname{Inf}\{\theta\left((f-h)^+\right):h\in ReL^\infty(m)\}.$$

ii) f bounded above $\Rightarrow \alpha^+(f)=0 \Rightarrow e^f \in L^\#$. iii) We have

$$\theta(f) = \operatorname*{Sup}_{t>0} \frac{1}{t}\alpha(tf) \qquad \forall\, f \in ReL(m) \text{ with } \alpha^+(-f)=0.$$

Proof: i) To prove the left estimation let $0<t<1$ and $n \in \mathbb{N}$ with $n<\frac{1}{t}\leqq n+1$. Then

$$\frac{1}{t}\alpha\big((tf-1)^+\big) \geqq n\alpha\big((tf-1)^+\big) \geqq \frac{n}{n+1}\alpha\big((n+1)(tf-1)^+\big) \geqq \Big(1-\frac{1}{n+1}\Big)\alpha\Big(\frac{1}{t}(tf-1)^+\Big)$$

$$\geqq (1-t)\alpha\Big(\big(f-\frac{1}{t}\big)^+\Big) \geqq (1-t)\operatorname{Inf}\{\alpha\big((f-h)^+\big): h \in ReL^\infty(m)\},$$

so that the assertion follows. The second estimation is immediate from IV.3.9. ii) The first implication is obvious while the second follows from IV.3.13 and i).

iii) Let $f \in ReL(m)$ with $\alpha^+(-f)=0$. Then $e^{-f} \in L^\#$ from ii) so that $-\infty<\frac{1}{t}\alpha(tf) \leqq \frac{1}{t}\theta(tf)=\theta(f)$ $\forall t>0$ from IV.3.9. Thus we have to prove that $c:=\operatorname{Sup}\{\frac{1}{t}\alpha(tf): t>0\}$ is $\geqq \theta(f)$. We can of course assume that c is finite. Note that $\alpha(tf) \leqq tc$ or $e^{-tc} \leqq e^{-\alpha(tf)} = a(e^{-tf})$ $\forall t>0$. 1) Because of $\alpha^+(-f)=0$ there exists a sequence $0<t(n)\downarrow 0$ such that $\frac{1}{t(n)}\alpha\big((-t(n)f-1)^+\big)\to 0$. From IV.3.6 we obtain functions $u_n \in H$ with

$$|u_n| \leqq e^{-(-t(n)f-1)^+}, \text{ that is } |u_n| \leqq 1 \text{ and } |u_n| \leqq e^{1+t(n)f},$$

$$\varphi(u_n) = e^{-\alpha((-t(n)f-1)^+)} > 0, \text{ and hence } (\varphi(u_n))^{\frac{1}{t(n)}} \to 1.$$

From the second line $\frac{1}{t(n)}\log\varphi(u_n)\to 0$, so that the estimation $\frac{1}{t(n)}\log\varphi(u_n) \leqq \frac{1}{t(n)}(\varphi(u_n)-1) \leqq 0$ shows that $\frac{1}{t(n)}(\varphi(u_n)-1)\to 0$. Therefore

$$D_n := \frac{1}{t(n)}\big(\varphi(u_n)e^{-t(n)c}-1\big) = \frac{1}{t(n)}\big(\varphi(u_n)-1\big)e^{-t(n)c} + \frac{e^{-t(n)c}-1}{t(n)} \to -c$$

And from the first line we see that

$$\frac{1}{t(n)}|u_n|\big(e^{-t(n)f}-1\big) \leqq ef^-.$$

This is obvious on $[f \geqq 0]$ while on $[f \leqq 0]$ we have

$$\frac{1}{t(n)}|u_n|\big(e^{-t(n)f}-1\big) \leqq \frac{e}{t(n)}\big(1-e^{t(n)f}\big) \leqq \frac{e}{t(n)}\big(-t(n)f\big) = e(-f) = ef^-.$$

2) Let us fix a function $V \in M$ with $\int f^- V dm < \infty$. We have to prove that $\int fVdm \leq c$. Now for $v \in H$ with $|v| \leq e^{-t(n)f}$ we obtain

$$\varphi(u_n)|\varphi(v)| = |\varphi(u_n v)| \leq \int |u_n v| Vdm \leq \int |u_n| e^{-t(n)f} Vdm.$$

It follows that

$$\varphi(u_n) e^{-t(n)c} \leq \varphi(u_n)\, a\left(e^{-t(n)f}\right) \leq \int |u_n| e^{-t(n)f} Vdm,$$

$$D_n \leq \frac{1}{t(n)} \int \left(|u_n| e^{-t(n)f} - 1\right) Vdm \leq \frac{1}{t(n)} \int |u_n| \left(e^{-t(n)f} - 1\right) Vdm.$$

Thus the function $F_n := ef^- - \frac{1}{t(n)} |u_n| \left(e^{-t(n)f} - 1\right) \geq 0$ has an integral $\int F_n Vdm \leq \leq e\int f^- Vdm - D_n$. Furthermore $F_n \to ef^- + f \geq 0$. Thus the Fatou theorem implies that $e\int f^- Vdm + \int fVdm \leq e\int f^- Vdm + c$ or $\int fVdm \leq c$. QED.

<u>3.6</u> <u>THEOREM</u>: i) Let $P \in E$ with $\alpha^+(-P) = 0$. Then $\alpha(P) = \theta(P) = \mathrm{Sup}\{\int PVdm : V \in M$ with $\int P^- Vdm < \infty\}$.

ii) Let $P \in E$ with $\alpha^+(\pm P) = 0$. Then $\alpha(P) = \int PVdm$ for all those $V \in M$ for which $\int PVdm$ exists in the extended sense.

Proof: Immediate from 3.5. QED.

<u>3.7</u> <u>COROLLARY</u>: Let $P \in \mathrm{ReL}(m)$ with $\alpha^+(\pm P) = 0$. Then $P \in E \Leftrightarrow$ the integral $\int PVdm$ has the same value $c \in [-\infty, \infty]$ for all those $V \in M$ for which it exists in the extended sense. In this case $\alpha(P) = c \in \mathbb{R}$.

<u>3.8</u> <u>SPECIAL CASE</u>: Let $P \in \mathrm{ReL}(m)$ and assume that there exists a sequence of functions $P_n \in \mathrm{ReL}^\infty(m)$ such that

$$\theta(|P - P_n|) = \sup_{V \in M} \int |P - P_n| Vdm \to 0.$$

Then $\alpha^+(\pm P) = 0$ and hence $e^{\pm P} \in L^\#$. Furthermore $\int |P| Vdm \leq \theta(|P|) < \infty \ \forall V \in M$. Therefore $P \in E \Leftrightarrow \int PVdm$ has the same value for all $V \in M \Leftrightarrow \alpha(P) = \int PVdm$ for all $V \in M$.

Proof: We have to prove that $\alpha^+(\pm P) = 0$. But for $\varepsilon = \pm 1$ we have $\theta((\varepsilon P - \varepsilon P_n)^+) \leq \leq \theta(|P - P_n|)$ so that 3.5.i) implies that $\alpha^+(\varepsilon P) = 0$. QED.

We conclude the section with a remarkable characterization of the Szegö situation.

3.9 THEOREM: For $F\in M$ the subsequent properties are equivalent.

i) $M = \{F\}$,

ii) $E = ReL^1(Fm)$,

iii) $ReL^\infty(m)\subset E$.

Proof: i) $\Rightarrow$ ii) For $P\in ReL^1(Fm)$ we have $e^{\pm P}\in L^o(Fm)=L^\#$ from IV.4.1 and hence $P\in E$ from 3.1. On the other hand $P\in E$ implies that $e^{\pm P}\in L^\#=L^o(Fm)$ and hence $P\in ReL^1(Fm)$. ii) $\Rightarrow$ iii) is trivial. iii) $\Rightarrow$ i) Let $V\in M$. Then 3.4 implies that $\int P(V-F)dm=0$ for all $P\in ReL^\infty(m)$ and hence that $V=F$. QED.

4. The basic Approximation Theorem

4.1 APPROXIMATION THEOREM: Assume that $P\in E^\infty:=E\cap ReL^\infty(m)$, so that $h:=P+iP^*\in H^\#$ and $\varphi(h)=\alpha(P)$. Then for each $\varepsilon>0$ there exists a function $h_\varepsilon\in H$ such that

i) $|h_\varepsilon|\leq|h|$ and $|\mathrm{Re}\, h_\varepsilon|\leq|\mathrm{Re}\, h|=|P|$,

ii) $|h_\varepsilon-h|\leq\varepsilon\mathrm{Max}(1,|h|^{1+\varepsilon})$,

iii) $\varphi(h_\varepsilon)$ is real and $|\varphi(h_\varepsilon)-\varphi(h)|\leq\varepsilon$.

Proof: 1) For $s=u+iv\in\mathbb{C}$ with $|u|<1$ we have $1-s^2=1-u^2+v^2-2iuv$ and hence $|1-s^2|\geq1-u^2+v^2\geq1-u^2>0$. It follows that

$$\left|\frac{s}{1-s^2}\right| \leq \frac{|s|}{1-u^2+v^2} \leq \frac{|s|}{1-u^2},$$

$$\left|\mathrm{Re}\,\frac{s}{1-s^2}\right| = \left|\mathrm{Re}\,\frac{s(1-\bar{s}^2)}{|1-s^2|^2}\right| = \frac{|u-u|s|^2|}{|1-s^2|^2} = \frac{|u|\,|1-|s|^2|}{|1-s^2|^2}$$

$$\leq \frac{|u|}{|1-s^2|} \leq \frac{|u|}{1-u^2+v^2} \leq \frac{|u|}{1-u^2}.$$

2) Let us fix $c\geq1$ with $|P|\leq c$. Then take $\sigma>0$ with $\sigma c<1$. From $\mathrm{Re}(c\pm h)\geq0$ we see that $e^{-t(c\pm h)}\in H$ $\forall t\geq0$ and hence $c\pm h\in H^+$. Thus $\frac{1}{1-\sigma c+\sigma(c\pm h)}=\frac{1}{1\pm\sigma h}\in H$ from V.4.1. Let us put

$$f_\sigma := \left(1-(\sigma c)^2\right)\frac{h}{1-(\sigma h)^2} = \frac{1-(\sigma c)^2}{2\sigma}\left(\frac{1}{1-\sigma h}-\frac{1}{1+\sigma h}\right).$$

Then $f_\sigma \in H$ and

$$\varphi(f_\sigma) = \frac{1-(\sigma c)^2}{2\sigma}\left(\frac{1}{1-\sigma\varphi(h)} - \frac{1}{1+\sigma\varphi(h)}\right) = \left(1-(\sigma c)^2\right)\frac{\varphi(h)}{1-(\sigma\varphi(h))^2}.$$

It is immediate from 1) that

$$|f_\sigma| \leqq \left(1-(\sigma c)^2\right)\frac{|h|}{1-(\sigma P)^2} \leqq |h|,$$

$$|\mathrm{Re} f_\sigma| \leqq \left(1-(\sigma c)^2\right)\frac{|P|}{1-(\sigma P)^2} \leqq |P|.$$

Furthermore $\varphi(f_\sigma)$ is real and $\varphi(f_\sigma)\to\varphi(h)$ for $\sigma\downarrow 0$. Also $f_\sigma\to h$ pointwise for $\sigma\downarrow 0$. But in order to prove ii) we need a more elaborate estimation of $|f_\sigma - h|$.

3) Let us fix $0<\varepsilon\leqq 1$. For $\sigma>0$ with $\sigma c<1$ then

$$f_\sigma - h = \frac{h(1-(\sigma c)^2)-h(1-(\sigma h)^2)}{1-(\sigma h)^2} = \frac{\sigma^2 h(h^2-c^2)}{1-(\sigma h)^2},$$

so that from 1) we obtain

$$|f_\sigma - h| \leqq \frac{2\sigma^2(\mathrm{Max}(c,|h|))^3}{1-(\sigma P)^2+(\sigma P^*)^2} \leqq 2\sigma^\varepsilon\left(\mathrm{Max}(c,|h|)\right)^{1+\varepsilon}\frac{\sigma^{2-\varepsilon}(c^2+(P^*)^2)^{\frac{2-\varepsilon}{2}}}{1-(\sigma c)^2+(\sigma P^*)^2}$$

$$\leqq 2\sigma^\varepsilon\left(\mathrm{Max}(c,|h|)\right)^{1+\varepsilon}\frac{(\sigma c)^{2-\varepsilon}+(\sigma P^*)^{2-\varepsilon}}{1-(\sigma c)^2+(\sigma P^*)^2}$$

$$\leqq \frac{2\sigma^\varepsilon(\mathrm{Max}(c,|h|))^{1+\varepsilon}}{1-(\sigma c)^2} + \frac{2\sigma^\varepsilon(\mathrm{Max}(c,|h|))^{1+\varepsilon}}{(1-(\sigma c)^2)^\varepsilon}\frac{(1-(\sigma c)^2)^\varepsilon(\sigma P^*)^{2-\varepsilon}}{1-(\sigma c)^2+(\sigma P^*)^2}$$

$$\leqq \frac{4\sigma^\varepsilon(\mathrm{Max}(c,|h|))^{1+\varepsilon}}{1-(\sigma c)^2} \leqq \frac{4\sigma^\varepsilon c^{1+\varepsilon}}{1-(\sigma c)^2}\,\mathrm{Max}(1,|h|^{1+\varepsilon}).$$

For the third estimation we used the calculus inequality

$$(u+v)^\alpha \leqq \mathrm{Max}(1,2^{\alpha-1})(u^\alpha+v^\alpha) \quad \forall\, u,v\geqq 0 \text{ and } \alpha>0,$$

while the fifth estimation is based on the inequality

$$uv \leqq t^p\frac{u^p}{p} + \frac{1}{t^q}\frac{v^q}{q} \quad \forall\, u,v\geqq 0 \text{ and } t>0 \text{ for conjugate } 1<p,q<\infty,$$

which we know from the proof of II.1.3, for the conjugate exponents $\frac{2}{\varepsilon}$

and $\frac{2}{2-\varepsilon}$. It is now clear that we can take $h_\varepsilon:=f_\sigma$ for $\sigma>0$ sufficiently small. QED.

4.2 REMARK: We cannot expect in 4.1 an estimation of the form $|h_\varepsilon-h| \leqq \varepsilon \operatorname{Max}(1,|h|)$ for $\varepsilon>0$ instead of ii). In fact, this would imply that $(1-\varepsilon)|h| \leqq |h_\varepsilon|+\varepsilon$ $\forall\varepsilon>0$ and hence the boundedness of the conjugate function P^*, a fact which is known to be not always true in the unit disk situation.

Let us turn to the most important consequences of the above approximation theorem. We have the chain of inclusions

$$E^\infty \subset \{P\in \operatorname{Re}L^\infty(m):\exists P_n\in \operatorname{Re}H \text{ with } |P_n| \leqq |P| \text{ and } P_n \to P\} \subset \overline{\operatorname{Re}H}^{\text{weak*}} \subset N^\perp = E^\infty,$$

where the first one is 4.1 and the last one is 3.4. Of course $N^\perp \subset \operatorname{Re}L^\infty(m)$ denotes the annihilator of $N\subset \operatorname{Re}L^1(m)$. Thus we obtain the subsequent theorem.

4.3 THEOREM: We have

$$E^\infty = \{P\in \operatorname{Re}L^\infty(m):\exists P_n\in \operatorname{Re}H \text{ with } |P_n| \leqq |P| \text{ and } P_n\to P\} = \overline{\operatorname{Re}H}^{\text{weak*}} = N^\perp.$$

4.4 CONSEQUENCE: We have

$$K \cap \operatorname{Re}L^1(m) = \mathbb{R}F + \bar{N}^{\operatorname{Re}L^1(m)},$$

whenever $F\in M$ is dominant over X.

Proof: The inclusion $\supset$ is obvious. In order to prove $=$ take a function $P\in \operatorname{Re}L^\infty(m)$ which annihilates the second member, which means that $\int PVdm=0$ $\forall V\in M$. From 4.3 we obtain a sequence of functions $h_n\in H$ with $|\operatorname{Re} h_n| \leqq |P|$ and $\operatorname{Re} h_n \to P$, and also with $\operatorname{Re}\varphi(h_n)=\int(\operatorname{Re} h_n)Fdm \to \int PFdm=0$. Thus for $f\in K\cap \operatorname{Re}L^1(m)$ we have

$$\int h_n f dm = \varphi(h_n)\int fdm, \quad \int(\operatorname{Re} h_n)\, fdm = (\operatorname{Re}\varphi(h_n))\int fdm,$$

and hence $\int Pfdm=0$. Since the second member is a closed linear subspace of $\operatorname{Re}L^1(m)$ the assertion follows. QED.

The next theorem is the complexified version of the last equality in theorem 4.3.

4.5 THEOREM: Assume that $F\in M$ is dominant over X. Then

$$H = N^{\perp} \cap (H_{\varphi}F)^{\perp} \quad \text{with } H_{\varphi}:=\{u\in H:\varphi(u)=0\},$$

where $N^{\perp}:=E^{\infty}+iE^{\infty}$ is the annihilator of $N\subset L^1(m)$ in the complex $L^{\infty}(m)$.

The proof uses the subsequent simple but fundamental remark to which we shall come back in the next section. Apart from this, the heart of the proof is the approximation theorem 4.1 as before.

4.6 REMARK: Let $h=P+iQ\in H$ with $\mathrm{Im}\varphi(h)=0$. Then

$$\int Q^2 V dm = \int P^2 V dm - (\varphi(h))^2 \leq \int P^2 V dm \quad \forall\, V\in M.$$

Proof of 4.6: $\int h^2 V dm = \varphi(h^2) = (\varphi(h))^2$ is real and ≥ 0. It follows that $\int h^2 V dm = \int (P^2-Q^2+2iPQ) V dm = \int P^2 V dm - \int Q^2 V dm$. QED.

Proof of 4.5 : The inclusion $\subset$ is obvious. In order to prove $\supset$ consider a function $h\in N^{\perp}\cap (H_{\varphi}F)^{\perp}\subset L^{\infty}(m)$ and put $h-\int hFdm=:P+iQ$. We have to prove that $h\in H$. 1) We have $P,Q\perp N$ and hence $P,Q\in E^{\infty}$. From 2.8 and 3.4 we know that

$$u:=P+iP^*\in H^{\#} \quad \text{with } \varphi(u) = \alpha(P) = \int PFdm = 0,$$

$$v:=Q+iQ^*\in H^{\#} \quad \text{with } \varphi(v) = \alpha(Q) = \int QFdm = 0.$$

And from 4.1 we obtain functions $u_n=P_n+iF_n\in H$ and $v_n=Q_n+iG_n\in H$ such that

$$|u_n| \leq |u|,\ |P_n| \leq |P|,\ u_n \to u,\ \varphi(u_n) \text{ real and } \to\varphi(u) = 0,$$

$$|v_n| \leq |v|,\ |Q_n| \leq |Q|,\ v_n \to v,\ \varphi(v_n) \text{ real and } \to\varphi(v) = 0.$$

2) From 4.6 we see that

$$\int |u_n|^2 Fdm \leq 2\int P_n^2 Fdm \leq 2\int P^2 Fdm,$$

$$\int |v_n|^2 Fdm \leq 2\int Q_n^2 Fdm \leq 2\int Q^2 Fdm,$$

so that $\int |u|^2 Fdm<\infty$ and $\int |v|^2 Fdm<\infty$ in view of the Fatou theorem.

3) Now from $h-\int hFdm \perp HF$ we conclude that

$$\int\big(h-\int hFdm - u_n\big)(u_n - iv_n)Fdm = -\varphi(u_n)\big(\varphi(u_n)-i\varphi(v_n)\big) \to 0.$$

On the other hand the integrand converges $\to(h-\int hFdm-u)(u-iv)$ pointwise and with the majorant $(|h|+|\int hFdm|+|u|)(|u|+|v|)\in L^1(Fm)$ in view of 2). It follows that

$$0 = \int\big(h-\int hFdm-u\big)(u-iv)Fdm = \int\big((P+iQ)-(P+iP^*)\big)\big((P+iP^*)-i(Q+iQ^*)\big)Fdm$$

$$= \int i(Q-P^*)\big((P+Q^*)-i(Q-P^*)\big)Fdm = \int(Q-P^*)^2Fdm + i\int(Q-P^*)(P+Q^*)Fdm.$$

Thus we obtain $Q=P^*$. Therefore $h=\int hFdm+P+iP^*=\int hFdm+u$ and hence $h\in H^\#\cap L^\infty(m)=H$. QED.

It is a pleasant consequence that the functions in H admit a simple characterization via multiplicativity under integration as follows.

<u>4.7</u> <u>COROLLARY</u>: Assume that $F\in M$ is dominant over X. Then for $f\in L^\infty(m)$ the subsequent properties are equivalent.

i) $f\in H$.

ii) $\int(f+u)^2Vdm = \big(\int(f+u)Vdm\big)^2$ for all $u\in H$ and $V\in M$.

iii) $\int f^2Vdm = (\int fVdm)^2$ for all $V\in M$ and $\int fuFdm = \int fFdm\int uFdm$ for all $u\in H$.

Proof: i) ⇒ ii) and ii) ⇒ iii) are obvious. So assume iii). The second assumption shows that $f\perp H_\varphi F$. Thus after 4.5 it remains to prove that $f\perp N$. But for $U,V\in M$ and $0\leqq t\leqq 1$ we have $(1-t)U+tV\in M$ and hence

$$(1-t)\int f^2Udm + t\int f^2Vdm = \big((1-t)\int fUdm + t\int fVdm\big)^2$$

$$= \big(\int fUdm + t\int f(V-U)dm\big)^2.$$

This implies at once that $\int f(V-U)dm=0$. Hence $f\perp N$. QED.

The next result is the reformulation of theorem 4.5 which corresponds to 4.4.

<u>4.8</u> <u>CONSEQUENCE</u>: We have

$$K = \overline{HF + N + iN}^{L^1(m)},$$

whenever $F\in M$ is dominant over X.

Proof: The inclusion $\supset$ is obvious. In order to prove $=$ take a function $h\in L^\infty(m)$ which annihilates the second member. This means that $h\perp HF$ and $h\perp N$. Then 4.5 implies that $h\in H$ and $\varphi(h)=\int hF dm=0$. It follows that h annihilates K. Since the second member is a closed linear subspace of $L^1(m)$ the assertion follows. QED.

We conclude with still another characterization of the Szegö situation.

4.9 THEOREM: For $F\in M$ the subsequent properties are equivalent.

i) $M = \{F\}$.

ii) $\overline{ReH}^{weak*} = ReL^\infty(m)$ (the weak*Dirichlet property).

iii) $K \cap ReL^1(m) = \mathbb{R}F$.

Proof: i) $\Leftrightarrow$ ii) is clear from 4.3, and i) $\Leftrightarrow$ iii) is clear from 4.4. QED.

4.10 REMARK: It is natural to ask whether the above characterization of the Szegö situation on the basis of 4.3 and 4.4 has a counterpart based upon 4.5 and 4.8. The answer is no: In the Szegö situation $M=\{F\}$ we obtain

$$H = (H_\varphi F)^\perp \quad \text{and} \quad K = \overline{HF}^{L^1(m)},$$

and these two properties are seen to be equivalent for each $F\in M$. But from them one cannot conclude that $M=\{F\}$ as the subsequent example will show.

4.11 EXAMPLE: We start from the usual $(H^\infty(D),\varphi_z)$ for $z\in D$ on (S,Baire,λ) as discussed in IV.3.15. We fix some $0<a<1$ and put $X:=S\cup\{a\}$ and $m:=\lambda+\delta_a \in$ $\in Pos(X,\text{Baire})$. Let us write $f^\circ:=f|S \in L(\lambda)$ for $f\in L(m)$. On (X,Baire,m) we consider the Hardy algebra situation (H,φ) defined to be

$$H:= \{f\in L^\infty(m): f^\circ \in H^\infty(D) \text{ and } f(a) = \langle f^\circ\lambda\rangle(a) = \int f^\circ P(a,\cdot)d\lambda\}$$

$$= \{(u,\langle u\lambda\rangle(a) = \int uP(a,\cdot)d\lambda): u\in H^\infty(D)\},$$

$$\varphi:\varphi(f) = \varphi_0(f^\circ) = \langle f^\circ\lambda\rangle(0) = \int f^\circ d\lambda \quad \forall\, f\in H.$$

We have the subsequent properties.

1) It is clear that $\chi_S \in M$. Furthermore $G:=(1-\frac{1-a}{1+a}P(a,\cdot),\frac{1-a}{1+a}) \in M$, where we know that $G \geqq 0$ from Section I.1 initial remark i). In fact, for $f= =(u,\langle u\lambda\rangle(a)) \in H$ we have

$$\int fGdm = \int u\left(1-\frac{1-a}{1+a}P(a,\cdot)\right)d\lambda + \langle u\lambda\rangle(a)\frac{1-a}{1+a} = \int ud\lambda = \int f^\circ d\lambda = \varphi(f).$$

Note that $G \in M$ is dominant.

2) $M = \{(1-t)\chi_S + tG: 0 \leqq t \leqq 1\}$. In order to see $\subset$ let $V \subset M$ so that $0 \leqq V \in L^1(m)$ and $\forall u \in H^\infty(D)$ we have

$$\varphi_o(u) = \varphi\big((u,\langle u\lambda\rangle(a))\big) = \int (u,\langle u\lambda\rangle(a))Vdm$$

$$= \int uV^\circ d\lambda + \langle u\lambda\rangle(a)V(a) = \int u\big(V^\circ + V(a)P(a,\cdot)\big)d\lambda.$$

Hence $V^\circ + V(a)P(a,\cdot)=1$. It follows that $V(a)P(a,\cdot) \leqq 1$ and hence $0 \leqq t:= =V(a)\frac{1+a}{1-a} \leqq 1$. Then $V^\circ = 1-V(a)P(a,\cdot) = (1-t)+t(1-\frac{1-a}{1+a}P(a,\cdot)) = ((1-t)\chi_S+tG)^\circ$ and hence $V=(1-t)\chi_S+tG$ from the definitions.

3) $L^\# = \{f \in L(m): f^\circ \in L^\#(H^\infty(D)) = L^O(\lambda)\}$. Furthermore $H^\# = \{f \in L(m): f^\circ \in H^\#(D)$ and $f(a)=\varphi_a(f^\circ)\}$ and $\varphi(f)=\varphi_o(f^\circ)$ $\forall f \in H^\#$. This is immediate.

4) One computes that

$$G^\circ = a\,\frac{(1-Z)^2}{(a-Z)(1-aZ)}.$$

Hence $1/G^\circ \in H^\#(D)$ and $\varphi_a(1/G^\circ)=0$ in view of V.4.4 and V.4.10. Thus 3) shows that $(1/G^\circ,0) \in H^\#$.

5) For $f \in K$ we have $f/G \in H^\#$ and $\varphi(f/G)=\int fdm$. In fact, for $u \in H^\infty(D)$ we have

$$\varphi_o(u)\int fdm = \varphi\big((u,\langle u\lambda\rangle(a))\big)\int fdm = \int (u,\langle u\lambda\rangle(a))fdm$$

$$= \int uf^\circ d\lambda + \langle u\lambda\rangle(a)f(a) = \int u\big(f^\circ + f(a)P(a,\cdot)\big)d\lambda,$$

so that $f^\circ + f(a)P(a,\cdot) \in K(H^\infty(D),\varphi_o) = \overline{H^\infty(D)}^{L^1(m)} \subset H^\#(D)$ in view of 4.8 and $L^1(\lambda) \subset L^O(\lambda) = L^\#(H^\infty(D))$. Thus 4) implies that

$$\frac{f^\circ+f(a)P(a,\cdot)}{G^\circ}\in H^{\#}(D) \text{ with } \varphi_a\left(\frac{f^\circ+f(a)P(a,\cdot)}{G^\circ}\right)=0.$$

Also 4) implies that

$$\frac{P(a,\cdot)}{G^\circ}=\frac{1+a}{1-a}\,\frac{1-G^\circ}{G^\circ}\in H^{\#}(D) \text{ with } \varphi_a\left(\frac{P(a,\cdot)}{G^\circ}\right)=-\frac{1+a}{1-a},$$

so that we conclude that $(f/G)^\circ=f^\circ/G^\circ\in H^{\#}(D)$ with $\varphi_a((f/G)^\circ)=\frac{1+a}{1-a}f(a)=$ $=f(a)/G(a)$. Thus 3) shows that $f/G\in H^{\#}$. Now $\int|f/G|Gdm=\int|f|dm<\infty$ so that IV.3.4 implies that $\varphi(f/G)=\int(f/G)Gdm=\int fdm$.

6) We have $K=\overline{HG}^{L^1(m)}$. In order to see $\subset$ let $f\in K$ so that $f/G\in H^{\#}$ after 5). Thus there are functions $f_n\in H$ with $|f_n|\leqq 1$, $f_n\to 1$ and $f_nf/G\in H$. It follows that $f_nf\in HG$ and $f_nf\to f$ in $L^1(m)$-norm so that $f\in L^1(m)$-norm closure (HG). Thus we have fulfilled the promise of 4.10.

7) Moreover the present situation furnishes examples of nonconstant real-valued $h\in H^{\#}$ and dominant $G\in M$ such that $\int|h|^pGdm<\infty$ not only for p=1 but also for certain p>1, namely for $1\leqq p<\frac{3}{2}$. This implies that $L^0(Gm)\not\subset L^{\#}$ since otherwise $|h|\in L^1(Gm)\Rightarrow e^{|h|}\in L^0(Gm)\subset L^{\#}$ would enforce that h=const after V.2.3. One example is $h=(1/G^\circ,0)\in H^{\#}$ after 4) since

$$\int|h|^pGdm=\int\left(\frac{1}{G^\circ}\right)^{p-1}d\lambda=\frac{1}{a^{p-1}}\int\left(\left|\frac{a-z}{1-z}\right|\right)^{2p-2}d\lambda$$
$$=\frac{1}{2\pi a^{p-1}}\int_{-\pi}^{\pi}\left(\frac{|a-e^{it}|}{|1-e^{it}|}\right)^{2p-2}dt<\infty \text{ for } 1\leqq p<\frac{3}{2}.$$

Another example is $h=\left(\frac{1+z}{1-z}\right)^2$ which is $\in H^{\#}$ in view of V.4.4 and V.4.10 as above.

5. The Marcel Riesz and Kolmogorov Estimations

We return to the conjugation operator $E\to \text{Re}L(m):P\mapsto P^*$. In the present section we start to extend the classical Marcel Riesz and Kolmogorov estimations. The discussion will be continued in Section 7 under additional assumptions on the Hardy algebra situation (H,φ).

We start with the Kolmogorov estimation. Let us fix $0<\tau<1$. Recall from V.5.6 that for $h\in H^+$ we have

$$\cos\frac{\tau\pi}{2}\int |h|^{\tau}Vdm \leq \mathrm{Re}(\varphi(h))^{\tau} \quad \forall\, V\in M,$$

$$\cos\frac{\tau\pi}{2}\,\theta\,(|h|^{\tau}) \leq \mathrm{Re}(\varphi(h))^{\tau}.$$

In particular if $0\leq P\in E$ then $P+iP^*\in H^+$ from the definitions and $\varphi(P+iP^*)=$ $=\alpha(P)=\theta(P)$ from 2.8 and 3.3.i). It follows that

$$\cos\frac{\tau\pi}{2}\,\theta\big(|P+iP^*|^{\tau}\big) \leq \big(\alpha(P)\big)^{\tau} = \big(\theta(P)\big)^{\tau} \quad \forall\, 0\leq P\in E.$$

Our aim is to remove the restriction $P\geq 0$ at least for bounded functions $P\in E^{\infty}$.

5.1 THEOREM: For $0<\tau<1$ we have

$$\cos\frac{\tau\pi}{2}\,\theta\big(|P+iP^*|^{\tau}\big) \leq 2^{1-\tau}\big(\theta(|P|)\big)^{\tau} \quad \forall\, P\in E^{\infty}.$$

Proof: We make essential use of the fundamental fact IV.2.5. Let $f\in \mathrm{Re}H$ with $f\geq |P|$. Then $f\pm P\in E^{\infty}$ and ≥ 0, and hence $\alpha(f)\pm\alpha(P)=\alpha(f\pm P)\geq 0$. For $h^{\pm}:=(f\pm P)+i(f\pm P)^*=(f+if^*)\pm(P+iP^*)$ we know that

$$\cos\frac{\tau\pi}{2}\,\theta\big(|h^{\pm}|^{\tau}\big) \leq \big(\alpha(f)\pm\alpha(P)\big)^{\tau}.$$

Now $P+iP^*=\frac{1}{2}(h^+-h^-)$. With repeated use of the calculus inequality (*) from the proof of 4.1 we thus obtain

$$|P+iP^*|^{\tau} \leq \frac{1}{2^{\tau}}(|h^+|+|h^-|)^{\tau} \leq \frac{1}{2^{\tau}}(|h^+|^{\tau}+|h^-|^{\tau}),$$

$$\cos\frac{\tau\pi}{2}\,\theta\big(|P+iP^*|^{\tau}\big) \leq \frac{1}{2^{\tau}}\cos\frac{\tau\pi}{2}\big(\theta(|h^+|^{\tau})+\theta(|h^-|^{\tau})\big)$$

$$\leq \frac{1}{2^{\tau}}\Big[\big(\alpha(f)+\alpha(P)\big)^{\tau}+\big(\alpha(f)-\alpha(P)\big)^{\tau}\Big]$$

$$\leq \frac{1}{2^{\tau}}\Big[2^{\frac{1}{\tau}-1}\Big[\big(\alpha(f)+\alpha(P)\big)+\big(\alpha(f)-\alpha(P)\big)\Big]\Big]^{\tau}=2^{1-\tau}\big(\alpha(f)\big)^{\tau}.$$

But from IV.2.5 and IV.3.10 we see that

$$\mathrm{Inf}\{\alpha(f): f\in \mathrm{Re}H \text{ with } f\geq |P|\} = \mathrm{Inf}\{\mathrm{Re}\varphi(u): u\in H \text{ with } \mathrm{Re}\,u\geq |P|\}$$

$$= \alpha^{o}(|P|) = \theta(|P|).$$

The assertion follows. QED.

Let us turn to the Marcel Riesz estimation. For $1\leq p<\infty$ and $V\in M$ the estimation in question assumes the subsequent forms which for each $0\leq R\leq\infty$ will be seen to be equivalent.

$$(\infty)\qquad \|P^*\|_{L^p(Vm)} \leq R\,\|P\|_{L^p(Vm)} \qquad \forall\, P\in E^\infty,$$

$$(0)\qquad \|\operatorname{Im} h\|_{L^p(Vm)} \leq R\,\|\operatorname{Re} h\|_{L^p(Vm)} \qquad \forall\, h\in H \text{ with } \operatorname{Im}\varphi(h) = 0,$$

$$(\#)\qquad \|\operatorname{Im} h\|_{L^p(Vm)} \leq R\,\|\operatorname{Re} h\|_{L^p(Vm)} \qquad \forall\, h\in H^\# \text{ with } \operatorname{Im}\varphi(h) = 0 \text{ and } \int|h|^p V dm<\infty.$$

The implications $(\infty)\Rightarrow(0)$ and $(\#)\Rightarrow(0)$ are obvious. The implication $(0)\Rightarrow(\infty)$ is an immediate consequence from the basic approximation theorem 4.1 combined with the Fatou theorem, and the implication $(0)\Rightarrow(\#)$ is clear after the definition of $H^\#$. We define $0\leq R(p,V)\leq\infty$ to be the smallest constant $0\leq R\leq\infty$ for which the above estimations $(\infty)(0)(\#)$ hold true.

5.2 REMARK: i) For each $V\in M$ we have either $R(p,V)=0$ $\forall 1\leq p<\infty$ or $R(p,V)\geq 1$ $\forall 1\leq p<\infty$. In fact, if $u|[V>0]=\text{const}=\varphi(u)$ $\forall u\in H$ then $R(p,V)=0$ $\forall 1\leq p<\infty$. Otherwise take some $u\in H$ with $\varphi(u)=0$ and $u|[V>0]\neq 0$ and apply (0) to u and iu. ii) For each $V\in M$ we have $R(2,V)\leq 1$ from 4.6 and hence $R(2,V)=0$ or $=1$.

5.3 REMARK: i) If $1\leq p<\infty$ and $V\in M$ is dominant such that there exists some nonconstant real-valued $h\in H^\#$ with $\int|h|^p V dm<\infty$ then $R(p,V)=\infty$. In fact, if $R(p,V)<\infty$ and $h\in H^\#$ is real-valued with $\int|h|^p V dm<\infty$ then $(\#)$ applied to $i(h-\operatorname{Re}\varphi(h))$ enforces that $h=\operatorname{Re}\varphi(h)=\text{const}$. ii) In the example 4.11 we thus have $R(p,G)=\infty$ $\forall 1\leq p<\frac{3}{2}$.

5.4 PROPOSITION: For each $V\in M$ we have $R(np,V)\leq 6nR(p,V)$ $\forall 1\leq p<\infty$ and $n\in\mathbb{N}$. In particular $R(2n,V)\leq 6n<\infty$ $\forall n\in\mathbb{N}$ (this estimation will be improved in 5.14).

Proof: We can assume that $1\leq R(p,V)<\infty$. i) We fix $h=P+iQ\in H$ with $\varphi(h)=0$. Then

$$|Q|^n\leq|h|^n=|i^{1-n}h^n|\leq|\operatorname{Re}(i^{1-n}h^n)|+|\operatorname{Im}(i^{1-n}h^n)|,$$

$$\||Q|^n\|_{L^p(Vm)}\leq\|\operatorname{Re}(i^{1-n}h^n)\|_{L^p(Vm)}+\|\operatorname{Im}(i^{1-n}h^n)\|_{L^p(Vm)}\leq$$
$$\leq 2R(p,V)\|\operatorname{Re}(i^{1-n}h^n)\|_{L^p(Vm)}.$$

Now we have

$$i^{1-n}h^n = \sum_{\ell=0}^{n} i^{1-\ell}\binom{n}{\ell}P^\ell Q^{n-\ell},$$

$$\mathrm{Re}\left(i^{1-n}h^n\right) = \sum_{\substack{k\geq 0\\ 2k+1\leq n}} (-1)^k\binom{n}{2k+1}P^{2k+1}Q^{n-1-2k}.$$

Thus with A and B the $L^{np}(Vm)$-norms of P and Q we obtain

$$B^n = \left(\int|Q|^{np}Vdm\right)^{\frac{1}{p}} \leq 2R(p,V) \sum_{\substack{k\geq 0\\ 2k+1\leq n}} \binom{n}{2k+1}\left(\int|P|^{(2k+1)p}|Q|^{(n-1-2k)p}Vdm\right)^{\frac{1}{p}}$$

$$\leq 2R(p,V) \sum_{\substack{k\geq 0\\ 2k+1\leq n}} \binom{n}{2k+1}\left(\int|P|^{np}Vdm\right)^{\frac{2k+1}{np}}\left(\int|Q|^{np}Vdm\right)^{\frac{n-1-2k}{np}}$$

$$= 2R(p,V) \sum_{\substack{k\geq 0\\ 2k+1\leq n}} \binom{n}{2k+1}A^{2k+1}B^{n-1-2k} = R(p,V)\left((B+A)^n-(B-A)^n\right).$$

We claim that $B\leq 3nR(p,V)A$. For the proof we can assume that $B>A\geq 0$ and $n\geq 2$. Let us put $0\leq x:=\frac{A}{B}<1$ and $0<S:=\frac{n-1}{2n}\,\frac{1}{R(p,V)}<1$. Then

$$1 \leq R(p,V)\left((1+x)^n-(1-x)^n\right) = R(p,V)n\int_{-x}^{x}(1+t)^{n-1}dt$$

$$\leq R(p,V)n\int_{-x}^{x}e^{(n-1)t}dt = R(p,V)\frac{n}{n-1}\left(e^{(n-1)x}-e^{-(n-1)x}\right),$$

$$\left(e^{(n-1)x}\right)^2-1 \geq 2Se^{(n-1)x} \quad \text{or} \quad e^{(n-1)x} \geq S+\sqrt{1+S^2},$$

$$(n-1)x \geq \log\left(S+\sqrt{1+S^2}\right) = \int_0^S\frac{1}{\sqrt{1+t^2}}dt \geq \frac{S}{\sqrt{2}} \geq \frac{2S}{3} = \frac{n-1}{3n}\,\frac{1}{R(p,V)},$$

so that $x>0$ and $\frac{B}{A}=\frac{1}{x}\leq 3nR(p,V)$ as claimed. ii) Let $h=P+iQ\in H$ with real $a:=\varphi(h)=\int PVdm$. Then

$$|a| \leq \int|P|Vdm \leq \|P\|_{L^{np}(Vm)},$$

so that from i) applied to $h-a=(P-a)+iQ$ we obtain

$$\|Q\|_{L^{np}(Vm)} \leq 3nR(p,V)\,\|P-a\|_{L^{np}(Vm)} \leq 6nR(p,V)\|P\|_{L^{np}(Vm)}.$$

It follows that $R(np,V) \leq 6nR(p,V)$. QED.

5.5 COROLLARY: Let $P \in E^\infty$ and $h:=P+iP^* \in H^\#$. For each $1 \leq p < \infty$ then

i) $\theta(|h|^p) = \mathrm{Sup}\{\int |h|^p V dm : V \in M\} < \infty$.

ii) If the $h_\varepsilon \in H$ $\forall \varepsilon > 0$ are as in the approximation theorem 4.1 then $\theta(|h_\varepsilon - h|^p) = O(\varepsilon^p)$ for $\varepsilon \downarrow 0$.

Proof: i) follows from 5.4 for the exponents $p=2n$ $(n=1,2\ldots)$ and is then clear for all $1 \leq p < \infty$. ii) Fix $1 \leq p < s < \infty$. For $0 < \varepsilon \leq \frac{s}{p} - 1$ then $|h_\varepsilon - h|^p \leq$ $\leq \varepsilon^p (\mathrm{Max}(1,|h|))^{(1+\varepsilon)p} \leq \varepsilon^p (\mathrm{Max}(1,|h|))^s = \varepsilon^p \mathrm{Max}(1,|h|^s)$ and hence $\theta(|h_\varepsilon - h|^p) \leq$ $\leq \varepsilon^p \theta(\mathrm{Max}(1,|h|^s))$. The assertion follows from i). QED.

We turn to the central results of the section.

5.6 THEOREM: Assume that H contains nonconstant inner functions. For each $V \in M$ then

$$R(p,V) \geq \mathrm{Max}(\tan\frac{\pi}{2p}, \cot\frac{\pi}{2p}) \qquad \forall\ 1 \leq p < \infty.$$

In particular $R(1,V)=\infty$.

5.7 THEOREM: For each $V \in MJ$ we have

$$R(p,V) \leq \mathrm{Max}(\tan\frac{\pi}{2p}, \cot\frac{\pi}{2p}) \qquad \forall\ 1 \leq p < \infty.$$

We combine these results with V.6.9 and Section IV.4 to obtain the subsequent particular theorem.

5.8 THEOREM: Assume the Szegö situation $M=\{F\}$ and $H \neq \mathbb{C}$. Then

$$R(p,F) = \mathrm{Max}(\tan\frac{\pi}{2p}, \cot\frac{\pi}{2p}) \qquad \forall\ 1 \leq p < \infty.$$

Proof of 5.6: We fix an inner function $u \in H$ with $\varphi(u)=0$ which exists after V.3.5.i). Then $h:=\frac{1-u}{1+u} \in H^+$ with $\mathrm{Re}\, h=0$ and $\varphi(h)=1$ after V.4.2. We have $\int |h| V dm = \infty$ $\forall\, V \in M$ since $\int |h| V dm < \infty$ implies after IV.3.4 that $\varphi(h) = \int h V dm$ and hence $\mathrm{Re}\varphi(h)=0$ which contradicts $\varphi(h)=1$. For $1 \leq p < \infty$ and $0 < \tau < 1$ the main branch $h^{\frac{\tau}{p}}$ is $\in H^+$ with $\varphi(h^{\frac{\tau}{p}})=1$ after V.4.9. We have

$$h = |h| e^{i\frac{\pi}{2}\operatorname{sgn}(\frac{h}{i})} \quad \text{and hence } h^{\frac{\tau}{p}} = |h|^{\frac{\tau}{p}} e^{i\frac{\tau\pi}{2p}\operatorname{sgn}(\frac{h}{i})}.$$

And the estimation V.5.6 shows that $\int |h^{\frac{\tau}{p}}|^p V dm = \int |h|^\tau V dm < \infty \ \forall\, V \in M$. From the Fatou lemma we have $\int |h|^\tau V dm \to \infty$ for $\tau \uparrow 1$. i) For the tan estimation apply (#) to $h^{\frac{\tau}{p}}$ to obtain

$$(\sin \tfrac{\tau\pi}{2p})(\int |h|^\tau V dm)^{\frac{1}{p}} \leqq R(p,V)(\cos \tfrac{\tau\pi}{2p})(\int |h|^\tau V dm)^{\frac{1}{p}} \quad \forall V \in M,$$

and hence $\sin \frac{\pi}{2p} \leqq R(p,V) \cos \frac{\pi}{2p}$ for $\tau \uparrow 1$. ii) For the cotan estimation apply (#) to $i(h^{\frac{\tau}{p}} - 1)$ to obtain

$$\|(\cos \tfrac{\tau\pi}{2p})|h|^{\frac{\tau}{p}} - 1\|_{L^p(Vm)} \leqq R(p,V)(\sin \tfrac{\tau\pi}{2p}) \| \, |h|^{\frac{\tau}{p}} \|_{L^p(Vm)},$$

$$(\cos \tfrac{\tau\pi}{2p})(\int |h|^\tau V dm)^{\frac{1}{p}} \leqq 1 + R(p,V)(\sin \tfrac{\tau\pi}{2p})(\int |h|^\tau V dm)^{\frac{1}{p}} \quad \forall\, V \in M,$$

and hence $\cos \frac{\pi}{2p} \leqq R(p,V) \sin \frac{\pi}{2p}$ for $\tau \uparrow 1$. QED.

The proof of 5.7 requires several lemmata. We define $C_*^\infty(¢)$ to consist of the functions $\in C^\infty(¢)$ with compact support. In the same sense we define $C_*(¢)$, $L_*^\infty(¢)$, $ca_*(¢)$,.... We start with the subsequent fundamental representation formula which has a standard proof based on the Green formula A.3.2.

<u>5.9</u> <u>LEMMA</u>: For $f \in C_*^2(¢)$ we have

$$f(z) = \frac{1}{2\pi} \int \log|z-t| \Delta f(t) dL(t) \qquad \forall\, z \in ¢.$$

The function $G \in \operatorname{ReC}(¢)$ is defined to be subharmonic iff

$$\int G(z) \Delta f(z) dL(z) \geqq 0 \quad \forall\, 0 \leqq f \in C_*^\infty(¢),$$

that is iff the distributional derivative ΔG is a Radon measure $\geqq 0$ on ¢. In the case $G \in \operatorname{ReC}^2(¢)$ this means that the usual derivative ΔG is $\geqq 0$ on ¢.

<u>5.10</u> <u>LEMMA</u>: Assume that $G \in \operatorname{ReC}(¢)$ is subharmonic. For each $V \in MJ$ then

$$G(\varphi(u)) \leqq \int G(u) V dm \qquad \forall\, u \in H.$$

Proof of 5.10: i) We first consider a subharmonic function $G\in\mathrm{Re}C^2(\mathbb{C})$. We fix $R>0$ and $F\in C_*^2(\mathbb{C})$ with $F(z)=1\ \forall|z|\leq R$. From 5.9 applied to $GF\in C_*^2(\mathbb{C})$ we obtain

$$G(z) = \frac{1}{2\pi}\int_{V(R)} \log|z-t|\Delta G(t)dL(t) + \frac{1}{2\pi}\int_{\mathbb{C}-V(R)} \log|z-t|\Delta(GF)(t)dL(t) \quad \forall z\in V(R),$$

with $V(R):=\{z\in\mathbb{C}: |z|<R\}$. The last term represents a harmonic function on $V(R)$ and hence is $=\mathrm{Re}\,h(z)$ for some $h\in\mathrm{Hol}(V(R))$. Let now $u\in H$ with $|u|\leq c<R$. Then $|\varphi(u)|\leq c<R$ as well, and it is clear that $h(u)\in H$ with $\varphi(h(u))= h(\varphi(u))$. Furthermore we have the estimation

$$\int_{V(R)} |\log|z-t||\,\Delta G(t)dL(t) \leq \text{const} \int_{V(R)} |\log|z-t||dL(t)$$

$$\leq \text{const} \int_{V(2R)} |\log|t||dL(t) \quad \forall z\in V(R),$$

so that we can apply the Fubini theorem to obtain for $V\in MJ$ the desired estimation

$$\int G(u)Vdm = \frac{1}{2\pi}\int \Big(\int_{V(R)} \log|u-t|\Delta G(t)dL(t)\Big)Vdm + \mathrm{Re}\int h(u)Vdm$$

$$= \frac{1}{2\pi}\int_{V(R)} \Big(\int\log|u-t|Vdm\Big)\Delta G(t)dL(t) + \mathrm{Re}\varphi(h(u))$$

$$\geq \frac{1}{2\pi}\int_{V(R)} \log|\varphi(u)-t|\Delta G(t)dL(t) + \mathrm{Re}\,h(\varphi(u)) = G(\varphi(u)).$$

ii) Consider now a subharmonic function $G\in\mathrm{Re}C(\mathbb{C})$. We fix some $0\leq F\in C_*^\infty(\mathbb{C})$ with $\int F(z)dL(z)=1$ and put $F_\varepsilon: F_\varepsilon(z)=\varepsilon^{-2}F(\frac{z}{\varepsilon})\ \forall z\in\mathbb{C}$. We form the convolution

$$G_\varepsilon = G*F_\varepsilon: G_\varepsilon(z) = \int G(z-t)F_\varepsilon(t)dL(t) = \int G(t)F_\varepsilon(z-t)dL(t) \quad \forall z\in\mathbb{C}.$$

Then $G_\varepsilon\in C^\infty(\mathbb{C})$ and is subharmonic since

$$\Delta G_\varepsilon(z) = \int G(t)\Delta F_\varepsilon(z-t)dL(t) \geq 0 \quad \forall z\in\mathbb{C}.$$

Moreover $G_\varepsilon\to G$ uniformly on each compact set $\subset\mathbb{C}$ for $\varepsilon\downarrow 0$. Thus for $V\in MJ$ and $u\in H$ we have $G_\varepsilon(\varphi(u))\leq\int G_\varepsilon(u)Vdm$ from i) and hence $G(\varphi(u))\leq\int G(u)Vdm$. QED.

For $1<p<\infty$ we define

$$A_p := \mathrm{Max}\Big(\tan\frac{\pi}{2p}, \operatorname{cotan}\frac{\pi}{2p}\Big) = \begin{Bmatrix} \tan\frac{\pi}{2p} & \text{for } 1<p\leq 2 \\ \operatorname{cotan}\frac{\pi}{2p} & \text{for } 2\leq p<\infty \end{Bmatrix} \geq 1.$$

In view of 5.10 the proof of 5.7 will be complete once we have established the subsequent lemma.

5.11 LEMMA: For each $1<p<\infty$ there exists a subharmonic function $G \in ReC(\mathbb{C})$ such that

$$(*) \qquad G(z) \leq A_p^p |x|^p - |y|^p \quad \forall\, z=x+iy \in \mathbb{C},$$

and $G(z) \geq 0 \ \forall z \in \mathbb{R}$.

Proof of 5.11 ⇒ 5.7: We can assume that $1<p<\infty$. For $V \in MJ$ and $u \in H$ with $Im\varphi(u)=0$ we obtain from 5.11 and 5.10

$$(A_p^p \| Re\, u \|_{L^p(Vm)})^p - (\| Im\, u \|_{L^p(Vm)})^p$$

$$= \int (A_p^p |Re\, u|^p - |Im\, u|^p) Vdm \geq \int G(u) Vdm \geq G(\varphi(u)) \geq 0. \text{ QED.}$$

The proof of 5.11 depends on certain sophisticated calculus inequalities. For $1<p<\infty$ we define

$$B_p := \left\{ \begin{array}{ll} \dfrac{(\sin\frac{\pi}{2p})^{p-1}}{\cos\frac{\pi}{2p}} & \text{for } 1<p\leq 2 \\[2ex] \dfrac{(\cos\frac{\pi}{2p})^{p-1}}{\sin\frac{\pi}{2p}} & \text{for } 2\leq p<\infty \end{array} \right\} > 0.$$

The inequalities in question are as follows.

5.12 LEMMA: i) In the case $1<p\leq 2$ we have

$$B_p \cos pt \leq A_p^p (\cos t)^p - (\sin t)^p \qquad \forall\ 0\leq t\leq \tfrac{\pi}{2}.$$

ii) In the case $2\leq p<\infty$ we have

$$-B_p \cos p(t-\tfrac{\pi}{2}) \leq A_p^p (\cos t)^p - (\sin t)^p \qquad \forall\ \tfrac{\pi}{2}-\tfrac{\pi}{p}\leq t\leq \tfrac{\pi}{2}.$$

This implies that

$$B_p \leq A_p^p (\cos t)^p - (\sin t)^p \qquad \forall\ 0\leq t\leq \tfrac{\pi}{2}-\tfrac{\pi}{p},$$

since the estimation is true for $t=\frac{\pi}{2}-\frac{\pi}{p}$ and the second member is monotone decreasing in $0 \leqq t \leqq \frac{\pi}{2}$.

Proof of 5.12: For p=2 we have $A_p=1$ and $B_p=1$ so that both times the assertion reads $\cos 2t \leqq (\cos t)^2-(\sin t)^2 \ \forall\, 0 \leqq t \leqq \frac{\pi}{2}$ and hence is obvious. Thus we can assume that $p \neq 2$. Moreover it suffices to prove the inequalities in question in the respective open intervals.

i) In the case $1<p<2$ we consider the function

$$F: F(t) = \frac{(\sin t)^p + B_p \cos pt}{(\cos t)^p} \qquad \forall\, 0<t<\frac{\pi}{2}.$$

It is C^∞ with $F(\frac{\pi}{2p}) = (\tan\frac{\pi}{2p})^p = A_p^p$. The assertion is $F(t) \leqq F(\frac{\pi}{2p})\ \forall\, 0<t<\frac{\pi}{2}$. One computes that

$$F'(t) = p\,\frac{(\sin t)^{p-1}}{(\cos t)^{p+1}}\,(1-B_p f(t)) \quad \text{with } f: f(t) = \frac{\sin(p-1)t}{(\sin t)^{p-1}},$$

$$f'(t) = (p-1)\,\frac{\sin(2-p)t}{(\sin t)^p} \quad \forall\, 0<t<\frac{\pi}{2}.$$

We see that $f'(t)>0$ in $0<t<\frac{\pi}{2}$. Thus f strictly increases and hence $1-B_p f$ strictly decreases in $]0,\frac{\pi}{2}[$. Now $1-B_p f(\frac{\pi}{2p})=0$. Thus $1-B_p f$ and hence F' are >0 in $]0,\frac{\pi}{2p}[$ and <0 in $]\frac{\pi}{2p},\frac{\pi}{2}[$, so that F strictly increases in $]0,\frac{\pi}{2p}]$ and strictly decreases in $[\frac{\pi}{2p},\frac{\pi}{2}[$. The assertion follows.

ii) In the case $2<p<\infty$ we consider the function

$$F: F(t) = \frac{(\sin t)^p - B_p \cos p(t-\frac{\pi}{2})}{(\cos t)^p} \qquad \forall\, \frac{\pi}{2}-\frac{\pi}{p}<t<\frac{\pi}{2}.$$

It is C^∞ with $F(\frac{\pi}{2}-\frac{\pi}{2p}) = (\operatorname{cotan}\frac{\pi}{2p})^p = A_p^p$. The assertion is $F(t) \leqq F(\frac{\pi}{2}-\frac{\pi}{2p})$ $\forall \frac{\pi}{2}-\frac{\pi}{p}<t<\frac{\pi}{2}$. One computes that

$$F'(t) = p\,\frac{(\sin t)^{p-1}}{(\cos t)^{p+1}}\,(1-B_p f(t)) \quad \text{with } f: f(t) = \frac{\cos(p-1)(t-\frac{\pi}{2})}{(\sin t)^{p-1}},$$

$$f'(t) = -(p-1)\,\frac{\sin(p-2)(t-\frac{\pi}{2})}{(\sin t)^p} \quad \forall\, \frac{\pi}{2}-\frac{\pi}{p}<t<\frac{\pi}{2}.$$

We see that $f'(t)>0$ in $\frac{\pi}{2}-\frac{\pi}{p}<t<\frac{\pi}{2}$. Thus f strictly increases and hence $1-B_p f$ strictly decreases in $]\frac{\pi}{2}-\frac{\pi}{p},\frac{\pi}{2}[$. Now $1-B_p f(\frac{\pi}{2}-\frac{\pi}{2p})=0$. Thus $1-B_p f$

and hence F' are >0 in $]\frac{\pi}{2}-\frac{\pi}{p},\frac{\pi}{2}-\frac{\pi}{2p}[$ and <0 in $]\frac{\pi}{2}-\frac{\pi}{2p},\frac{\pi}{2}[$, so that F strictly increases in $]\frac{\pi}{2}-\frac{\pi}{p},\frac{\pi}{2}-\frac{\pi}{2p}]$ and strictly decreases in $[\frac{\pi}{2}-\frac{\pi}{2p},\frac{\pi}{2}[$. The assertion follows. QED.

Proof of 5.12 $\Rightarrow$ 5.11: i) In the case $1<p\leq 2$ we define $G:\mathbb{C}\rightarrow\mathbb{R}$ for $z=|z|e^{it}\in\mathbb{C}$ as follows: if $\operatorname{Re} z\geq 0$ and $|t|\leq\frac{\pi}{2}$ then

$$G(z):=\ B_p|z|^p\cos pt\ =\ B_p\operatorname{Re}(z^p)\qquad\text{(the main branch)},$$

and if $\operatorname{Re} z\leq 0$ then $G(z):=G(-\bar{z})$. Note that if $\operatorname{Re} z=0$ then $z=-\bar{z}$ so that $G(z)$ is well-defined. We obtain the subsequent properties. 1) G is continuous. 2) G is harmonic in the open halfplanes $\Delta:\operatorname{Re} z>0$ and $-\Delta:\ \operatorname{Re} z<0$. 3) For all $z=x+iy\in\mathbb{C}$ we have $G(x+iy)=G(-x+iy)=G(x-iy)$. 4) A standard application of the Green formula A.3.2 combined with 1) - 3) leads to

$$\int G(z)\Delta f(z)dL(z)\ =\ 2pB_p\ \sin\frac{p\pi}{2}\int_{-\infty}^{\infty}|t|^{p-1}f(it)dt\qquad\forall\ f\in C_*^2(\mathbb{C}).$$

Thus G is subharmonic. 5) In order to prove (*) we can in view of 3) restrict ourselves to $z=x+iy=|z|e^{it}$ with $x,y\geq 0$ and hence $0\leq t\leq\frac{\pi}{2}$. But then 5.12 shows that

$$G(z)\ =\ B_p|z|^p\cos pt\leq\ A_p^p(|z|\cos t)^p\ -\ (|z|\sin t)^p\ =\ A_p^p|x|^p-|y|^p.$$

6) For $z\in\mathbb{R}$ we have $G(z)=G(|z|)=B_p|z|^p\geq 0$.

ii) In the case $2\leq p<\infty$ we define $G:\mathbb{C}\rightarrow\mathbb{R}$ for $z=|z|e^{it}\in\mathbb{C}$ as follows: if $\operatorname{Im} z=\operatorname{Re}\frac{z}{i}\geq 0$ and $0\leq t\leq\pi$ then

$$G(z):=\begin{cases}-B_p|z|^p\cos p(t-\frac{\pi}{2})=-B_p\operatorname{Re}(\frac{z}{i})^p\text{(the main branch)} & \text{for } |t-\frac{\pi}{2}|\leq\frac{\pi}{p}\\ B_p|z|^p & \text{for } |t-\frac{\pi}{2}|\geq\frac{\pi}{p}\end{cases},$$

and if $\operatorname{Im} z\leq 0$ then $G(z):=G(\bar{z})$. Note that $G(z)$ is well-defined in all cases. We obtain the subsequent properties. 1) G is continuous. 2) G is harmonic in the open double cone $U:z=|z|e^{it}\neq 0$ with $||t|-\frac{\pi}{2}|<\frac{\pi}{p}$ and is C^∞ with $\Delta G(z)=p^2B_p|z|^{p-2}$ in the open double cone $V:z=|z|e^{it}\neq 0$ with $\frac{\pi}{p}<||t|-\frac{\pi}{2}|\leq\frac{\pi}{2}$. 3) For all $z=x+iy\in\mathbb{C}$ we have $G(x+iy)=G(x-iy)=G(-x+iy)$.
4) A standard application of the Green formula A.3.2 combined with 1)-3)

Leads to

$$\int G(z)\Delta f(z)dL(z) = p^2 B_p \int_V |z|^{p-2} f(z)dL(z) \quad \forall\ f\in C_*^2(\mathbb{C}).$$

Thus G is subharmonic. 5) In order to prove (*) we can in view of 3) restrict ourselves to $z = x + iy = |z|e^{it}$ with $x,y \geq 0$ and hence $0 \leq t \leq \frac{\pi}{2}$. But then 5.12 shows that

$$\frac{\pi}{2} - \frac{\pi}{p} \leq t \leq \frac{\pi}{2}: \ G(z) = -B_p |z|^p \cos p(t-\frac{\pi}{2}) \leq$$

$$\leq A_p^p(|z|\cos t)^p - (|z|\sin t)^p = A_p^p|x|^p - |y|^p,$$

$$0 \leq t \leq \frac{\pi}{2} - \frac{\pi}{p}: \ G(z) = B_p |z|^p$$

$$\leq A_p^p(|z|\cos t)^p - (|z|\sin t)^p = A_p^p|x|^p - |y|^p.$$

6) For $z\in\mathbb{R}$ we have $G(z) = B_p|z|^p \geq 0$. QED.

At this point the proof of 5.7 is complete. The subsequent results are important by-products of the above discussion.

<u>5.13</u> <u>THEOREM</u>: Assume that $V\in M$. i) In the case $1 < p \leq 2$ we have

$$\|\mathrm{Im}\, h \pm A_p\, \varphi(h)\|_{L^p(Vm)} \leq A_p \|\mathrm{Re}\, h\|_{L^p(Vm)} \quad \text{and hence}$$

$$\|\mathrm{Im}\, h\|_{L^p(Vm)} \leq A_p \|\mathrm{Re}\, h\|_{L^p(Vm)}$$

$$\forall\ h\in H^+ \text{ with } \mathrm{Im}\,\varphi(h) = 0 \text{ and } \int |h|^p V dm < \infty.$$

ii) In the case $2 \leq p < \infty$ we have

$$\|\mathrm{Re}\, h\|_{L^p(Vm)} \leq \|A_p \mathrm{Im}\, h \pm \varphi(h)\|_{L^p(Vm)}$$

$$\forall h\in H^+ \text{ with } \mathrm{Im}\,\varphi(h) = 0 \text{ and } \int |h|^p V dm < \infty.$$

In 5.13.i) the restriction to H^+ instead of $H^\#$ is essential as is evident from 5.3.ii). We shall come back to this question in Section 7.

Proof: i) In the case $1 < p \leq 2$ we know from 5.12.i) that

$$B_p \cos pt \leq A_p^p (\cos t)^p - |\sin t|^p \qquad \forall\ |t| \leq \frac{\pi}{2}.$$

For $h \in H^+$ with $\int |h|^p V dm < \infty$ we have $h^p \in H^\#$ with $\varphi(h^p) = (\varphi(h))^p$ after V.4.9 and $\varphi(h^p) = \int h^p V dm$ after IV.3.4. Thus $B_p \mathrm{Re}(h^p) \leqq A_p^p (\mathrm{Re}\, h)^p - |\mathrm{Im}\, h|^p$ implies that

$$\int |\mathrm{Im}\, h|^p V dm \leqq A_p^p \int |\mathrm{Re}\, h|^p V dm - B_p \mathrm{Re}(\varphi(h))^p.$$

We apply this to $h^o := h + i\varepsilon A_p \varphi(h)$ with $\varepsilon = \pm 1$ under the assumption $\mathrm{Im}\varphi(h) = 0$. We see that

$$\varphi(h^o) = \varphi(h)\big(1 + i\varepsilon A_p\big) = \frac{\varphi(h)}{\cos\frac{\pi}{2p}} \exp\big(i\frac{\varepsilon\pi}{2p}\big),$$

$$\big(\varphi(h^o)\big)^p = \Big(\frac{\varphi(h)}{\cos\frac{\pi}{2p}}\Big)^p \exp\big(i\,\frac{\varepsilon\pi}{2}\big) \text{ and hence } \mathrm{Re}\big(\varphi(h^o)\big)^p = 0,$$

so that the result follows. ii) In the case $2 \leqq p < \infty$ we know from 5.12.ii) that

$$-B_p \cos p(t - \tfrac{\pi}{2}) \leqq A_p^p |\cos t|^p - (\sin t)^p \quad \forall\ 0 \leqq t \leqq \pi,$$

$$-B_p \cos pt \leqq A_p^p |\sin t|^p - (\cos t)^p \quad \forall |t| \leqq \tfrac{\pi}{2}.$$

For $h \in H^+$ with $\int |h|^p V dm < \infty$ we obtain as above

$$\int |\mathrm{Re}\, h|^p V dm \leqq A_p^p \int |\mathrm{Im}\, h|^p V dm + B_p \mathrm{Re}\big(\varphi(h)\big)^p.$$

We apply this to $h^o := h + i\varepsilon \frac{1}{A_p} \varphi(h)$ with $\varepsilon = \pm 1$ under the assumption $\mathrm{Im}\varphi(h) = 0$. As above we see that $\mathrm{Re}(\varphi(h^o))^p = 0$, so that the result follows. QED.

<u>5.14</u> <u>THEOREM</u>: Assume that $V \in M$ and $p = 2n\ (n = 1, 2, \ldots)$. Then

$$\|\mathrm{Im}\, h \pm A_p \varphi(h)\|_{L^p(Vm)} \leqq A_p \|\mathrm{Re}\, h\|_{L^p(Vm)} \quad \text{and hence}$$

$$\|\mathrm{Im}\, h\|_{L^p(Vm)} \leqq A_p \|\mathrm{Re}\, h\|_{L^p(Vm)}$$

$$\forall\ h \in H^\# \text{ with } \mathrm{Im}\varphi(h) = 0 \text{ and } \int |h|^p V dm < \infty.$$

Thus $R(p, V) \leqq A_p = \mathrm{cotan}\frac{\pi}{2p}$.

Proof: We know from 5.12.ii) that

$$-(-1)^n B_p \cos pt \leqq A_p^p (\cos t)^p - (\sin t)^p \quad \forall\ |t| \leqq \pi.$$

Thus for $h \in H^{\#}$ with $\int |h|^p V dm < \infty$ we obtain

$$\int (\mathrm{Im}\, h)^p V dm \leqq A_p^p \int (\mathrm{Re}\, h)^p V dm + (-1)^n B_p \mathrm{Re}(\varphi(h))^p.$$

We apply this to $h^o := h + i\varepsilon A_p \varphi(h)$ with $\varepsilon = \pm 1$ under the assumption $\mathrm{Im}\varphi(h) = 0$. We see that

$$\varphi(h^o) = \varphi(h)\left(1 + i\varepsilon A_p\right) = \frac{i\varepsilon\varphi(h)}{\sin\frac{\pi}{2p}} \exp\left(-i\frac{\varepsilon\pi}{2p}\right),$$

$$(\varphi(h^o))^p = (-1)^n \left(\frac{\varphi(h)}{\sin\frac{\pi}{2p}}\right)^p \exp\left(-i\frac{\varepsilon\pi}{2}\right) \text{ and hence } \mathrm{Re}(\varphi(h^o))^p = 0,$$

so that the result follows. QED.

6. Special Situations

In the first part of the section we discuss several properties of an individual function $F \in M$ as to their mutual dependence.

6.1 REMARK: For $F \in M$ the subsequent properties are equivalent.

i) $M \subset F(\mathrm{ReL}^\infty(m))$, that is for each $V \in M$ there exists a constant $c > 0$ such that $V \leqq cF$.

i') There exists a constant $c > 0$ such that $V \leqq cF \ \forall\, V \in M$.

ii) F is an internal point of the convex set $M \subset \mathrm{ReL}^1(m)$, that is to each $V \in M$ there exists an $\varepsilon > 0$ such that $F - \varepsilon(V - F) \in M$.

ii') There exists an $\varepsilon > 0$ such that $F - \varepsilon(V - F) \in M \ \forall\, V \in M$.

iii) $N = \{c(V - F) : V \in M \text{ and } c > 0\}$.

The functions $F \in M$ which possess the equivalent properties i) - iii) in 6.1 are called internal functions. From ii) we see that in the case $\dim N < \infty$ internal functions always exist. Note that the proof of 6.1 will not use the reducedness of (H, φ).

Proof: i) ⇒ i') Assume that the assertion is not true. Then there exist functions $V_n \in M$ such that $V_n \leqq n2^n F$ is false $(n = 1, 2 \ldots)$. Now $V := \sum_{n=1}^{\infty} \frac{1}{2^n} V_n \in M$ and hence $V \leqq cF$ for some $c > 0$. It follows that $V_n \leqq 2^n V \leqq c2^n F \leqq n2^n F$ for all sufficiently large n. So we obtain a contradiction.

i') ⇒ ii') For $V \in M$ and $\varepsilon>0$ the relation $F-\varepsilon(V-F) \in M$ is equivalent to $F-\varepsilon(V-F) \geqq 0$ or $V \leqq (1+\frac{1}{\varepsilon})F$. Thus the implication is clear. ii') ⇒ ii) is trivial. ii) ⇒ iii) We have to prove the inclusion $\subset$. Let $f \in N$, that is $f = c(U-V)$ with $U,V \in M$ and $c>0$. From ii) we have an $\varepsilon>0$ such that $W:= =F-\varepsilon(V-F)=(1+\varepsilon)F-\varepsilon V \in M$. It follows that

$$f = c(U-V) = c(U+\frac{W-(1+\varepsilon)F}{\varepsilon}) = c\frac{1+\varepsilon}{\varepsilon}\left(\frac{W+\varepsilon U}{1+\varepsilon} - F\right).$$

iii) ⇒ i) Let $V \in M$. Then $F-V \in N$ and hence $F-V=c(U-F)$ for some $U \in M$ and $c>0$. It follows that $F-V \geqq -cF$ or $V \leqq (1+c)F$. QED.

6.2 REMARK: For $F \in M$ consider the subsequent properties.

o) $L^{o}(Fm) \subset L^{\#}$.

1) If $0 \leqq f_n \in ReL^1(Fm)$ and $\int f_n F dm \to 0$ then $\alpha(f_n) \to 0$.

∞) If $0 \leqq f_n \in ReL^{\infty}(m)$ and $\int f_n F dm \to 0$ then $\alpha(f_n) \to 0$.

1↓) If $0 \leqq f_n \in ReL^1(Fm)$ and $f_n \downarrow 0$ then $\alpha(f_n) \to 0$.

∞↓) If $0 \leqq f_n \in ReL^{\infty}(m)$ and $f_n \downarrow 0$ then $\alpha(f_n) \to 0$.

Then 1) and ∞) are equivalent, and are equivalent to o) plus $F>0$ on X and equivalent to 1↓) plus $F>0$ on X. Of course ∞) ⇒ ∞↓), but we do not claim the converse even if $F>0$ (note that ∞↓) does not depend on F!).

The functions $F \in M$ which possess the equivalent properties 1) and ∞), o) plus $F>0$ and 1↓) plus $F>0$ in 6.2 are called enveloped functions.

Proof: i) We start with the equivalence 1) ⇔ ∞). The implication 1) ⇒ ∞) is trivial. For the converse conclude from IV.3.12 that $0 \leqq f_n \uparrow f \in ReL(m)$ implies that $\alpha(f_n) \uparrow \alpha(f)$. It is obvious that this remark proves 1) ⇐ ∞). ii) We next prove that ∞) implies $F>0$ on X. If $B \in \Sigma$ with $\int \chi_B F dm=0$ then ∞) implies that $\alpha(\chi_B)=0$. From IV.3.6 we obtain a function $u \in H$ with $|u| \leqq \exp(-\chi_B)$ and $\varphi(u)=\exp(-\alpha(\chi_B))=1$. For each $V \in M$ it follows that $1=\int uV dm \leqq \int \exp(-\chi_B)\, V dm = 1-(1-\frac{1}{e})\int \chi_B V dm$ and hence $\int \chi_B V dm=0$. Hence $m(B)=0$ since (H,φ) is reduced.

iii) We deduce 1) from o) and $F>0$. Let $0 \leqq f_n \in ReL^1(Fm)$ with $\int f_n F dm \to 0$. If $\alpha(f_n) \to 0$ is false then after transition to a subsequence we have $\alpha(f_n) \to \lambda > 0$. And we can assume that $\int f_n F dm \leqq \frac{1}{n2^n}$. It follows that $G:=\sum_{n=1}^{\infty} n f_n \in L^1(Fm)$

(and in view of $F>0$ on X is well-defined even modulo m) so that $e^G \in L^{\#}$ after 0). Hence for $\varepsilon>0$ we obtain $f_n \leq \frac{G}{n} \leq \varepsilon + \frac{1}{n}(G-\varepsilon n)^+ \leq \varepsilon + (G-\varepsilon n)^+$ and hence $\alpha(f_n) \leq \varepsilon + \alpha((G-\varepsilon n)^+)$. From IV.3.13 it follows that $\lambda \leq \varepsilon$ $\forall \varepsilon>0$ and hence a contradiction. iv) Since 1) $\Rightarrow$ 1↓) is trivial it remains to prove that 1↓) $\Rightarrow$ o). Let $0 \leq f \in ReL^1(Fm)$ and put $f_n := Min(f,n)$ $\forall n \geq 1$. Then $f-f_n = (f-f_n)^+ \downarrow 0$ so that $\alpha((f-f_n)^+) \to 0$. From IV.3.13 we conclude that $e^f \in L^{\#}$. QED.

In order to illustrate condition ∞↓) above we insert the next result. We shall come back to this context in Chapter VIII (see also IV.4.5).

6.3 REMARK: Consider the subsequent conditions.

i) M is compact in $\sigma(ReL^1(m), ReL^{\infty}(m))$.

ii) If $0 \leq f_n \in ReL^{\infty}(m)$ and $f_n \downarrow 0$ then $\theta(f_n) \to 0$.

iii) = ∞↓) If $0 \leq f_n \in ReL^{\infty}(m)$ and $f_n \downarrow 0$ then $\alpha(f_n) \to 0$.

Then i) $\Rightarrow$ ii) $\Rightarrow$ iii) (let us announce that also ii) $\Rightarrow$ i) as it will be seen in VIII.3.1).

Proof: i) $\Rightarrow$ ii) The functions $f_n^{\sim}: V \mapsto \int f_n V dm$ are $\sigma(ReL^1(m), ReL^{\infty}(m))$ continuous real-valued functions on M with supnorm $\|f_n^{\sim}\| = \theta(f_n)$. Since $f_n^{\sim} \downarrow 0$ the Dini theorem implies that $\theta(f_n) \to 0$. ii) $\Rightarrow$ iii) is obvious from IV.3.9. QED.

For $0 \leq F \in ReL^1(m)$ and $1 \leq p < \infty$ let us now define

$$R^p(Fm) := \overline{Re\,H}^{ReL^p(Fm)} := \{f \in ReL(m) : \exists f_n \in Re\,H \text{ with } \int |f-f_n|^p F dm \to 0\},$$

so that likewise

$$R^p(Fm) = \overline{E^{\infty}}^{ReL^p(Fm)}$$

The final result of the first part of the present section then reads as follows.

6.4 PROPOSITION: For $F \in M$ consider the subsequent properties.

i) F is internal.

ii) F is enveloped.

iii) $R^1(Fm) \subset E$ and $\alpha(f) = \int fFdm \ \forall f \in R^1(Fm)$.

iv) $R^1(Fm) \subset E$.

v) $R^1(Fm) \cap ReL^\infty(m) \subset E^\infty$ and hence $=E^\infty$.

vi) $N \subset \overline{N \cap F(ReL^\infty(m))}^{ReL^1(m)}$.

Then i) $\Rightarrow$ ii) $\Rightarrow$ iii) $\Rightarrow$ iv) $\Rightarrow$ v) $\Leftrightarrow$ vi). Hence if $N \cap F(ReL^\infty(m))$ is $L^1(m)$-norm closed then i) - vi) are equivalent.

Proof: i) $\Rightarrow$ ii) If $V \leqq cF \ \forall V \in M$ then from IV.3.9 we obtain $\alpha(f) \leqq \theta(f) \leqq$ $\leqq c\int fFdm$ for all $0 \leqq f \in ReL(m)$. Thus condition 1) in 6.2 is obvious. ii) $\Rightarrow$ iii) Let $f \in R^1(Fm)$. Since $F>0$ on X we have a sequence $f_n \in Re\,H$ such that $f_n \to f$ and $|f_n| \leqq G$ with $G \in ReL^1(Fm)$ and hence $e^G \in L^O(Fm) \subset L^\#$. From 2.4.i) we see that $f \in E$ and $\alpha(f) = \int fFdm$. iii) $\Rightarrow$ iv) and iv) $\Rightarrow$ v) are trivial. Thus it remains to prove the equivalence v) $\Leftrightarrow$ vi). Now observe that both v) and vi) imply that $F>0$ on X. For vi) this is obvious, and to deduce it from v) note that the characteristic function χ_B of $B:=[F=0]$ is in $R^1(Fm) \cap ReL^\infty(m) \subset E^\infty$ so that $\int \chi_B V dm = \int \chi_B F dm = 0 \ \forall V \in M$ and hence $m(B) = 0$ since (H,φ) is reduced. Therefore we can prove the equivalence v) $\Leftrightarrow$ vi) under the assumption that $F>0$ on X. Now we have

$$N \cap F(ReL^\infty(m)) = \{f \in K \cap F(ReL^\infty(m)) : \int fdm = 0\}$$

$$= \{f \in F(ReL^\infty(m)) : f \perp H\} = \{f \in F(ReL^\infty(m)) : f \perp Re\,h\}$$

$$= \{f \in F(ReL^\infty(m)) : f \perp R^1(Fm)\},$$

$$(\tfrac{1}{F}N) \cap ReL^\infty(m) = \{f \in ReL^\infty(m) : f \perp FR^1(Fm)\}.$$

Since $FR^1(Fm) \subset ReL^1(m)$ is a closed linear subspace it follows from the bipolar theorem that

$$FR^1(Fm) = \{f \in ReL^1(m) : f \perp (\tfrac{1}{F}N) \cap ReL^\infty(m)\},$$

$$R^1(Fm) = \{f \in ReL(m) : fF \in ReL^1(m) \text{ and } f \perp N \cap F(ReL^\infty(m))\},$$

$$R^1(Fm) \cap ReL^\infty(m) = \{f \in ReL^\infty(m) : f \perp N \cap F(ReL^\infty(m))\}.$$

We compare the last equation with

$$E^{\infty} = \{f \in ReL^{\infty}(m) : f \perp N\}.$$

It follows from the bipolar theorem that $R^1(Fm) \cap ReL^{\infty}(m) = E^{\infty}$ iff $N \cap F(ReL^{\infty}(m))$ and N have the same $L^1(m)$-norm closure. Thus the equivalence v) $\Leftrightarrow$ vi) becomes obvious. QED.

The second part of the section centers around the condition $\dim N < \infty$. We start with a remarkable lemma which is unrelated to the abstract Hardy algebra situation. It is true in both the real and the complex case.

6.5 LEMMA: Let $1 \leqq p < \infty$ and $T \subset L^p(m)$ be a closed linear subspace such that $T \subset FL^{\infty}(m)$ for some $F \in L^p(m)$. Then T is finite-dimensional.

Proof: i) We can assume that $F>0$ since we can take $|F|+1$ instead of F. Then we can pass over to $\frac{1}{F}T \subset L^p(F^p m)$ and hence assume that $F=1$. The desired result then reads as follows: If a linear subspace $T \subset L^{\infty}(m)$ is $L^p(m)$-norm closed for some $1 \leqq p < \infty$ then $\dim T < \infty$. We shall prove this version. ii) It is clear that T is $L^{\infty}(m)$-norm closed. Thus from the closed graph theorem we obtain a constant $c>0$ such that $\|f\|_{L^{\infty}(m)} \leqq c\|f\|_{L^p(m)}$ $\forall f \in T$. We claim that for each $1 \leqq s < \infty$ there exists $c(s)>0$ such that $\|f\|_{L^{\infty}(m)} \leqq$ $\leqq c(s)\|f\|_{L^s(m)}$ $\forall f \in T$. This is obvious for $p<s<\infty$. For $1 \leqq s < p$ we have

$$\|f\|^p_{L^p(m)} = \int |f|^p dm \leqq \|f\|^{p-s}_{L^{\infty}(m)} \int |f|^s dm \leqq c^{p-s}\|f\|^{p-s}_{L^p(m)} \|f\|^s_{L^s(m)},$$

$$\|f\|_{L^p(m)} \leqq c^{\frac{p}{s}-1} \|f\|_{L^s(m)} \qquad \forall f \in T,$$

and hence the assertion as well. iii) Now we deduce from $\|f\|_{L^{\infty}(m)} \leqq$ $\leqq c\|f\|_{L^2(m)}$ $\forall f \in T$ that T must be finite-dimensional. Take an $n \leqq \dim T$ and orthonormal $f_1,\ldots,f_n \in T$. For $t_1,\ldots,t_n \in \mathbb{C}$ then

$$\left|\sum_{\ell=1}^{n} t_\ell f_\ell\right| \leqq c\left\|\sum_{\ell=1}^{n} t_\ell f_\ell\right\|_{L^2(m)} = c\left(\sum_{\ell=1}^{n} |t_\ell|^2\right)^{\frac{1}{2}}.$$

If representative functions for $f_1,\ldots,f_n$ are chosen then the above inequality is true outside a fixed m-null set $N \in \Sigma$ for all $t_1,\ldots,t_n \in \mathbb{C}$ with rational components and hence for all $t_1,\ldots,t_n \in \mathbb{C}$. Now for $x \in X-N$ take $t_\ell := \overline{f_\ell(x)}$ $(\ell=1,\ldots,n)$ to obtain

$$\sum_{\ell=1}^{n} |f_\ell(x)|^2 \leq c^2.$$

After integration it follows that $n \leq c^2 m(X)$. Thus $\dim T < \infty$. QED.

6.6 PROPOSITION: Let $F \in M$. Then the subsequent properties are equivalent.

i) $\overline{N}^{\mathrm{ReL}^1(m)} \subset F(\mathrm{ReL}^\infty(m))$.

ii) $N \cap F(\mathrm{ReL}^\infty(m))$ is $L^1(m)$-norm closed and F fulfills the equivalent conditions i) - vi) in 6.4.

iii) $\dim N < \infty$ and F is internal.

Proof: i) $\Rightarrow$ iii) follows from 6.5 and iii) $\Rightarrow$ ii) $\Rightarrow$ i) are obvious. QED.

The subsequent results do not depend on lemma 6.5.

6.7 THEOREM: Let $0 < F \in \mathrm{ReL}(m)$ be such that $\overline{N}^{\mathrm{ReL}^1(m)} \subset F(\mathrm{ReL}^\infty(m))$ (we do not assume that F be in $L^1(m)$ and hence cannot conclude that N must be finite-dimensional). Then

i) $E \cap \frac{1}{F}\overline{N}^{\mathrm{ReL}^1(m)} = \{0\}$,

ii) $\mathrm{Re}\, H + \frac{1}{F}\overline{N}^{\mathrm{ReL}^1(m)}$ is weak*dense in $\mathrm{ReL}^\infty(m)$.

Proof: i) Assume that $f \in E \cap \frac{1}{F}\overline{N}^{\mathrm{ReL}^1(m)}$. Then f is bounded so that $f \perp N$ implies that $f \perp \overline{N}^{\mathrm{ReL}^1(m)}$. In particular $f \perp Ff \in \overline{N}^{\mathrm{ReL}^1(m)}$. It follows that $\int Ff^2 dm = 0$ and hence $f = 0$. ii) Assume that $f \in \mathrm{ReL}^1(m)$ annihilates $\mathrm{Re}\, H$ and $\frac{1}{F}\overline{N}^{\mathrm{ReL}^1(m)}$. Then $f \perp E^\infty$ and hence $f \in \overline{N}^{\mathrm{ReL}^1(m)}$ from 4.3 and the bipolar theorem. Hence $f \perp \frac{1}{F}f$ which is a bounded function. It follows that $\int \frac{1}{F}f^2 dm = 0$ and hence $f = 0$. QED.

6.8 COROLLARY: Assume that $\dim N < \infty$. Let $0 < F \in \mathrm{ReL}(m)$ be such that $N \subset F(\mathrm{ReL}^\infty(m))$. Then

i) $E \cap \frac{1}{F} N = \{0\}$,

ii) $E^\infty \oplus \frac{1}{F} N = \mathrm{ReL}^\infty(m)$.

6.9 PROPOSITION: Assume that $\dim N < \infty$ and that $F \in M$ is internal. Then

i) $R^p(Fm) = E \cap ReL^p(Fm)$ for $1 \leq p < \infty$,

ii) $R^p(Fm) \oplus \frac{1}{F} N = ReL^p(Fm)$ for $1 \leq p < \infty$,

iii) $K = (H^{\#} \cap L^1(Fm))F + N + iN$.

Proof: From 6.8.ii) we see that $R^p(Fm) + \frac{1}{F} N = ReL^p(Fm)$. And $R^p(Fm) \subset \subset E \cap ReL^p(Fm)$ since $R^1(Fm) \subset E$ from 6.4. Combine this with $(E \cap ReL^p(Fm)) \cap \cap \frac{1}{F} N = \{0\}$ from 6.8.i) to deduce the directness of the sum in ii) and the relation i). iii) From 4.8 we have

$$K = \overline{HF}^{L^1(m)} + N + iN = \overline{H}^{L^1(Fm)} F + N + iN,$$

so that $\overline{H}^{L^1(Fm)} = H^{\#} \cap L^1(Fm)$ is to be shown. But this is contained in the subsequent remark which will be separated for future reference. QED.

6.10 REMARK: Let $F \in M$ be enveloped and $1 \leq p < \infty$. From $L^p(Fm) \subset L^0(Fm) \subset L^{\#}$ we see that $H^{\#} \cap L^p(Fm)$ is $L^p(Fm)$-norm closed and hence is the $L^p(Fm)$-norm closure of H. Likewise for $H^{\#}_{\varphi} := \{u \in H^{\#} : \varphi(u) = 0\}$ we see that $H^{\#}_{\varphi} \cap L^p(Fm)$ is $L^p(Fm)$-norm closed and hence is the $L^p(Fm)$-norm closure of H_{φ} (see 4.5).

7. Return to the Marcel Riesz and Kolmogorov Estimations

We shall assume that $\dim N < \infty$ and use 6.9 to extend the estimations obtained in Section 5. We start with the Kolmogorov estimation.

7.1 THEOREM: Assume that $\dim N < \infty$. For $0 < \tau < 1$ then

$$\cos \frac{\tau\pi}{2}\, \theta(|P + iP^*|^{\tau}) \leq 2^{1-\tau} (\theta(|P|))^{\tau} \quad \forall\, P \in E.$$

Proof: Let $P \in E$ with $\theta(|P|) < \infty$. We fix an internal $F \in M$ and a constant $c > 0$ such that $V \leq cF\ \forall\, V \in M$. Then $P \in E \cap ReL^1(Fm) = R^1(Fm)$ so that there exists a sequence of functions $P_n \in E^{\infty}$ with $P_n \to P$ and $|P_n| \leq$ some $f \in ReL^1(Fm)$. From 5.1 we have

$$\cos \frac{\tau\pi}{2}\, \theta(|P_p^* - P_q^*|^{\tau}) \leq 2^{1-\tau} (\theta(|P_p - P_q|))^{\tau} \leq 2^{1-\tau} c^{\tau} \|P_p - P_q\|^{\tau}_{L^1(Fm)} .$$

Hence there exists a subsequence $1 \leq n(1)<\ldots<n(\ell)<n(\ell+1)<\ldots$ such that $P^*_{n(\ell)} \to$ some $Q \in \mathrm{ReL}(m)$ for $\ell\to\infty$ and $|P^*_{n(\ell)}| \leq$ some $g \in \mathrm{ReL}(m)$ with $\int g^\tau V dm < \infty$ for all $V \in M$. Once more from 5.1 we obtain

$$\cos\frac{\tau\pi}{2}\int|P_{n(\ell)} + iP^*_{n(\ell)}|^\tau V dm \leq 2^{1-\tau}(\theta(|P_{n(\ell)}|))^\tau$$

$$\leq 2^{1-\tau}(\theta(|P-P_{n(\ell)}|)+\theta(|P|))^\tau \leq 2^{1-\tau}(c\|P-P_{n(\ell)}\|_{L^1(Fm)}+\theta(|P|))^\tau,$$

$$\cos\frac{\tau\pi}{2}\int|P+iQ|^\tau V dm \leq 2^{1-\tau}(\theta(|P|))^\tau \quad \forall\ V\in M,$$

and hence $\cos\frac{\tau\pi}{2}\,\theta(|P+iQ|^\tau) \leq 2^{1-\tau}(\theta(|P|))^\tau$. It remains to show that $Q = P^*$. But this is a routine application of IV.3.2-3.3 and 2.4.i) in view of $e^{|t|f} \in L^\circ(Fm) \subset L^\# \ \forall\ t \in \mathbb{R}$. QED.

Let us turn to the Marcel Riesz estimation. It is clear after the above proof that it can be extended as follows.

<u>7.2</u> <u>THEOREM</u>: Assume that $\dim N<\infty$ and that $F \in M$ is internal. For $1 \leq p<\infty$ then

$$\|P^*\|_{L^p(Fm)} \leq R(p,F)\|P\|_{L^p(Fm)} \qquad \forall\ P \in E.$$

The last aim is the subsequent theorem.

<u>7.3</u> <u>THEOREM</u>: Assume that $\dim N<\infty$ and that $F \in M$ is internal. Then $R(p,F) < \infty \quad \forall\ 1<p<\infty$.

In view of 5.4 we can restrict ourselves to the case $1<p\leq 2$. Thus we have to remove the positivity restriction in 5.13.i). To do this we need the subsequent ad-hoc lemma.

<u>7.4</u> <u>LEMMA</u>: Let $1 \leq p <\infty$ and $\mathrm{ReL}^p(m) = T + R$, where $T \subset \mathrm{ReL}^p(m)$ is a closed linear subspace with $1 \in T$ and $R \subset \mathrm{ReL}^\infty(m)$ is a finite-dimensional linear subspace with $T \cap R = \{0\}$. Then there exists a constant $c>0$ such that

$$\mathrm{Inf}\{\|f\|_{L^p(m)} : |P| \leq f \in T\cap\mathrm{ReL}^\infty(m)\} \leq c\|P\|_{L^p(m)} \qquad \forall\ P \in \mathrm{ReL}^\infty(m).$$

Proof of 7.4: From the closed graph theorem we obtain a constant $a>0$ such that

$$\|f\|_{L^p(m)},\ \|u\|_{L^p(m)} \leqq a\,\|f+u\|_{L^p(m)} \qquad \forall\ f\in T \text{ and } u\in R.$$

Furthermore there exists a constant $b>0$ with

$$\|u\|_{L^\infty(m)} \leqq b\,\|u\|_{L^p(m)} \qquad \forall\ u\in R.$$

Let now $P\in \mathrm{ReL}^\infty(m)$. Then $|P| = f+u$ with $f\in T$ and $u\in R$. Thus $f\in T\cap \mathrm{ReL}^\infty(m)$ and hence $|P| \leqq f+\|u\|_{L^\infty(m)} \in T\cap \mathrm{ReL}^\infty(m)$. It follows that the infimum in question is

$$\leqq \|f+\|u\|_{L^\infty(m)}\|_{L^p(m)} \leqq \|f\|_{L^p(m)} + b\|u\|_{L^p(m)}(m(X))^{\frac{1}{p}} \leqq (a+ba(m(X))^{\frac{1}{p}})\|P\|_{L^p(m)}.$$

QED.

Proof of 7.3:i) We apply 7.4 to the direct sum decomposition $\mathrm{ReL}^p(Fm) = R^p(Fm)+\frac{1}{F}N$ from 6.9.ii) and take into account 6.9.i). It follows that there exists a constant $c>0$ such that

$$\mathrm{Inf}\,\{\|f\|_{L^p(Fm)} : |P| \leqq f\in E^\infty\} \leqq c\|P\|_{L^p(Fm)} \qquad \forall\ P\in \mathrm{ReL}^\infty(m).$$

ii) We assume that $1<p\leqq 2$ and deduce the assertion from 5.13. Let $P\in E^\infty$. For $|P|\leqq f\in E^\infty$ we have $0\leqq f\pm P\in E^\infty$. Thus $h^\pm := (f\pm P)+i(f\pm P)^*\in H^+$ with $\varphi(h^\pm) = \alpha(f\pm P)\geqq 0$ and $\int |h^\pm|^p F dm<\infty$ after 5.5.i). From 5.13.i) we obtain

$$\|f^*\pm P^*\|_{L^p(Fm)} \leqq (\tan\frac{\pi}{2p})\|f\pm P\|_{L^p(Fm)},$$

$$\|P^*\|_{L^p(Fm)} = \|\tfrac{1}{2}(f^*+P^*) - \tfrac{1}{2}(f^*-P^*)\|_{L^p(Fm)}$$

$$\leqq \tfrac{1}{2}(\tan\frac{\pi}{2p})\left(\|f+P\|_{L^p(Fm)} + \|f-P\|_{L^p(Fm)}\right) \leqq 2(\tan\frac{\pi}{2p})\|f\|_{L^p(Fm)},$$

so that from i) the assertion follows with $R(p,F)\leqq 2c(\tan\frac{\pi}{2p})$. QED.

Notes

For the classical conjugate function theory we refer to ZYGMUND [1968] Chapter VII, KATZNELSON [1968] Chapter III and DUREN [1970] Chapter 4.

Substantial portions of the theory have been extended to the Dirichlet algebra situation in DEVINATZ [1966] and to the Szegö situation in LUMER [1965]. In retrospect our theorem 3.9 makes clear that the transition beyond the Szegö situation is a very serious step. The transition was initiated in LUMER [1965] and worked out in GAMELIN-LUMER [1968] and LUMER [1968]. Their version of the theory is reproduced in part in GAMELIN [1969]. But the abstract conjugation theory did not attain its full coherence until the definition of the present text came up. The composite functions exp(t(P+iP*)) for conjugable $P \in \mathrm{ReL}(m)$ and $t \in \mathbb{R}$ have been in the literature ever since, but their rôle had to be converted to become the definition. The definition in question and the version of the fundamentals presented in Sections 1-3 are in KÖNIG [1967c] (exept the present form of 3.6 which is new). Let us once more emphasize the close relation to the function class H^+ dealt with in Sections V.4-5: if $0 \leqq P \in E$ then $h:=P+iP^* \in H^+$ from the definitions.

In Section 5 there are some brand-new results on the Marcel Riesz estimation. PICHORIDES [1972] obtained the sharp estimation 5.8 in the unit disk situation, for $1<p\leqq 2$ with the present method and for $2<p<\infty$ via the usual duality argument based on the weak*Dirichlet property 4.9.ii). YABUTA [1977] transferred the method to the abstract Hardy algebra situation to obtain the tan estimation in 5.6 and the case $1<p\leqq 2$ in 5.7 which combine to 5.8 as before, and also the essence of 5.13.i). The cotan estimation in 5.6 and the case $2<p<\infty$ in 5.7, and also 5.13.ii) and the sharp estimation 5.14 are new results due to König. Moreover YABUTA [1977] discovered that the example 4.11 from KÖNIG [1965] leads to the counter-example 5.3.ii). In 5.4 another idea from YABUTA [1977] combines with a classical method of proof due to Bochner. For further results see KÖNIG [1978].

The approximation theorem 4.1 is a result of the Function Algebra Seminar in Seattle 1970. The present version and its proof are due to König after an oral communication from Brian Cole of a somewhat weaker result. The annihilator theorem 4.5 is (in the version of 4.8) one of the main results in KÖNIG [1967c], where the proof is quite different. As an example of a partial result in the literature we quote GAMELIN [1969] theorem IV.4.5. Theorem 4.9 is the main achievement of the simultaneous papers HOFFMAN - ROSSI [1965] and KÖNIG [1965].

The considerations in Section 6 on special situations are close to the respective sections in GAMELIN [1969] which in turn are based upon AHERN-SARASON [1967a], GLICKSBERG [1968] and GAMELIN-LUMER [1968]. But there are some new results. For the finite-dimension lemma 6.5 we refer to GROTHENDIECK [1954].

Chapter VII

Analytic Disks and Isomorphisms with the Unit Disk Situation

The first section develops the basic tool for the sequel: It is proved that a Hardy algebra situation (H,φ) satisfies a certain form of the famous invariant subspace theorem iff it is a reduced Szegö situation. Hereafter then a reduced Szegö situation (H,φ) is assumed. The remainder of the chapter centers around the question whether such an (H,φ) can be proved to be isomorphic to the unit disk situation $(H^\infty(D),\varphi_o)$ in some sense or other. We consider two sets of properties which are all true in the unit disk situation and prove two equivalence theorems to the effect that the properties within either set are equivalent. The theorems are the maximality theorem due to Muhly and the analytic disk theorem due to Wermer. A class of examples shows that the two sets of properties are independent. Then it is proved that an (H,φ) which possesses either set of properties is isomorphic to $(H^\infty(D),\varphi_o)$ in the sharp sense of an isomorphism of the entire structure which comes from a map $(X,\Sigma,Fm)\rightarrow(S,\text{Baire},\lambda)$ between the basic measure spaces. Here $M=\{F\}$. We also prove that the weakest possible kind of isomorphism, that is a purely algebraic isomorphism of the algebras H and $H^\infty(D)$, implies that (H,φ) possesses the above sets of properties and hence that (H,φ) is isomorphic to the unit disk situation $(H^\infty(D),\varphi_o)$ in the sharp sense described above.

1. The Invariant Subspace Theorem

Let (H,φ) be a Hardy algebra situation. Consider a weak* closed linear subspace $T\subset L^\infty(m)$ which is invariant in the sense that $HT\subset T$. We define

$$T_\varphi := \overline{\text{Lin}\, H_\varphi T}^{\text{weak*}},$$

where $H_\varphi:=\{u\in H:\varphi(u)=0\}$ as introduced in VI.4.5 (in the case $T=H$ the notation is in accordance with the former one). Then $T_\varphi\subset L^\infty(m)$ is an invariant weak* closed linear subspace as well, and $T_\varphi\subset T$. It can happen that $T_\varphi=T$. If $T_\varphi\neq T$ then T is defined to be simply invariant.

There are obvious examples of simply invariant subspaces: If $P\in L^\infty(m)$

is of modulus $|P|=1$ then $T:=PH$ is an invariant weak* closed linear subspace of $L^\infty(m)$ (the weak* closedness is obvious from the Krein-Smulian Consequence IV.3.14) and $T_\varphi=PH_\varphi$, so that T is simply invariant since $P\notin T_\varphi$. In the opposite direction we claim the celebrated invariant subspace theorem.

1.1 INVARIANT SUBSPACE THEOREM: The subsequent assertions are equivalent.

i) (H,φ) is a reduced Szegö situation.

ii) Each simply invariant weak* closed linear subspace $T\subset L^\infty(m)$ is of the form $T=PH$ for some $P\in T$ of modulus $|P|=1$.

In this case for each simply invariant $T\subset L^\infty(m)$ the function $P\in T$ with $|P|=1$ such that $T=PH$ is of course unique up to a constant factor of modulus one.

Proof: i)$\Rightarrow$ii) We have $M=\{F\}$ with $F>0$ on X. The easiest proof is via an Hilbert space argument in $L^2(Fm)$. 1) We claim that $\overline{T}^{L^2(Fm)}\cap L^\infty(m)=T$. The inclusion $\supset$ is trivial. In order to prove $\subset$ let $f\in\overline{T}^{L^2(Fm)}$. Then there exist functions $f_n\in T$ such that $f_n\to f$ pointwise and $|f_n|\leqq$ some $G\in L^2(Fm)\subset L^0(Fm)=L^\#$. Take functions $u_\ell\in H$ with $|u_\ell|\leqq 1$, $u_\ell\to 1$ and $|u_\ell|G\leqq c_\ell$. Then $u_\ell f_n\in T$ and $|u_\ell f_n|\leqq|u_\ell|G\leqq c_\ell$ so that for $n\to\infty$ we obtain $u_\ell f\in T$. For $\ell\to\infty$ we have $u_\ell f\to f$. Thus if $f\in L^\infty(m)$ then $|u_\ell f|\leqq|f|\leqq \text{const}<\infty$ and hence $f\in T$. 2) We have $\overline{T_\varphi}^{L^2(Fm)}\cap L^\infty(m)=T_\varphi$ as well. It follows that $\overline{T_\varphi}^{L^2(Fm)}\neq\overline{T}^{L^2(Fm)}$. 3) Take a function $P\in\overline{T}^{L^2(Fm)}$ of $L^2(Fm)$-norm one which is orthogonal to $\overline{T_\varphi}^{L^2(Fm)}$ in the Hilbert space sense. Thus $\int\overline{P}uf\,Fdm=0$ for all $u\in H_\varphi$ and $f\in T$ and hence $f\in\overline{T}^{L^2(Fm)}$. In particular $\int|P|^2uFdm=0$ $\forall u\in H_\varphi$. It follows that $|P|^2F\in M$ and hence $|P|^2F=F$ or $|P|=1$. In particular $P\in T$ from 1) and hence $PH\subset T$. 4) To see the converse observe for $f\in T$ that the above equation $\int\overline{P}uf\,Fdm=0$ $\forall u\in H_\varphi$ means that $\overline{P}f\in(H_\varphi F)^\perp$ so that VI.4.5 implies that $\overline{P}f\in H$ or $f\in PH$. Thus $T\subset PH$ and hence $T=PH$.

ii)$\Rightarrow$i): 1) We prove that (H,φ) must be reduced. From IV.1.10 we know that $\chi_{Y(X)}\in H$. Hence $T:=\chi_{Y(X)}H$ is an invariant weak* closed linear subspace of $L^\infty(m)$ (the weak* closedness is clear from IV.3.14). Also $T_\varphi=\chi_{Y(X)}H_\varphi$ and hence $T_\varphi\neq T$ since $\chi_{Y(X)}\notin T_\varphi$ in view of $\varphi(\chi_{Y(X)})=1$. But there are no functions $P\in T$ of modulus $|P|=1$ except when $m(Y(X))=1$. Therefore

(H,φ) must be reduced. 2) Let $f\in L^{\infty}(m)$ such that $|f|\geqq$ some $\delta>0$. Then IV.3.14 shows as above that $T:=fH$ is an invariant weak* closed linear subspace of $L^{\infty}(m)$ and that $T_{\varphi}=fH_{\varphi}$. Thus $T_{\varphi}\neq T$ since $f\notin T_{\varphi}$. From ii) now $T=PH$ for some $P\in T$ with $|P|=1$. It follows that $\frac{1}{P}f=\bar{P}f$ is an invertible element of H so that $a(|f|)a(|\frac{1}{f}|)=1$ after V.1.3. We take $f=e^{th}$ for $h\in ReL^{\infty}(m)$ and $t\in\mathbb{R}$ to obtain $a(e^{th})a(e^{-th})=1$. It follows that $ReL^{\infty}(m)\subset E$ from VI.2.1 so that (H,φ) is Szegö after VI.3.9. QED.

1.2 COROLLARY: Let (H,φ) be a reduced Szegö situation with $M=\{F\}$ and $1\leqq p<\infty$. Let $T\subset L^{p}(Fm)$ be a closed linear subspace which is invariant in the sense that $HT\subset T$, and define

$$T_{\varphi}^{p}:=\overline{\operatorname{Lin} H_{\varphi}T}^{L^{p}(Fm)}.$$

Then $T=P(H^{\#}\cap L^{p}(Fm))$ for some $P\in T$ of modulus $|P|=1$ iff T is simply invariant in the sense that $T_{\varphi}^{p}\neq T$.

Proof of 1.2: 1) It is clear that $T_{\varphi}^{p}\subset L^{p}(Fm)$ is an invariant closed linear subspace as well, and that $T_{\varphi}^{p}\subset T$. 2) If $T=P(H^{\#}\cap L^{p}(Fm))$ for some $P\in T$ with $|P|=1$ then VI.6.10 shows that $T_{\varphi}^{p}=P(H_{\varphi}^{\#}\cap L^{p}(Fm))$. Thus $T_{\varphi}^{p}\neq T$ since $P\notin T_{\varphi}^{p}$. So let us assume that $T_{\varphi}^{p}\neq T$ and prove the opposite direction. 3) We have $\overline{T\cap L^{\infty}(m)}^{L^{p}(Fm)}=T$. Here $\subset$ is obvious and $\supset$ results from $L^{p}(Fm)\subset$ $\subset L^{0}(Fm)=L^{\#}$. Likewise $\overline{T_{\varphi}^{p}\cap L^{\infty}(m)}^{L^{p}(Fm)}=T_{\varphi}^{p}$. It follows that $T_{\varphi}^{p}\cap L^{\infty}(m)\neq$ $\neq T\cap L^{\infty}(m)$. 4) From IV.3.14 we see that $B:=T\cap L^{\infty}(m)$ is an invariant weak* closed linear subspace of $L^{\infty}(m)$. Furthermore $H_{\varphi}B\subset H_{\varphi}T\subset T_{\varphi}^{p}$ and hence $H_{\varphi}B\subset$ $\subset T_{\varphi}^{p}\cap L^{\infty}(m)$ so that $B_{\varphi}\subset T_{\varphi}^{p}\cap L^{\infty}(m)$ since $T_{\varphi}^{p}\cap L^{\infty}(m)$ is weak* closed as well. Thus $B_{\varphi}\neq B$. From 1.1 we obtain a function $P\in B$ of modulus $|P|=1$ such that $B=PH$. Then VI.6.10 implies that $T=\bar{B}^{L^{p}(Fm)}=P\bar{H}^{L^{p}(Fm)}=P(H^{\#}\cap L^{p}(Fm))$. QED.

2. The Maximality Theorem

For the remainder of the chapter we fix a reduced Szegö situation (H,φ) with $M=\{F\}$. We could of course assume that $F=1$ as it is often done. But this is not advisable in particular for the benefit of the next section.

In the present section we want to prove the subsequent theorem. It is a surprise because the properties involved look quite different.

2.1 MAXIMALITY THEOREM: The subsequent properties are equivalent.

i) H is a maximal proper weak* closed subalgebra of $L^{\infty}(m)$.

ii) If $T \subset L^{\infty}(m)$ is an invariant weak* closed linear subspace such that $fT \subset T$ for some $f \in L^{\infty}(m)$ which is $\notin H$ then $T = \chi_V L^{\infty}(m)$ for some $V \in \Sigma$.

iii) For each nonzero $u \in H$ we have $m([u=0])=0$.

iv) H contains no zero divisors.

The proof will be after the scheme i)⇒ii)⇒iii)⇒iv)⇒i). The most complicated part is iv)⇒i), and i)⇒ii) involves a bit of work which deserves independent interest. iii)⇒iv) is obvious. And ii)⇒iii) is simple: Fix a nonzero $a \in H$ and assume that $A:=[a=0]$ has $m(A)>0$. Then $T:=\{f \in L^{\infty}(m): fa \in H\}$ is an invariant weak* closed linear subspace of $L^{\infty}(m)$. We have $\chi_A L^{\infty}(m) \subset T$ and hence $\chi_A T \subset T$. Since $\chi_A \notin H$ in view of the assumption it follows from ii) that $T = \chi_V L^{\infty}(m)$ for some $V \in \Sigma$. But since $H \subset T$ it must then be $T = L^{\infty}(m)$. In particular $\bar{a} \in T$ or $|a|^2 \in H$ so that $|a|^2 = \text{const}$ and hence $a=0$. So we arrive at a contradiction which proves ii)⇒iii).

Let us turn to the preparation for the step i)⇒ii). The subsequent consideration is unrelated to the abstract Hardy algebra situation.

For a weak* closed linear subspace $T \subset L^{\infty}(m)$ we introduce the transporter $\tau(T) := \{f \in L^{\infty}(m): fT \subset T\}$. It is seen that $\tau(T)$ is a weak* closed complex subalgebra of $L^{\infty}(m)$ which contains the constants. In particular for the weak* closed linear subspaces $T := \chi_V L^{\infty}(m)$ with $V \in \Sigma$ we have $\tau(T) = L^{\infty}(m)$. It is not hard to prove the converse.

2.2 LEMMA: Let $T \subset L^{\infty}(m)$ be a weak* closed linear subspace such that $\tau(T) = L^{\infty}(m)$. Then $T = \chi_V L^{\infty}(m)$ for some $V \in \Sigma$.

Proof: The class of subsets $\Delta := \{D \in \Sigma : \chi_D \in T\}$ has the subsequent properties. 1) If $D \in \Delta$ and $U \in \Sigma$ is contained in D then $U \in \Delta$. This follows from $\chi_U = \chi_U \chi_D \in T$. 2) If $U, V \in \Delta$ then $U \cup V \in \Delta$. This follows from $\chi_{U \cup V} = \chi_U + \chi_V - \chi_{U \cap V}$ in view of 1). 3) If $D_n \in \Delta$ for $n \geq 1$ and $D_n \uparrow D$ then $D \in \Delta$. 4) If $f \in T$ then $[f \neq 0] \in \Delta$. In fact, we have $|f|^2/(\frac{1}{n}+|f|^2) = (\bar{f}/(\frac{1}{n}+|f|^2))f \in T$ for $n \geq 1$, and these functions are between 0 and 1 and $\to \chi_{[f \neq 0]}$ for $n \to \infty$. Now we can prove the assertion: From 2) and 3) we obtain a set $V\in\Delta$ such that $m(V)$ is maximum. Then $D\subset V$ (modulo an m-null set) for all $D\in\Delta$. From 4) in particular $[f\neq 0]\subset V$ or $f\in\chi_V L^\infty(m)$ for all $f\in T$. It follows that $T\subset\chi_V L^\infty(m)$ and hence $T=\chi_V L^\infty(m)$ since $\supset$ is trivial. QED.

Proof of 2.1.i) $\Rightarrow$ ii): The transporter $\tau(T)\subset L^\infty(m)$ contains H since T is invariant and is assumed to be $\neq$ H. Hence $\tau(T)=L^\infty(m)$ after i). Then the assertion follows from 2.2. QED.

The proof of iv) $\Rightarrow$ i) requires several steps. Let us isolate the subsequent lemma.

2.3 LEMMA: Let $B\subset L^\infty(m)$ be a weak* closed subalgebra such that $H\subsetneqq B$. Define $\Delta:=\{D\in\Sigma:\chi_D\in B\}$ so that Δ is a σ-algebra $\subset\Sigma$ which in particular contains the m-null sets. Then for each nonzero $f\in L(m)$ there exists a set $D\in\Delta$ such that $f\chi_D\neq 0$ and $f\chi_{X-D}\neq 0$.

Proof: 1) Define $f\in L(m)$ to be Δ-measurable iff some representative function and hence each representative function of f is Δ-measurable. Let $L(m|\Delta)$ consist of the Δ-measurable $f\in L(m)$. 2) Assume that $D\in\Delta$ has $m(D)>0$ and is minimal in the sense that each $U\in\Delta$ contained in D fulfills $m(U)=0$ or $m(D-U)=0$. Then each $f\in L(m|\Delta)$ is constant on D. In fact, for complex numbers $a\neq b$ the sets $D\cap[|f-a|<\frac{1}{2}|a-b|]$ and $D\cap[|f-b|<\frac{1}{2}|a-b|]$ are in Δ and hence cannot both have positive m-measure, so a and b cannot both be in the value carrier $\omega(f|D)$. 3) Each real-valued function $f\in B$ is Δ-measurable. To see this let $F:x\mapsto F(x)$ be a fixed representative function of f which is real-valued and bounded $|F(x)|\leqq R$ $\forall x\in X$. Define Ω to consist of the functions $h:\mathbb{R}\to\mathbb{C}$ such that $h(F)\bmod m$ is in B. Then i) Ω contains the polynomials. ii) If $h:\mathbb{R}\to\mathbb{C}$ and $h_n\in\Omega$ are such that $h_n\to h$ and $|h_n|\leqq$ some common constant both on $[-R,R]$ then $h\in\Omega$. Therefore iii) Ω contains the continuous functions and hence the Baire functions $h:\mathbb{R}\to\mathbb{C}$ bounded on $[-R,R]$. Now for a Baire set $A\subset\mathbb{R}$ we have $\chi_A(F)=\chi_{[F\in A]}$ and hence $\chi_A(F)\bmod m=$ $=\chi_{[f\in A]}\in B$ so that $[f\in A]\in\Delta$ as claimed.

4) There exist sets $U\in\Delta$ with $0<m(U)<m(X)$. Otherwise X were minimal in the sense of 2) so that after 2) and 3) all real-valued functions $f\in B$ were constant. Fix now $f=P+iQ\in B$. For $t\in\mathbb{R}$ consider $e^{t(P+iP^*)}\in H^\times$ and $f_t:=$ $:=e^{tf}e^{-t(P+iP^*)}=e^{it(Q-P^*)}\in B^\times$. We have $|f_t|=1$ so that $\bar{f}_t=1/f_t\in B$ and hence $\mathrm{Re}f_t$, $\mathrm{Im}f_t\in B$. Thus failure of the assertion would imply that $f_t=\mathrm{const}=c(t)$

and hence that $e^{tf}=c(t)e^{t(P+iP^*)}\in H$ for all $t\in\mathbb{R}$, so that $f\in H$. But then $B\subset H$ in contradiction to our basic assumption.

5) To each $D\in\Delta$ with $m(D)>0$ there exists a subset $U\in\Delta$ such that $0<m(U)<m(D)$, that is there are no sets $D\in\Delta$ which are minimal in the sense of 2). In view of 4) we can assume that $m(D)<m(X)$ so that $\chi_D\notin H$. We have $F_n:=\exp(-n(1-\chi_D))\in B^\times$ for $n\geqq 1$. From V.1.6 we obtain $f_n\in H^\times$ with $|f_n|=F_n$ which are determined up to a constant factor of modulus one. Thus $F_n=f_nb_n$ where $b_n\in B^\times$ of modulus $|b_n|=1$. Now assume that D were minimal. Then $b_n|D=\text{const}$ after 2) and 3) so that we can assume that $b_n=1$ and hence $f_n=F_n=1$ on D. But on X-D we have $|f_n|=F_n=e^{-n}$. It follows that $f_n\to\chi_D$ for $n\to\infty$ and hence that $\chi_D\in H$. But this is a contradiction.

6) We can now prove the assertion of the lemma. Fix a nonzero function $f\in L(m)$ and define $\Delta(f):=\{D\in\Delta: f\chi_D=f\}$. We list some immediate properties. i) $\Delta(f)\subset\Delta$. ii) $\Delta(f)\neq\emptyset$ since $X\in\Delta(f)$. iii) For each $D\in\Delta(f)$ we have $m(D)>0$ since $f\neq 0$. iv) If $U,V\in\Delta(f)$ then $U\cap V\in\Delta(f)$. v) If $D_n\in\Delta(f)$ for $n\geqq 1$ and $D_n\downarrow D$ then $D\in\Delta(f)$. Now we assume the assertion to be false for f and deduce a contradiction as follows. From iv) and v) we obtain a set $D\in\Delta(f)$ such that $m(D)$ is minimum. Of course $m(D)>0$. We show that D must be minimal in the sense of 2) in contradiction to 5). In fact, if $U\in\Delta$ is contained in D and $m(U)<m(D)$ then $U\notin\Delta(f)$ and hence $f\chi_U\neq f$ or $f\chi_{X-U}\neq 0$. Thus $f\chi_U=0$ from our assumption. But then $f\chi_{D-U}=f\chi_D-f\chi_U=f$ and hence $D-U\in\Delta(f)$. It follows that $m(D-U)=m(D)$ so that $m(U)=0$. Thus D is indeed minimal in the sense of 2) which is the desired contradiction. QED.

Proof of 2.1.iv) $\Rightarrow$ i): Let $B\subset L^\infty(m)$ be a weak* closed subalgebra such that $H\subsetneqq B\subsetneqq L^\infty(m)$. We shall see that 2.3 provides us with lots of zero divisors in H. Define $T\subset L^\infty(m)$ to consist of the functions $f\in L^\infty(m)$ such that fF annihilates B. Then we have the subsequent facts.

1) T contains functions $\neq 0$.

2) $BT\subset T$.

3) $T\subset H_\varphi\subset H$.

To see 1) take a nonzero $h\in L^1(m)$ which annihilates B. Then $\frac{h}{F}\in L^1(Fm)\subset L^0(Fm)\subset L^\#$. Hence there are functions $u_\ell\in H$ with $|u_\ell|\leqq 1$, $u_\ell\to 1$ and $u_\ell\frac{h}{F}\in L^\infty(m)$. It follows that all these $u_\ell\frac{h}{F}$ are in T. 2) is obvious since B is an algebra. To prove 3) we invoke the fundamental theorem VI.4.8 in the form VI.6.9.iii). If $f\in T$ then fF annihilates B and hence H so that $fF\in K$ and $\int fFdm=0$. It follows that $f\in H^{\#}\cap L^{\infty}(m)=H$ and $\varphi(f)=\int fFdm=0$ which proves 3).

Now the time for 2.3 has come. For a nonzero $f\in T$ we obtain a set $D\in\Delta$ such that $f\chi_D$ and $f\chi_{X-D}$ are nonzero members of T as well. Thus we arrive at nonzero functions in H the product of which is =0. And it is clear that we can iterate the procedure and thus obtain lots of zero divisors in H indeed. QED.

2.4 RETURN to the UNIT DISK: The algebra $H^{\infty}(D)\subset L^1(\lambda)$ satisfies iii) as we know from I.3.8. And it satisfies iv) since it is algebraically isomorphic to $Hol^{\infty}(D)$ which contains no zero divisors as a consequence of the basic facts of analytic function theory. The properties i) and ii) are less immediate but are well-known properties of $H^{\infty}(D)$ as well.

3. The Analytic Disk Theorem

The analytic disk theorem is a beautiful answer to certain questions which are natural in view of the unit disk situation. One question is on H_{φ} which appears to be the simplest nontrivial invariant weak* closed linear subspace of $L^{\infty}(m)$: is H_{φ} always simply invariant? In the unit disk situation $(H,\varphi)=(H^{\infty}(D),\varphi_o)$ this is true. The representation after the invariant subspace theorem 1.1 is immediate but comes from the depths of analytic function theory: it is $H_{\varphi}=ZH$ with Z the coordinate function. The other question is whether H always admits nonzero weak* continuous multiplicative linear functionals other than φ and how to describe these functionals? In the unit disk situation $(H,\varphi)=(H^{\infty}(D),\varphi_o)$ the answer is clear from II.3.6 and IV.1.12: the functionals in question are the φ_a for the points $a\in D$. Each time (H,φ_a) is a reduced Szegö situation, and we see that $H_{\varphi_a}=\frac{Z-a}{1-\bar{a}Z}H$.

The analytic disk theorem discloses the intimate connection between the above questions. It depends upon a fundamental construction which links (H,φ) to the unit disk situation $(H^{\infty}(D),\varphi_o)$. Let us start with this construction.

3.1 THEOREM: Assume that H_{φ} is simply invariant and fix an inner function $I\in H_{\varphi}$ such that $H_{\varphi}=IH$.

i) For each $u \in H^{\#}$ there exists a unique sequence of complex numbers $a_n (n=0,1,2,...)$ such that

$$u - \sum_{\ell=0}^{n} a_\ell I^\ell \in I^{n+1} H^{\#} \quad \text{for all } n \geq 0,$$

called the Taylor coefficients of the function u. If $u \in H^{\#} \cap L^1(Fm)$ then $a_n = \int u \bar{I}^n Fdm \ \forall n \geq 0$.

ii) For each $u \in H^{\#}$ the power series expansion

$$\hat{u} : \hat{u}(z) = \sum_{n=0}^{\infty} a_n z^n \quad \text{converges for all } z \in D,$$

and defines a function $\hat{u} \in Hol^{\#}(D)$. If $u \in H^{\#} \cap L^1(Fm)$ then

$$\hat{u}(z) = \int u \frac{I}{I-z} Fdm = \int u \frac{1-|z|^2}{|I-z|^2} Fdm \quad \forall \ z \in D.$$

iii) If $u \in H^{\#} \cap L^p(Fm)$ for some $1 \leq p \leq \infty$ then $\hat{u} \in Hol^p(D)$ and

$$N_p \hat{u} \leq ||u||_{L^p(Fm)} \qquad \text{(see Section I.1).}$$

iv) For $u, v \in H^{\#}$ and $c \in \mathbb{C}$ we have $(u+v)^\wedge = \hat{u} + \hat{v}$, $(cu)^\wedge = c\hat{u}$ and $(uv)^\wedge = \hat{u}\hat{v}$. Furthermore $\hat{1} = 1$ and $\hat{I} = Z$.

v) For $a \in D$ we have

$$\{u \in H^{\#} : \hat{u}(a) = 0\} = \frac{I-a}{1-\bar{a}I} H^{\#}.$$

Proof: 1) We have $H^{\#}_\varphi = IH^{\#}$. The inclusion $\supset$ is trivial. To see $\subset$ let $f \in H^{\#}_\varphi$ and take functions $u_\ell \in H$ with $|u_\ell| \leq 1$, $u_\ell \to 1$ and $u_\ell f \in H$. Then $u_\ell f \in H_\varphi$ and hence $u_\ell \bar{I} f \in H$. It follows that $\bar{I} f \in H^{\#}$ or $f \in IH^{\#}$. 2) For $u \in H^{\#}$ we obtain from 1) a sequence of functions $u_n \in H^{\#} (n=0,1,2,...)$ such that $u = u_0$ and $u_n = \varphi(u_n) + I u_{n+1} \ \forall n \geq 0$. Then $I^n u_n = I^n \varphi(u_n) + I^{n+1} u_{n+1} \ \forall n \geq 0$ so that we obtain the existence assertion in i) with $a_n := \varphi(u_n)$. The other assertions in i) are then immediate. 3) If $u, v \in H^{\#}$ and uv have the Taylor coefficients a_n, b_n and c_n then

$$c_n = \sum_{\ell=0}^{n} a_\ell b_{n-\ell} \quad \forall \ n \geq 0.$$

This follows upon multiplication of the relations which define the Taylor coefficients. For u+v and cu the correspondence is obvious. 4) Define a function $u \in H^{\#}$ to be honest iff its power series expansion in ii) converges for all $z \in D$ and thus defines a function $\hat{u} \in Hol(D)$. Then the honest functións $\in H^{\#}$ form a subalgebra which contains the functions 1 and

I and on which the relations claimed in iv) are fulfilled.

5) For $u\in H^{\#}\cap L^1(Fm)$ we have $|a_n|\leq\int|u|Fdm$ so that u is an honest function. And for $z\in D$ we have

$$\int u\frac{I}{I-z}Fdm=\int u\frac{1}{1-\bar{I}z}Fdm=\int u\Big(\sum_{n=0}^{\infty}(\bar{I}z)^n\Big)Fdm=\sum_{n=0}^{\infty}a_nz^n=\hat{u}(z),$$
$$\int u\frac{I\bar{z}}{1-I\bar{z}}Fdm=\int u\Big(\sum_{n=1}^{\infty}(I\bar{z})^n\Big)Fdm=\sum_{n=1}^{\infty}\big(\int uI^nFdm\big)\bar{z}^n=0,$$
$$\int u\frac{1-|z|^2}{|I-z|^2}Fdm=\int u\Big(\frac{I}{I-z}+\frac{I\bar{z}}{1-I\bar{z}}\Big)Fdm=\hat{u}(z),$$

as claimed in ii). From this one obtains iii) as in I.1.2 with the aid of the relations $\int\frac{1-|z|^2}{|I-z|^2}Fdm=1\ \forall z\in D$ from the above for the constant function $=1$ and $\int_S P(z,s)d\lambda(s)=1\ \forall z\in D$ from I.1.1.

6) We next prove v) for $H^{\#}\cap L^1(Fm)$ instead of $H^{\#}$. Here $h:=\frac{I-a}{1-\bar{a}I}$ is an inner function, and from $(1-\bar{a}I)h=I-a$ we see that $(1-\bar{a}Z)\hat{h}=Z-a$ or $\hat{h}=\frac{Z-a}{1-\bar{a}Z}$ so that $\hat{h}(a)=0$. Thus we have $\supset$. In order to prove $\subset$ let $u\in H^{\#}\cap L^1(Fm)$ with $\hat{u}(a)=0$. For all $v\in H$ then $0=\hat{v}(a)\hat{u}(a)=(vu)^\wedge(a)=\int v\frac{uI}{I-a}Fdm$. Hence $\frac{uI}{I-a}F\in K$ with $\int\frac{uI}{I-a}Fdm=0$. Thus from VI.6.9.iii) we see that $\frac{uI}{I-a}\in H^{\#}\cap L^1(Fm)$ with $\varphi(\frac{uI}{I-a})=0$. From 1) we obtain $\frac{u}{I-a}\in H^{\#}\cap L^1(Fm)$ and hence $\frac{u}{h}=(1-\bar{a}I)\frac{u}{I-a}\in H^{\#}\cap L^1(Fm)$. 7) From 6) we deduce that a function $u\in H^{\#}\cap L^1(Fm)$ which is invertible in $H^{\#}$ satisfies $\hat{u}(z)\neq 0$ for all $z\in D$.

8) We can now prove that all functions $f\in H^{\#}$ are honest. To see this write $f=\frac{u}{v}$ with $u,v\in H$ and $v\in(H^{\#})^{\times}$ after V.1.10. Let u,v and f have the Taylor coefficients a_n,b_n and c_n. From 7) we know that $\hat{v}(z)\neq 0\ \forall z\in D$. Thus $\hat{u}/\hat{v}$ is an holomorphic function in D, and 3) tells us that its power series expansion in the origin has the coefficients c_n. It follows that $f\in H^{\#}$ is an honest function. 9) To see that $\hat{f}\in Hol^{\#}(D)$ take functions $u_\ell\in H$ with $|u_\ell|\leq 1$, $u_\ell\to 1$ and $u_\ell f\in H$. Then $\hat{u}_\ell\in Hol^{\infty}(D)$ with $|\hat{u}_\ell|\leq 1$ and $\hat{u}_\ell\hat{f}=(u_\ell f)^\wedge\in Hol^{\infty}(D)$, and we see that $\hat{u}_\ell(z)=\int u_\ell\frac{1-|z|^2}{|I-z|^2}Fdm\to 1\ \forall z\in D$. Thus $\hat{f}\in Hol^{\#}(D)$ as claimed.
10) It remains to prove v). The inclusion $\supset$ is obvious. To see $\subset$ let

$f\in H^{\#}$ with $\hat{f}(a)=0$ and write $f=\frac{u}{v}$ with $u,v\in H$ and $v\in(H^{\#})^{\times}$ as above. Then $\hat{u}(a)=(fv)^{\wedge}(a)=\hat{f}(a)\hat{v}(a)=0$ and hence $u=\frac{I-a}{1-\bar{a}I}w$ with $w\in H$ after 6). It follows that $f=\frac{I-a}{1-\bar{a}I}\frac{w}{v}\in\frac{I-a}{1-\bar{a}I}H^{\#}$. QED.

For the moment we prefer to look at the individual points $z\in D$. Then we obtain the subsequent consequence.

3.2 CONSEQUENCE: Assume that H_{φ} is simply invariant and fix an inner function $I\in H_{\varphi}$ such that $H_{\varphi}=IH$. For each $z\in D$ then

i) $\varphi_z:u\mapsto\hat{u}(z)$ is a nonzero weak* continuous multiplicative linear functional on H. In particular $\varphi_0=\varphi$. Furthermore $\varphi_z(I)=z$ so that different points $z\in D$ produce different functionals φ_z.

ii) The Hardy algebra situation (H,φ_z) is a reduced Szegö situation with $M_z=\left\{\frac{1-|z|^2}{|I-z|^2}F\right\}$.

iii) The extension of φ_z to $H^{\#}$ in the sense of IV.3.3 continues to be $\varphi_z:u\to\hat{u}(z)$.

iv) We have $H_{\varphi_z}=\frac{I-z}{1-\bar{z}I}H$.

Proof: Most of the assertions are immediate from 3.1. For ii) observe that Re H is weak* dense in Re $L^{\infty}(m)$ after VI.4.9. And iii) is truc because V.1.10 tells us that the extension of φ_z to $H^{\#}$ in the sense of IV.3.3 is the unique multiplicative extension. QED.

Now we come to the main result of the section.

3.3 ANALYTIC DISK THEOREM: i) H_{φ} is simply invariant iff H possesses nonzero weak* continuous multiplicative linear functionals other than φ.

ii) Assume that this is true and fix an inner function $I\in H_{\varphi}$ such that $H_{\varphi}=IH$. Then the functionals in question are the φ_z for the points $z\in D$ as obtained above.

iii) On the set $\Sigma*(H)$ of all nonzero weak* continuous multiplicative linear functionals on H introduce the Gelfand topology, that is the weakest topology in which for each $u\in H$ the function $\Sigma*(H)\to\mathbb{C}:\psi\mapsto\psi(u)$ is continuous. Then the map $D\to\Sigma*(H):z\mapsto\varphi_z$ is a homeomorphism.

Proof: 1) Assume that $\psi:H\to\mathbb{C}$ is a nonzero weak* continuous multiplicative linear functional $\neq\varphi$. Then $H_{\psi}:=\{u\in H:\psi(u)=0\}\subset H$ is an invariant weak* closed linear subspace of $L^\infty(m)$. We claim that H_ψ is simply invariant. In fact, take a function $u\in H$ such that $\psi(u)=0$ and $\varphi(u)\neq 0$. Then $u\in H_\psi$ but $u\notin H_\varphi$ so that u is not in $(H_\psi)_\varphi\subset H_\varphi$. Thus $(H_\psi)_\varphi\neq H_\psi$. From 1.1 we obtain a function $Q\in H_\psi$ with $|Q|=1$ such that $H_\psi=QH$. Then Q is a nonconstant inner function since otherwise we had $H_\psi=H$ or $\psi=0$. Thus $b:=\varphi(Q)=$ $=\int QFdm$ is $\in D$. We introduce the function

$$G:=\frac{|Q-b|^2}{1-|b|^2}=\frac{1}{1-|b|^2}\left(1-\bar{b}Q-b\bar{Q}+|b|^2\right).$$

From $1-|b|\leq|Q-b|\leq 1+|b|$ we see that

$$0<\frac{1-|b|}{1+|b|}\leq G\leq\frac{1+|b|}{1-|b|}<\infty.$$

We claim that $\psi(u)=\int uG\,Fdm$ for all $u\in H$. To see this note that

$$\int GFdm=\frac{1}{1-|b|^2}\left(1-\bar{b}b-b\bar{b}+|b|^2\right)=1,$$

$$QG\in H \text{ with } \varphi(QG)=\int QGFdm=\frac{1}{1-|b|^2}\left(b-\bar{b}b^2-b+|b|^2b\right)=0.$$

Now for $u\in H$ we have $u=\psi(u)+Qv$ with $v\in H$ and hence $\int uGFdm=\int(\psi(u)+Qv)GFdm=$ $=\psi(u)+\varphi(v)\varphi(QG)=\psi(u)$ as claimed. It follows that the Hardy algebra situation (H,ψ) is a reduced Szegö situation with representative function $GF\in\operatorname{Re}L^1(m)$.

2) With $\psi:H\to\mathbb{C}$ as before we apply 1) to (H,ψ) and to the functional $\varphi\neq\psi$. It follows that there exists a function $I\in H_\varphi$ with $|I|=1$ such that $H_\varphi=IH$. Let us fix such a function $I\in H_\varphi$. Then $a:=\psi(I)$ is $\in D$ and $\varphi:H\to\mathbb{C}$ has the representative function $\frac{|I-a|^2}{1-|a|^2}GF\in\operatorname{Re}L^1(m)$. This function must of course be $=F$. It follows that $G=\frac{1-|a|^2}{|I-a|^2}$ which means that $\psi=\varphi_a$. Thus we have proved i) and ii).

3) For the proof of iii) let us identify $D\cong\Sigma^*(H)$ via the map $z\mapsto\varphi_z$. The Gelfand topology τ is then the weakest topology on D such that the functions $z\cong\varphi_z\mapsto\varphi_z(u)=\hat{u}(z)$, that is the functions $\hat{u}$ are continuous $\forall u\in H$. We have to prove that $\tau=\nu|D$ where ν is the absolute value topology on $\mathbb{C}$. Now the functions $\hat{u}$ are holomorphic on D and hence continuous in $\nu|D$ $\forall u\in H$ whence $\tau\subset\nu|D$. On the other hand the identity $\hat{I}=Z:(D,\tau)\to(\mathbb{C},\nu)$ is continuous which means that $\nu|D\subset\tau$. Thus $\tau=\nu|D$. QED.

4. The Isomorphism Theorem

In the present section we want to construct isomorphisms between (H,φ) and the unit disk situation $(H^\infty(D),\varphi_o)$ and to characterize those reduced Szegö situations (H,φ) for which this can be done. Of course we must explain what an isomorphism is to be. It is clear that it should comprehend at least an algebraic isomorphism between H and $H^\infty(D)$. Then H cannot contain zero divisors so that the equivalent conditions i)-iv) of the maximality theorem 2.1 must be satisfied. Furthermore if the notion of an isomorphism is not too weak then it should comprehend that H_φ be simply invariant so that the equivalent conditions of the analytic disk theorem 3.3.i) must be satisfied. The result of the present section will be that under these two assumptions there is an isomorphism between (H,φ) and $(H^\infty(D),\varphi_o)$ in a sense which is as sharp as one could wish. For a fixed inner function $I\in H_\varphi$ with $H_\varphi=IH$ such an isomorphism comes from the evaluation map $H^\#\to Hol^\#(D):u\mapsto\hat{u}$ established in 3.1. However, we want to obtain the isomorphism as the substitution map which comes from a map $(X,\Sigma,Fm)\to(S,Baire,\lambda)$ between the basic measure spaces. Now the above I is not a measurable map $(X,\Sigma)\to(S,Baire)$ but an equivalence class modulo m of such maps. But nevertheless we can obtain a well-defined substitution map $L(m)\leftarrow L(\lambda):foI\leftarrow\!\shortmid f$. It will turn out to be bijective and its restriction $H^\#\leftarrow H^\#(D):foI\leftarrow\!\shortmid f$ to be inverse to the above evalution map $H^\#\to Hol^\#(D):u\mapsto\hat{u}$.

We start with the definition and the basic properties of the desired substitution map.

<u>4.1</u> <u>PROPOSITION</u>: Assume that $I\in H_\varphi$ is an inner function. Then

i) $\lambda(B)=(Fm)([I\in B])$ for all Baire sets $B\subset S$.

ii) For each $f\in L(\lambda)$ there is a unique $foI\in L(m)$ such that $foI=RoJ$ modulo m for all representative Baire functions $R:S\to\mathbb{C}$ of f and all representative functions $J:x\mapsto J(x)\in S$ of I. The map $L(\lambda)\to L(m):f\mapsto foI$ is of course an algebraic homomorphism and homomorphic with respect to complex conjugation. Furthermore it is injective.

iii) For $f_n,f\in L(\lambda)$ pointwise convergence $f_n\to f$ in the $L(\lambda)$ sense implies pointwise convergence $f_noI\to foI$ in the $L(m)$ sense.

iv) We have $\int_S fd\lambda=\int(foI)Fdm$ for all $f\in L^\infty(\lambda)$. Hence the same is true for all $0\leqq f\in Re\,L(\lambda)$.

v) For $1\leq p\leq\infty$ we have $\|f\|_{L^p(\lambda)} = \|foI\|_{L^p(Fm)}$ for all $f\in L(\lambda)$ (of course with the value ∞ included). For $f\in L(\lambda)$ therefore $f\in L^p(\lambda)\Leftrightarrow foI\in L^p(Fm)$.

vi) For $f\in L(\lambda)$ we have $f\in H^\#(D)\Leftrightarrow foI\in H^\#$.

Proof: 1) For a Baire function $R:S\to\mathbb{C}$ the substitution of the representative functions $J:x\mapsto J(x)\in S$ of I leads to functions RoJ which are equivalent modulo m so that $RoI:=RoJ \bmod m\in L(m)$ is well-defined. 2) We claim that $\int_S Rd\lambda=\int(RoI)Fdm$ for all $R\in B(S,\text{Baire})$. In fact, the set of all functions $R\in B(S,\text{Baire})$ for which this is true is a linear subspace which is closed under bounded pointwise convergence and contains the functions $R=z^n$ $\forall n\in\mathbb{Z}$. Hence it must be $=B(S,\text{Baire})$. It follows that the relation is true for all $0\leq R\in\text{Re}\,L(S)$. 3) For $R=\chi_B$ with $B\subset S$ a Baire set we obtain i). 4) For $R\in L(B)$ it follows from 2) applied to $|R|$ that R is a null function modulo λ iff $RoI=0$. Thus for $f\in L(\lambda)$ the substitution $foI:=RoI\in L(m)$ is well-defined as claimed in ii), and we have $foI=0$ iff $f=0$. 5) iii) is immediate from i). Thus we have i)-v). It remains to prove vi).

6) For $f=z^n$ with $n\geq 0$ we have $foI=I^n\in H$. Hence from I.3.3.iii) and the above iii) we see that $f\in H^\infty(D)\Rightarrow foI\in H$. 7) For $f\in H^\#(D)$ take functions $u_\ell\in H^\infty(D)$ with $|u_\ell|\leq 1, u_\ell\to 1$ and $u_\ell f\in H^\infty(D)$. Then $u_\ell oI\in H$ with $|u_\ell oI|\leq 1, u_\ell oI\to 1$ and $(u_\ell oI)(foI)=(u_\ell f)oI\in H$. It follows that $foI\in H^\#$. 8) Assume that $f\in L(\lambda)$ is such that $foI\in H$. Then of course $f\in L^\infty(\lambda)$. We claim that $f\in H^\infty(D)$. To see this take an $h\in L^1(\lambda)$ which annihilates $H^\infty(D)$. From VI.6.9.iii) then $h\in H^\# \cap L^1(\lambda)$ with $\int_S hd\lambda=0$. Hence $hoI\in H^\#\cap L^1(Fm)$ after 7) with $\varphi(hoI) = \int(hoI)Fdm = \int_S hd\lambda=0$. It follows that $\int_S hfd\lambda=\int((hf)oI)Fdm=\int(hoI)(foI)Fdm= =\varphi(hoI)\varphi(foI)=0$. Hence $f\in H^\infty(D)$. 9) Assume that $f\in L(\lambda)$ is such that $foI\in H^\#$. Then $foI\in L^\#=L^O(Fm)$ and hence $f\in L^O(\lambda)$ in view of iv). Since $L^O(\lambda)$ is$=L^\#(D):=$ the class $L^\#$ for the unit disk situation there are functions $u_\ell\in H^\infty(D)$ with $|u_\ell|\leq 1$, $u_\ell\to 1$ and $u_\ell f\in L^\infty(\lambda)$. We have $(u_\ell f)oI=(u_\ell oI)(foI)\in H^\#\cap L^\infty(m)=H$ and hence $u_\ell f\in H^\infty(D)$ from 8). It follows that $f\in H^\#(D)$. QED.

Now let $I\in H_\varphi$ be an inner function such that $H_\varphi=IH$. Then the substitution map $H^\#(D)\to H^\#:f\mapsto foI$ can be composed with the evaluation map $H^\#\to Hol^\#(D):u\mapsto\hat{u}$ from 3.1. The result is as follows.

<u>4.2</u> <u>PROPOSITION</u>: Assume that $I\in H_\varphi$ is an inner function such that $H_\varphi=IH$. Then the composition

$$H^{\#}(D) \to H^{\#} \to Hol^{\#}(D): f \mapsto f\circ I \mapsto (f\circ I)^{\wedge}$$

is the identity map in the sense that $f\in H^{\#}(D)$ is the boundary function of $(f\circ I)^{\wedge}\in Hol^{\#}(D)$. An equivalent formulation is

$$\varphi_z(f) = \varphi_z(f\circ I) \qquad \forall f\in H^{\#}(D) \text{ and } z\in D,$$

where the first φ_z is relative to the unit disk situation after IV.3.15 while the second φ_z is in the sense of 3.2.i).

Proof:i) Let $U\in H^{\infty}(D)$ and $u=\langle U\lambda\rangle\in Hol^{\infty}(D)$. For $z\in D$ then

$$u(z) = \int_S P(z,\cdot)U d\lambda = \int \bigl((P(z,\cdot)U)\circ I\bigr) Fdm$$

$$= \int \bigl(P(z,\cdot)\circ I\bigr)\bigl(U\circ I\bigr) Fdm = \int (U\circ I)\, \frac{1-|z|^2}{|I-z|^2}\, Fdm = (U\circ I)^{\wedge}(z).$$

ii) For $f\in H^{\#}(D)$ write $f=\frac{U}{V}$ with $U,V\in H^{\infty}(D)$ and V invertible in $H^{\#}(D)$. Then $U=Vf$ and $U\circ I=(V\circ I)(f\circ I)$. From $\varphi_z(U)=\varphi_z(V)\varphi_z(f)$ and $\varphi_z(U\circ I)=\varphi_z(V\circ I)\varphi_z(f\circ I)$ as well as $\varphi_z(U)=\varphi_z(U\circ I)$ and $\varphi_z(V)=\varphi_z(V\circ I)\neq 0$ from i) it follows that $\varphi_z(f)=\varphi_z(f\circ I)$ $\forall z\in D$. QED.

For the subsequent consequence observe that $H^{\#}(D)\cap L^p(\lambda)=H^p(D)$ for all $1\leq p\leq\infty$. For $1\leq p\leq\infty$ this follows from IV.6.10 combined with I.3.3.

4.3 CONSEQUENCE: Assume that $I\in H_{\varphi}$ is an inner function such that $H_{\varphi}=IH$. Then the evaluation map $u\mapsto\hat{u}$ is surjective as a map $H^{\#}\to Hol^{\#}(D)$ and as a map $H^{\#}\cap L^p(Fm)\to Hol^p(D)$ for all $1\leq p\leq\infty$, hence in particular as a map $H\to Hol^{\infty}(D)$.

Proof: Immediate from 4.2 and 4.1.v) combined with the above remark. QED.

Let us keep in mind that we want an isomorphism between (H,φ) and $(H^{\infty}(D),\varphi_0)$ via substitution map and evaluation map. In the next remark we formulate the decisive bijectivity property of these maps in several equivalent versions.

4.4 REMARK: Assume that $I\in H_{\varphi}$ is an inner function such that $H_{\varphi}=IH$. Then the subsequent nine properties are equivalent.

The substitution map $f\mapsto f\circ I$ is surjective and hence bijective as a map

1#) $L(\lambda)\to L(m)$, 2#) $H^{\#}(D)\to H^{\#}$,

1) $L^{p}(\lambda)\to L^{p}(Fm)$ $\forall 1\leq p\leq\infty$, 2) $H^{p}(D)=H^{\#}(D)\cap L^{p}(\lambda)\to H^{\#}\cap L^{p}(Fm)$ $\forall 1\leq p\leq\infty$,

1∞) $L^{\infty}(\lambda)\to L^{\infty}(m)$, 2∞) $H^{\infty}(D)\to H$.

The evaluation map $u\mapsto\hat{u}$ is injective and hence bijective as a map

3#) $H^{\#}\to Hol^{\#}(D)$,

3) $H^{\#}\cap L^{p}(Fm)\to Hol^{p}(D)$ $\forall 1\leq p\leq\infty$,

3∞) $H\to Hol^{\infty}(D)$.

The proof will be after the scheme

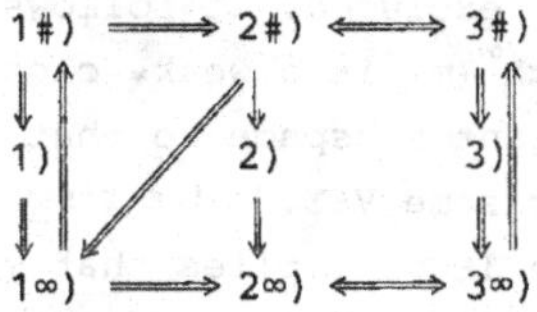

The implications 1#) ⇒ 1) ⇒ 1∞) and 2#) ⇒ 2) ⇒ 2∞) are immediate after 4.1.v), and the implications 1#) ⇒ 2#) and 1∞) ⇒ 2∞) are immediate after 4.1.vi). The implications 3#) ⇒ 3) ⇒ 3∞) are obvious. And 3∞) ⇒ 3#) is immediate after V.1.10. Furthermore 2#) ⇔ 3#) is a direct consequence of 4.2, and 2∞) ⇔ 3∞) is seen alike. It remains to prove 2#)⇒1∞) and 1∞) ⇒ 1#).

Proof of 2#)⇒1∞): Let $P\in\text{Re}\,L^{\infty}(m)$. Then $P+iP^{*}\in H^{\#}$ after VI.3.9 and VI.2.8. Hence there exists an $f=u+iv\in H^{\#}(D)$ such that $f\circ I=u\circ I+i(v\circ I)=P+iP^{*}$. It follows that $u\circ I=P$ and of course $u\in\text{Re}\,L^{\infty}(\lambda)$ from 4.1.v). The assertion follows. Proof of 1∞)⇒1#): Each function $P\in L(m)$ can be written $P=\frac{U}{V}$ with $U,V\in L^{\infty}(m)$ and $m([V=0])=0$. We have functions $u,v\in L^{\infty}(\lambda)$ such that $u\circ I=U$ and $v\circ I=V$. From 4.1.i) we see that $\lambda([v=0])=(Fm)([I\in[v=0]])=(Fm)([V=0])=0$. Hence $v\in(L(\lambda))^{\times}$ and $\frac{1}{v}\circ I=\frac{1}{V}$. The assertion follows. QED.

We are now arrived at the main result.

<u>4.5</u> <u>ISOMORPHISM THEOREM</u>: Assume that $H\neq\mathbb{C}$. Then the subsequent properties are equivalent.

i) The invariant subspace theorem is true in its classical form: An invariant weak* closed linear subspace $T\subset L^{\infty}(m)$ is of the form $T=PH$ for some $P\in T$ of modulus $|P|=1$ or of the form $T=\chi_{V}L^{\infty}(m)$ for some $V\in\Sigma$.

ii) H_φ is simply invariant, and H fulfills the equivalent conditions i)-iv) in the maximality theorem 2.1.

iii) H_φ is simply invariant, and for some (and hence for each) inner function $I \in H_\varphi$ such that $H_\varphi = IH$ the substitution map $f \mapsto f \circ I$ is surjective and hence bijective in the sense of 4.4.

In this case the substitution isomorphism $H^{\#}(D) \to H^{\#} : f \mapsto f \circ I$ satisfies $\varphi_z(f) = \varphi_z(f \circ I)$ for all $f \in H^{\#}(D)$ and $z \in D$ and hence in particular $\varphi_0(f) = \varphi(f \circ I)$ for all $f \in H^{\#}(D)$.

Proof: i)⇒ii) 1) If H_φ were not simply invariant then it were of the form $H_\varphi = \chi_V L^\infty(m)$ for some $V \in \Sigma$. Then $\chi_V \in H_\varphi$ which enforces that $\chi_V = 0$. Thus $H_\varphi = 0$ or $H = \mathbb{C}$ which has been excluded. It follows that H_φ is simply invariant. 2) Assume that $B \subset L^\infty(m)$ is a weak* closed subalgebra which contains H. Then B is an invariant subspace so that either $B = PH$ for some $P \in B$ with $|P| = 1$ or $B = \chi_V L^\infty(m)$ for some $V \in \Sigma$. In the first case $P^2 \in B = PH$ or $P \in H$ so that $B = H$. In the second case $1 \in H \subset B$ implies that $\chi_V = 1$ and hence that $B = L^\infty(m)$. Thus 2.1.i) is fulfilled.

ii)⇒iii) Let $I \in H_\varphi$ be an inner function such that $H_\varphi = IH$. Then $T := \{u \in H : \hat{u} = 0\}$ consists of the functions $u \in H$ with Taylor coefficients $a_n = 0$ $\forall n \geqq 0$. Thus T is the intersection of the $I^n H$ $\forall n \geqq 1$ and hence an invariant weak* closed linear subspace of $L^\infty(m)$. Also $\bar{I}T \subset T$ so that ii) implies that $T = \chi_V L^\infty(m)$ for some $V \in \Sigma$. But then $\chi_V \in T \subset IH$ enforces that $\chi_V = 0$ so that $T = 0$. Thus condition 3∞) in 4.4 is satisfied.

iii)⇒i) Let $I \in H_\varphi$ be an inner function such that $H_\varphi = IH$. 1) We claim that the linear combinations of the functions I^n with $n \in \mathbb{Z}$ are weak* dense in $L^\infty(m)$. To see this take a function $P \in L^1(m)$ with $\int I^n P dm = 0$ $\forall n \in \mathbb{Z}$. Then $\frac{P}{F} \in L^1(Fm)$ so that from 1) in 4.4 we obtain a function $u \in L^1(\lambda)$ with $u \circ I = \frac{P}{F}$. It follows that $\int\limits_S z^n u d\lambda = \int (z \circ I)^n (u \circ I) F dm = \int I^n P dm = 0$ $\forall n \in \mathbb{Z}$. Thus $u = 0$ and hence $P = 0$. 2) Let $T \subset L^\infty(m)$ be an invariant weak* closed linear subspace. Then $H_\varphi T = IT$ so that $T_\varphi = IT$. If $T_\varphi \neq T$ then $T = PH$ for some $P \in T$ with $|P| = 1$ after the invariant subspace theorem 1.1. Assume now that $T_\varphi = T$. Then $IT = T$ and hence $\bar{I}T = T$ so that $I, \bar{I} \in \tau(T)$ (see Section 2). Thus 1) implies that $\tau(T) = L^\infty(m)$. Then 2.2 shows that $T = \chi_V L^\infty(m)$ for some $V \in \Sigma$. QED.

In the next section we shall prove that another equivalent condition is the existence of a purely algebraic isomorphism between H and $H^\infty(D)$. If such an isomorphism exists then H contains no zero divisors, which is condition 2.1.iv). Thus the fact to be proved is that H_φ must be simply invariant. This is not at all obvious.

5. Complements on the simple Invariance of H_φ

The first aim is the subsequent illustrative result.

5.1 THEOREM: Assume that H_φ is not simply invariant. Let $S,T\subset L^\infty(m)$ be invariant weak* closed linear subspaces such that $S\subset T$ and $\dim\frac{T}{S}<\infty$. Then $S=T$ or $S=T_\varphi$.

The finite-dimension hypothesis will be introduced into the proof via an algebraic lemma which seems to be mathematical folklore.

5.2 COMMUTATIVE OPERATOR ALGEBRA LEMMA: Let V be a nontrivial finite-dimensional complex vector space and $A\subset L(V)$ be a commutative operator algebra on V which contains the identity operator. Then there exists a nonzero vector $a\in V$ which is an eigenvector to each $P\in A$.

Proof of 5.2: We want a one-dimensional linear subspace of V which is invariant under each $P\in A$. It suffices to prove that in case $\dim V>1$ there exists a nontrivial proper common invariant linear subspace. The assertion then follows upon successive application. Now we are done if $\dim A=1$. So take an operator $R\in A$ which is not a multiple of the identity I and fix an eigenvalue $c\in\mathbb{C}$ of R. Then the nullspace $U\subset V$ of $R-cI$ is nontrivial and $\neq V$. It remains to show that U is invariant under each $P\in A$. In fact, for $x\in U$ we have $(R-cI)Px=P(R-cI)x=0$ which means that $Px\in U$. QED.

5.3 REMARK: Assume that H_φ is not simply invariant. For each invariant weak* closed linear subspace $T\subset L^\infty(m)$ then $(T_\varphi)_\varphi=T_\varphi$.

Proof of 5.3: The assumption is that $H_\varphi=\overline{\mathrm{Lin}H_\varphi H_\varphi}^{\text{weak*}}$. Thus $H_\varphi H_\varphi T\subset H_\varphi T_\varphi\subset (T_\varphi)_\varphi$ implies that $H_\varphi T\subset (T_\varphi)_\varphi$ and hence that $T_\varphi\subset (T_\varphi)_\varphi$. QED.

Proof of 5.1: i) It suffices to prove $T_\varphi\subset S$. In the case $T_\varphi=T$ then $S=T$, and in the case $T_\varphi\neq T$ the invariant subspace theorem 1.1 implies that $\dim(T/T_\varphi)=1$ so that indeed $S=T$ or $S=T_\varphi$. ii) The assertion $T_\varphi\subset S$ will be proved by induction after $\dim(T/S)=:n$. The case $n=0$ is trivial.

So let us turn to the induction step $0\leq n-1\Rightarrow n$. To do this it suffices to construct an invariant weak* closed linear subspace $U\subset L^\infty(m)$ such that $S\subset U\subset T$ with $\dim(U/S)=1$ and $H_\varphi U\subset S$. Then $U_\varphi\subset S$. From the induction hypothesis it will follow that $T_\varphi\subset U$ so that $T_\varphi=(T_\varphi)_\varphi\subset U_\varphi\subset S$ after 5.3, which is the assertion. iii) In order to construct the intermediate linear subspace U we consider the n-dimensional quotient space T/S. Each $u\in H$ defines a linear operator $\langle u\rangle\in L(T/S)$ via $\langle u\rangle:[f]\mapsto[uf]$ $\forall f\in T$. It is obvious that $\langle u+v\rangle=\langle u\rangle+\langle v\rangle$, $\langle cu\rangle=c\langle u\rangle$ and $\langle uv\rangle=\langle u\rangle\langle v\rangle$ for $u,v\in H$ and $c\in\mathbb{C}$, as well as $\langle 1\rangle=$ identity. Thus $A:=\{\langle u\rangle:u\in H\}\subset L(T/S)$ is as required in 5.2. It follows that there exists an $f\in T$, $f\notin S$ and a function $\varepsilon:H\to\mathbb{C}$ such that $\langle u\rangle[f]=\varepsilon(u)[f]$ or $uf-\varepsilon(u)f\in S$ for all $u\in H$. It is clear that $\varepsilon:H\to\mathbb{C}$ is unique and is a multiplicative linear functional with $\varepsilon(1)=1$. Furthermore ε is weak* continuous since its nullspace $\{u\in H:\varepsilon(u)=0\}=\{u\in H:uf\in S\}$ is weak* closed. It follows that $\varepsilon=\varphi$ after the analytic disk theorem 3.3.i) since H_φ is not simply invariant. Thus we have $uf-\varphi(u)f\in S$ for all $u\in H$. Now define $U:=S+\mathbb{C}f$. Then it is clear that U is a weak* closed linear subspace of $L^\infty(m)$ such that $S\subset U\subset T$ and $\dim(U/S)=1$. And the above tells us that U is invariant and that $H_\varphi U\subset S$. QED.

The other main result of the section requires a functional analytic lemma which is of interest in itself.

<u>5.4</u> <u>LEMMA</u>: Let $T\subset L^\infty(m)$ be a weak* closed linear subspace and $P\in L^\infty(m)$, so that $PT\subset L^\infty(m)$ is a linear subspace as well. If PT is $L^\infty(m)$-norm closed then PT is weak* closed.

Proof: i) The linear map $T\to PT:f\mapsto Pf$ is surjective and continuous in the $L^\infty(m)$-norms. Thus if PT is $L^\infty(m)$-norm closed then the open mapping theorem provides us with a constant $C>0$ such that to each $g\in PT$ there is an $f\in T$ with $g=Pf$ and

$$\|f\|_{L^\infty(m)}\leqq C\|g\|_{L^\infty(m)}.$$

ii) The proof that PT is weak* closed will be based on the Krein-Smulian consequence IV.3.14. Thus consider a sequence of functions $g_n\in PT$ such that $|g_n|\leqq b<\infty$ and $g_n\to g\in L^\infty(m)$ pointwise. The claim is that $g\in PT$. From i) we obtain functions $f_n\in T$ with $g_n=Pf_n$ and $|f_n|\leqq Cb<\infty$. After Banach-Alaoglu there exists a weak* adherence point $f\in L^\infty(m)$ of the f_n for $n\to\infty$. Then $f\in T$ since T is weak* closed. For each fixed $u\in L^1(m)$ there is a subsequence $1\leqq n(1)<\dots<n(\ell)<n(\ell+1)<\dots$ such that $\int uPf_{n(\ell)}dm\to\int uPf\,dm$ for $\ell\to\infty$.

But $\int uPf_{n(\ell)}dm=\int ug_{n(\ell)}dm\to\int ugdm$. It follows that $\int ugdm=\int uPfdm\ \ \forall u\in L^1(m)$ and hence that $g=Pf\in PT$. QED.

5.5 THEOREM: Assume that $(\tilde{H},\tilde{\varphi})$ on $(\tilde{X},\tilde{\Sigma},\tilde{m})$ is another reduced Szegö situation besides (H,φ) and that there exists an algebraic isomorphism $\sigma:H\to\tilde{H}$. If $\tilde{H}_{\tilde{\varphi}}$ is simply invariant then H_φ is simply invariant as well.

5.6 CONSEQUENCE: Assume that there exists an algebraic isomorphism between H and $H^\infty(D)$. Then H_φ is simply invariant. Hence (H,φ) satisfies the equivalent conditions i)-iii) in the isomorphism theorem 4.5 (the converse is trivial).

Proof of 5.5: i) We use the notation $\sigma:f\mapsto f^\sigma$ for $f\in H$. We show first that $\sigma:H\to\tilde{H}$ preserves L^∞-norms. In fact, a function $f\in H$ is invertible in H iff f^σ is invertible in $\tilde{H}$. Hence the spectrum of f relative to H coincides with the spectrum of f^σ relative to $\tilde{H}$. Since in both cases norm = = spectral radius we obtain $\|f\|_{L^\infty(m)}=\|f^\sigma\|_{L^\infty(\tilde{m})}$ $\forall f\in H$ as claimed. ii) Let $\tilde{I}\in\tilde{H}_{\tilde{\varphi}}$ be an inner function such that $\tilde{H}_{\tilde{\varphi}}=\tilde{I}\tilde{H}$, and let $I\in H$ such that $I^\sigma=\tilde{I}$. Then

$$\begin{array}{ccc}
\sigma:H & \longrightarrow & \tilde{H}\\
\cup & & \cup\\
IH & \longrightarrow & \tilde{I}\tilde{H}=\tilde{H}_{\tilde{\varphi}}\\
\cup & & \cup\\
I^2H & \longrightarrow & \tilde{I}^2\tilde{H},
\end{array}$$

where in each line we have an isomorphism which preserves L^∞-norms after i). The linear spaces in the second column are $L^\infty(\tilde{m})$-norm closed since $|\tilde{I}|=1$. Hence the linear spaces in the first column are $L^\infty(m)$-norm closed and therefore weak* closed after 5.4. And they are of course invariant under H. Furthermore the codimensions are all=1 in the second column and hence in the first column. If now H_φ were not simply invariant then 5.1 would force both IH and I^2H to be $=H_\varphi$ so that we would arrive at a contradiction. It follows that H_φ must be simply invariant. QED.

6. A Class of Examples

We construct a class of examples in order to show that the two sets of equivalent properties which dominate the present chapter are independent.

The measure space is $(X,\Sigma,m):=(S\times S,\mathrm{Baire},\lambda\times\lambda)$. We have the coordinate functions $Z_1,Z_2:Z_\ell(s)=s_\ell$ for $s=(s_1,s_2)\in X$ and the monomials $Z^p:=Z_1^{p_1}Z_2^{p_2}$ for $p=(p_1,p_2)\in\mathbb{Z}\times\mathbb{Z}$. It is an immediate Stone-Weierstraß consequence that the linear combinations of all these monomials are weak* dense in $L^\infty(m)$. For each subset $\Gamma\subset\mathbb{Z}\times\mathbb{Z}$ with $0\in\Gamma$ and $\Gamma+\Gamma\subset\Gamma$ we define

$$H(\Gamma) := \overline{\mathrm{Lin}\{Z^p:p\in\Gamma\}}^{\mathrm{weak}*} \subset L^\infty(m).$$

Thus $H(\Gamma)$ is a weak* closed complex subalgebra of $L^\infty(m)$ which contains the constants. We need two immediate properties. i) If $\Gamma\cup(-\Gamma)=\mathbb{Z}\times\mathbb{Z}$ then $H(\Gamma)+\overline{H(\Gamma)}$ is weak* dense in $L^\infty(m)$, that is $\mathrm{Re}\,H(\Gamma)$ is weak* dense in $\mathrm{Re}\,L^\infty(m)$. ii) If $\Gamma\cap(-\Gamma)=\{0\}$ then the functional $\varphi:u\mapsto\int u\,dm$ is multiplicative on $H(\Gamma)$. Therefore: If $\Gamma\cup(-\Gamma)=\mathbb{Z}\times\mathbb{Z}$ and $\Gamma\cap(-\Gamma)=\{0\}$ then $(H(\Gamma),\varphi)$ is a reduced Szegö situation with $M=\{1\}$. In the sequel we fix a subset $\Gamma\subset\mathbb{Z}\times\mathbb{Z}$ with all these properties.

<u>6.1</u> <u>PROPOSITION</u>: $(H(\Gamma),\varphi)$ does not fulfill the equivalent conditions i)-iii) in the isomorphism theorem 4.5. That is $(H(\Gamma),\varphi)$ cannot at the same time possess the equivalent properties i)-iv) in the maximality theorem 2.1 and have a simply invariant $(H(\Gamma))_\varphi$.

Proof: Assume that this is not true. Then there is an inner function $I\in(H(\Gamma))_\varphi$ such that $(H(\Gamma))_\varphi=IH(\Gamma)$ and that the substitution map $H^\infty(D)\to H(\Gamma):f\mapsto f\circ I$ is bijective. Also $\varphi_o(f)=\varphi(f\circ I)$ for all $f\in H^\infty(D)$. For $p\in\Gamma$ thus $Z^p=f_p\circ I$ with $f_p\in H^\infty(D)$. Then $f_o=1$ while for nonzero $p\in\Gamma$ we have $\varphi_o(f_p)=\varphi(Z^p)=0$ so that after I.3.6 there is a unique $N(p)\in\mathbb{N}$ such that $f_p=Z^{N(p)}h_p$ with $h_p\in H^\infty(D)$ and $\varphi_o(h_p)\neq 0$. Of course we put $N(0)=0$. For $p,q\in\Gamma$ then $f_{p+q}=f_pf_q$ implies that $N(p+q)=N(p)+N(q)$. For $p\in\mathbb{Z}\times\mathbb{Z}$ not in Γ we have $p\neq 0$ and $-p\in\Gamma$ so that we can define $N(p):=-N(-p)<0$. The function $N:\mathbb{Z}\times\mathbb{Z}\to\mathbb{Z}$ then satisfies $N(p+q)=N(p)+N(q)$ for all $p,q\in\mathbb{Z}\times\mathbb{Z}$ as a look at the different possibilities shows, and we have $N(p)\neq 0$ whenever $p\neq 0$. But this is impossible: we have $N(p)=a_1p_1+a_2p_2$ $\forall p\in\mathbb{Z}\times\mathbb{Z}$ for some $a_1,a_2\in\mathbb{Z}$, so that there are lots of nonzero $p\in\mathbb{Z}\times\mathbb{Z}$ with $N(p)=0$. This contradiction proves the assertion. QED.

After the above result it remains to show that the subset $\Gamma\subset\mathbb{Z}\times\mathbb{Z}$ can be chosen such that $(H(\Gamma))_\varphi$ is simply invariant, and also such that the equivalent conditions i)-iv) in the maximality theorem 2.1 are satisfied.

6.2 PROPOSITION: Assume that the nonzero $a\in\Gamma$ is such that $p-a\in\Gamma$ for all nonzero $p\in\Gamma$. Then $(H(\Gamma))_{\varphi}=Z^{a}H(\Gamma)$.

Proof: We have to prove the inclusion $\subset$. Fix an $f\in(H(\Gamma))_{\varphi}$. In order to prove $Z^{-a}f\in H(\Gamma)$ it suffices to show that

$$\int Z^{-a}fZ^{p}dm = \int Z^{-a}fdm \int Z^{p}dm \quad \forall\ p\in\Gamma.$$

For then $Z^{-a}f\in K\cap L^{\infty}(m)$ so that VI.6.9.iii) implies that $Z^{-a}f\in(H(\Gamma))^{\#}\cap$ $\cap L^{\infty}(m)=H(\Gamma)$. But the desired equation is obvious for $p=0$, and for nonzero $p\in\Gamma$ both sides are $=0$ as a consequence of the assumptions. QED.

6.3 EXAMPLE: Let Γ consist of the points $p=(p_1,p_2)\in\mathbb{Z}\times\mathbb{Z}$ with $p_1>0$ and of those with $p_1=0$ and $p_2\geqq 0$, that is of the $p\geqq 0$ in the lexicographic ordering of $\mathbb{Z}\times\mathbb{Z}$. Then we are in the situation of 6.2 with $a=(0,1)$. Thus $(H(\Gamma))_{\varphi}=Z_2H(\Gamma)$. It is not hard to see that $H(\Gamma)$ consists of the functions $f\in L^{\infty}(m)$ such that i) for λ-almost all $s\in S$ the section $f(\cdot,s)\in L^{\infty}(\lambda)$ is in $H^{\infty}(D)$ and ii) the function $s\mapsto\varphi_{o}(f(\cdot,s))=\int f(\cdot,s)d\lambda$ for λ-almost all $s\in S$ defines a member of $H^{\infty}(D)$.

The other task is somewhat more involved. We need a basic fact from commutative Gelfand theory combined with the subsequent restricted maximality theorem.

6.4 REMARK: Let (H,φ) be a reduced Szegö situation. Assume that $B\subset L^{\infty}(m)$ is an $L^{\infty}(m)$-norm closed complex subalgebra which contains H such that φ can be extended to a multiplicative linear functional on B. Then $B=H$.

Proof: Let $\phi:B\rightarrow\mathbb{C}$ be a multiplicative linear functional which extends φ. Then ϕ is $L^{\infty}(m)$-norm continuous with $\|\phi\|=1$. And let $\Lambda:L^{\infty}(m)\rightarrow\mathbb{C}$ be a bounded linear extension of ϕ with $\|\Lambda\|=1$. After A.1.5 then $\|\Lambda\|=\Lambda(1)=1$ implies that Λ is positive: $\Lambda(f)\geqq 0$ for all $0\leqq f\in\operatorname{Re}L^{\infty}(m)$, and hence that $\Lambda(f)$ is real for all $f\in\operatorname{Re}L^{\infty}(m)$. Thus for $f\in\operatorname{Re}L^{\infty}(m)$ we obtain

$$u\in H \text{ with } \operatorname{Re}u\geqq f \Rightarrow \operatorname{Re}\varphi(u) = \operatorname{Re}\Lambda(u) = \Lambda(\operatorname{Re}u)\geqq\Lambda(f),$$

and hence $\alpha^{o}(f)\geqq\Lambda(f)$ from IV.2.5. But $\alpha^{o}(f)=\int fFdm$ from IV.3.10 where $M=\{F\}$. It follows that $\Lambda(f)\leqq\int fFdm$ $\forall f\in\operatorname{Re}L^{\infty}(m)$ and hence that $\Lambda(f)=\int fFdm$ $\forall f\in L^{\infty}(m)$. In particular $\phi(f)=\int fFdm$ $\forall f\in B$ so that the integral $f\mapsto\int fFdm$ remains multiplicative on B. Thus $f\in B$ implies that $fF\in K$ and hence $f\in H^{\#}\cap$

$\cap L^\infty(m)=H$ after VI.6.9.iii). It follows that B=H. QED.

Now for $\Gamma\subset\mathbb{Z}\times\mathbb{Z}$ as above define $\bar{\Gamma}\subset\mathbb{Z}\times\mathbb{Z}$ to consist of the $p\in\mathbb{Z}\times\mathbb{Z}$ such that for each nonzero $q\in\Gamma$ we can achieve $p+nq\in\Gamma$ for some $n\in\mathbb{N}$. Then $\Gamma\subset\bar{\Gamma}$ and $\bar{\Gamma}+\bar{\Gamma}\subset\bar{\Gamma}$ so that $H(\bar{\Gamma})$ is defined. It is clear that $\bar{\Gamma}\cup(-\bar{\Gamma})=\mathbb{Z}\times\mathbb{Z}$ but we must expect that $\bar{\Gamma}\cap(-\bar{\Gamma})\neq\{0\}$.

<u>6.5</u> <u>PROPOSITION</u>: If $B\subset L^\infty(m)$ is a weak* closed subalgebra with $H(\Gamma)\subsetneqq B$ then $H(\bar{\Gamma})\subset B$. In particular if $\bar{\Gamma}=\mathbb{Z}\times\mathbb{Z}$ then $H(\Gamma)$ is a maximal proper weak* closed subalgebra of $L^\infty(m)$.

Proof: We have to prove that $z^p\in B$ for each $p\in\bar{\Gamma}$. This is clear for $p\in\Gamma$. So assume that $p\in\bar{\Gamma}$ and $p\notin\Gamma$ so that $p\neq0$ and $-p\in\Gamma$. In order to prove that $z^p\in B$ we show that $\psi(z^{-p})\neq0$ for each nonzero multiplicative linear functional $\psi:B\to\mathbb{C}$: then it follows from commutative Gelfand theory that $z^{-p}\in H(\Gamma)\subset B$ is invertible in B and hence that $z^p\in B$. So let us assume that there exists such a ψ with $\psi(z^{-p})=0$. Take then a nonzero $q\in\Gamma$ and choose an $n\in\mathbb{N}$ such that $p+nq\in\Gamma$. Then from $z^{nq}=z^{p+nq}z^{-p}$ we obtain $(\psi(z^q))^n=\psi(z^{p+nq})\psi(z^{-p})=0$ and hence $\psi(z^q)=0$. It follows that $\psi(z^q)=\varphi(z^q)$ for all $q\in\Gamma$ and hence $\psi|H(\Gamma)=\varphi$. But this is impossible in view of 6.4. Hence it must be $\psi(z^{-p})\neq0$ as claimed. QED.

<u>6.6</u> <u>EXAMPLE</u>: For $\alpha\in\mathbb{R}$ an irrational number define $\Gamma:=\{p=(p_1,p_2)\in\mathbb{Z}\times\mathbb{Z}: p_1+\alpha p_2\geqq0\}=\{0\}\cup\{p: p_1+\alpha p_2>0\}$. It is obvious that Γ fulfills our assumptions and that $\bar{\Gamma}=\mathbb{Z}\times\mathbb{Z}$. Thus 6.5 tells us that $(H(\Gamma),\varphi)$ is an example of the desired kind.

<u>Notes</u>

The evolution of the invariant subspace theorem started with BEURLING [1949] who discovered the L^2-version in the unit disk situation. The basic idea of the abstract proof which carried as far as to the Szegö situation is due to HELSON-LOWDENSLAGER [1958]. SRINIVASAN-WANG [1966] observed that the usual version of the invariant subspace theorem is in fact equivalent to the weak*Dirichlet condition. But their equivalence theorem suffered from the fact that it missed the decisive implication Szegö $\Rightarrow$ weak*Dirichlet. Also their proof of the equivalence theorem seems to us to be incomplete (in the diagram on p. 239 the arrow which leads to (vii) is not conclusive because of the definition $H^{\infty}:=H^{2}\cap L^{\infty}$). Whithin the Szegö situation MERRILL-LAL [1969] and NAKAZI [1975] extended the theorem to certain classes of non-simply invariant subspaces. A step beyond the Szegö situation is in AHERN-SARASON [1967a]. The invariant subspace theme is of an impressive width and depth as it is demonstrated in the treatises of HOFFMAN [1962a], HELSON [1964] and RADJAVI-ROSENTHAL [1973].

The maximality theorem 2.1 is due to MUHLY [1972]. The present proof is from SALINAS [1976]. It furnishes quite some information on the σ-algebras $\Delta\subset\Sigma$ which correspond to the weak* closed intermediate algebras $H\subsetneqq B\subsetneqq L^{\infty}(m)$ in case of a non-maximal H. Certain classes of these B are determined in NAKAZI [1976]. The maximality theme is often dealt with in terms of maximal proper closed subalgebras of C(X) with X a compact Hausdorff space. See WERMER [1961] where also the respective version of 6.6 can be found.

The analytic disk theorem 3.3 is due to WERMER [1960] (in the Dirichlet algebra situation). The problem to discover analytic structures in spectra is of wide interest and was in fact one of the main stimuli for the development of the theory. For the results beyond the Szegö situation we refer to GAMELIN [1969] Chapter VI.

The essence of the isomorphism theorem 4.5 is in MERRILL [1968]. The equivalence i)⇔iii) was earlier in MÜRMANN [1967]. The substitution map as in 4.1 appeared first in LUMER [1966b]. Theorem 5.1 is due to SALINAS [1976]. The permanence theorem 5.5 with 5.4 is due to König and appears here for the first time. Section 6 is from SALINAS [1976] except that the new result 6.1 replaces certain direct verifications. It seems that there was no prior explicit example of a reduced Szegö situation (H,φ) with H_{φ} not simply invariant (and nontrivial).

Chapter VIII

Weak Compactness of M

The present chapter is devoted to the reduced Hardy algebra situation (H,φ) on (X,Σ,m) in the special case that the set M of the representative functions is compact in the weak topology $\sigma(ReL^1(m),ReL^\infty(m))$. This assumption was already under short consideration in IV.4.5 and VI.6.3. As we have noted in IV.4 the case that M is weakly compact includes the important case that N is finite-dimensional.

Our main result will be 3.3. The proof requires two auxiliary means of independent interest. The first one is the Hewitt-Yosida decomposition of bounded linear functionals on $L^\infty(m)$ into weak*continuous and singular parts, while the second one is a particular notion of convergence for sequences of functions, adapted to our purposes in $L^\infty(m)$. These topics will be dealt with in Sections 1 and 2 where we fix a nontrivial finite positive measure space (X,Σ,m).

1. The Decomposition Theorem of Hewitt-Yosida

A functional $\tau\in(L^\infty(m))'$ is defined to be singular iff for each $\varepsilon>0$ there exists a set $E\in\Sigma$ with $m(E)<\varepsilon$ such that $\tau(f) = \tau(f\chi_E)$ $\forall f\in L^\infty(m)$. Note that for any $\tau\in(L^\infty(m))'$ the collection of subsets $E\in\Sigma$ such that $\tau(f) = \tau(f\chi_E)$ $\forall f\in L^\infty(m)$ is closed under intersection. Hence a functional $\tau\in(L^\infty(m))'$ is singular iff there exists a decreasing sequence of subsets $E(n)\in\Sigma$ such that $m(E(n))\to 0$ and $\tau(f) = \tau(f\chi_{E(n)})$ $\forall f\in L^\infty(m)$ $(n=1,2,\dots)$. Furthermore it is obvious that the singular functionals form a linear subspace of $(L^\infty(m))'$.

<u>1.1</u> <u>THEOREM</u>: Each functional $\lambda\in(L^\infty(m))'$ admits a unique decomposition

$$\lambda = Gm + \tau : \lambda(f) = \int fGdm + \tau(f) \qquad \forall f\in L^\infty(m) ,$$

where $G\in L^1(m)$ and $\tau\in(L^\infty(m))'$ is singular. In particular λ is positive (that is $\lambda(f)\geq 0$ whenever $f\geq 0$) iff $G\geq 0$ and τ is positive.

We remark that the above decomposition can be interpreted as follows. The Gelfand transformation is an isometric isomorphism of the commutative B*algebra $L^\infty(m)$ with $C(K)$ on the structure space K of $L^\infty(m)$. Hence the functionals $\lambda\in(L^\infty(m))'$ are in one-to-one correspondence with the measures $\tilde{\lambda}\in ca(K) = (C(K))'$, and positive functionals λ correspond to positive measures $\tilde{\lambda}$. In particular m itself has a representative $\tilde{m}\in Pos(K)$. Now it is not hard to see that the above decomposition of $\lambda\in(L^\infty(m))'$ corresponds to the Lebesgue decomposition of $\tilde{\lambda}\in ca(K)$ with respect to $\tilde{m}$ (see BARBEY [1975] p.524 for more details). This remark will not be used in the sequel.

The proof of 1.1 requires the implication iii) $\Rightarrow$ i) in the subsequent lemma.

<u>1.2 LEMMA</u>: For a positive $\tau\in(L^\infty(m))'$ the subsequent properties are equivalent.

i) τ is singular.

ii) $\operatorname{Inf}\{\int h dm + \tau(1-h) : h\in L^\infty(m) \text{ with } 0\leqq h\leqq 1\} = 0$.

iii) If $0\leqq F\in L^1(m)$ with $\int hF dm \leqq \tau(h)$ $\forall 0\leqq h\in L^\infty(m)$ then $F=0$.

Proof of 1.2: i)$\Rightarrow$iii) For $E\in\Sigma$ with $\tau(f)=\tau(f\chi_E)$ $\forall f\in L^\infty(m)$ we have $\int F(1-\chi_E)dm \leqq \tau(1-\chi_E) = 0$ and hence $\int F dm = \int F\chi_E dm$. From this the assertion is clear. ii)$\Rightarrow$i) Fix $\varepsilon>0$ and choose functions $h_n\in L^\infty(m)$ with $0\leqq h_n\leqq 1$ and

$$\int h_n dm \leqq 2^{-2n}\varepsilon \quad \text{and} \quad \tau(1-h_n) \leqq 2^{-n}\tau(1) \quad (n=1,2,\ldots).$$

Define $A(n):=[h_n\geqq 2^{-n}]\in\Sigma$ so that $2^{-n}\chi_{A(n)} \leqq h_n \leqq 2^{-n} + \chi_{A(n)}$. From the first relation we obtain $m(A(n)) \leqq 2^{-n}\varepsilon$ while the second one implies that $\tau(\chi_{A(n)}) \geqq \tau(h_n) - 2^{-n}\tau(1) \geqq \tau(1) - 2^{-n+1}\tau(1)$. Let now $A\in\Sigma$ be the union of the $A(n)$. Then $m(A)\leqq\varepsilon$. And in view of $\tau(1) \geqq \tau(\chi_A) \geqq \tau(\chi_{A(n)})$ the above inequality implies that $\tau(1) = \tau(\chi_A)$. From this it follows that $\tau(f) = \tau(f\chi_A)$ $\forall f\in L^\infty(m)$. We can of course assume that $f\geqq 0$. Then $f \leqq$ some positive c and hence $0 \leqq \tau(f-f\chi_A) = \tau(f(1-\chi_A)) \leqq \tau(c(1-\chi_A)) = 0$ so that indeed $\tau(f) = \tau(f\chi_A)$. Thus τ is singular as claimed.

iii)$\Rightarrow$ii) Define the functional

$$\sigma:\sigma(f) = \operatorname{Inf}\{\int h dm + \tau(f-h) : h\in L^\infty(m) \text{ with } 0\leqq h\leqq f\} \text{ for } 0\leqq f\in L^\infty(m).$$

We claim the subsequent properties.

1) $0 \leq \sigma(f) \leq \int f dm$ and $\leq \tau(f)$ for $0 \leq f \in L^\infty(m)$.

2) $\sigma(tf) = t\sigma(f)$ for $t \geq 0$ and $0 \leq f \in L^\infty(m)$.

3) $\sigma(f+g) = \sigma(f)+\sigma(g)$ for $0 \leq f,g \in L^\infty(m)$.

Here 1) and 2) are clear. In 3) the inequality $\leq$ is obvious. To prove $\geq$ note that $h \in L^\infty(m)$ with $0 \leq h \leq f+g$ can be written $h = u+v$ where $0 \leq u \leq f$ and $0 \leq v \leq g$, for example with $u = \text{Inf}(h,f)$. The details are then clear. Now extend σ to a positive linear functional $\sigma \in (L^\infty(m))'$. It follows that $|\sigma(f)| \leq \int |f| dm \quad \forall f \in L^\infty(m)$: in fact, for some complex c with $|c|=1$ we have $|\sigma(f)| = c\sigma(f) = \sigma(cf) = \sigma(\text{Re } cf) \leq \sigma(|f|) \leq \int |f| dm$. Hence via extension of σ to a linear functional $\in (L^1(m))'$ we obtain the representation $\sigma(f) = \int fF dm \quad \forall f \in L^\infty(m)$ for some $F \in L^\infty(m)$ of modulus $|F| \leq 1$. Of course F must be ≥ 0. Now in view of 1) the assumption iii) implies that $F=0$ or $\sigma=0$. Thus $\sigma(1)=0$ which is the assertion ii). QED.

Proof of 1.1: 1) The uniqueness assertion is clear from the definition of singularity. 2) We prove the existence of the decomposition for a positive functional $\lambda \in (L^\infty(m))'$. Define

$$T = \{0 \leq F \in L^1(m) : \int fF dm \leq \lambda(f) \text{ for all } 0 \leq f \in L^\infty(m)\}.$$

First note that $F,G \in T$ implies that $\text{Sup}(F,G) \in T$ in view of

$$\int f \,\text{Sup}(F,G) dm = \int f\chi_{[F \geq G]} F dm + \int f\chi_{[F<G]} G dm$$

$$\leq \lambda(f\chi_{[F \geq G]}) + \lambda(f\chi_{[F<G]}) = \lambda(f) \quad \text{for } 0 \leq f \in L^\infty(m).$$

Put $s := \text{Sup}\{\int F dm : F \in T\} \leq \lambda(1) < \infty$ and choose $F_n \in T$ such that $\int F_n dm \to s$. Then $G_n := \text{Sup}(F_1,\ldots,F_n) \in T$ and hence $G_n \uparrow G \in T$ with $\int G dm = s$ after the Beppo Levi theorem. Now $\tau(f) := \lambda(f) - \int fG dm \quad \forall f \in L^\infty(m)$ defines a positive linear functional $\tau \in (L^\infty(m))'$. In view of the above it satisfies 1.2.iii) and hence 1.2.i). Thus we have a decomposition $\lambda = Gm + \tau$ of λ as claimed. 3) To prove the existence of the decomposition for arbitrary $\lambda \in (L^\infty(m))'$ we consider $\text{Re}\lambda$ and $\text{Im}\lambda$ separately and thus can assume that $\lambda(f)$ is real-valued if $f \in \text{Re}L^\infty(m)$. Then define the positive part λ^+ of λ to be

$$\lambda^+(f) = \text{Sup}\{\lambda(h) : h \in L^\infty(m) \text{ with } 0 \leq h \leq f\} \qquad \text{for } 0 \leq f \in L^\infty(m).$$

As in the proof of 1.2.iii)⇒ii) it extends to a positive linear functional $\lambda^+\in(L^\infty(m))'$ which is such that $\lambda^- := \lambda^+ - \lambda\in(L^\infty(m))'$ is a positive functional as well. The result follows upon application of 2) to λ^+ and λ^-. QED.

2. Strict Convergence

The crucial step in the proof of the main result 3.3 is the interchange of two limit processes. An elegant way to master this difficulty is to use a sophisticated notion of convergence for sequences in $L^\infty(m)$ which we call strict convergence. The present section is to develop those properties of the strict convergence which are needed in the sequel.

Assume that $f_n, f\in L^\infty(m)$ $(n=1,2,\ldots)$. We say that the f_n converge to f stricty iff $f_n\to f$ pointwise and in addition

$$|f_1| + \sum_{n=2}^{\infty} |f_n - f_{n-1}| \leq \text{const} < \infty .$$

We note two immediate properties. i) If $f_n\to f$ strictly then

$$|f_n| \leq |f_1| + \sum_{l=2}^{\infty} |f_l - f_{l-1}| \leq \text{const} < \infty \quad (n=1,2,\ldots).$$

In particular it follows that $\int f_n G dm \to \int f G dm$ $\forall G\in L^1(m)$. ii) If $f_n\to f$ in $L^\infty(m)$-norm then there exist strictly convergent subsequences $f_{n(l)}\to f$. In fact, choose $1\leq n(1)<\ldots<n(l)<n(l+1)<\ldots$ such that

$$\sum_{l=2}^{\infty} \| f_{n(l)} - f_{n(l-1)} \| < \infty .$$

2.1 REMARK: Assume that $h\in L(m)$ with $\operatorname{Re} h \geq 0$ and let $0<t(n)\downarrow 0$. Then $\frac{1}{1+t(n)h}\to 1$ strictly.

Proof: It is sufficient to show that for $1=t(0)>t(1)>\ldots>t(n)>t(n+1)>$ $>\ldots>0$ we have

$$\left|\frac{1}{1+z}\right| + \sum_{l=1}^{\infty} \left|\frac{1}{1+t(l)z} - \frac{1}{1+t(l-1)z}\right| \leq \frac{\pi}{2} \qquad \forall\ z\in\mathbb{C} \text{ with } \operatorname{Re} z \geq 0.$$

To see this fix $z = u+iv$ with $u \geq 0$ and $R := |z| \geq 0$ and define $b: b(t) = 1/(1+tz)$ $\forall 0 \leq t \leq 1$. Then

$$b'(t) = -\frac{z}{(1+tz)^2}, \quad |b'(t)| = \frac{R}{1+2tu+t^2R^2} \leq \frac{R}{1+t^2R^2} \qquad \forall 0 \leq t \leq 1.$$

Thus

$$\left|\frac{1}{1+z}\right| + \sum_{l=1}^{n} \left|\frac{1}{1+t(l)z} - \frac{1}{1+t(l-1)z}\right| = \left|\frac{1}{1+z}\right| + \sum_{l=1}^{n} |b(t(l))-b(t(l-1))|$$

$$\leq \left|\frac{1}{1+z}\right| + \int_0^1 |b'(t)|\,dt \leq \frac{1}{\sqrt{(1+u)^2+v^2}} + \int_0^1 \frac{R}{1+t^2R^2}dt \leq \frac{1}{\sqrt{1+R^2}} + \int_0^R \frac{1}{1+s^2}ds =: B(R),$$

which is $\leq \frac{\pi}{2}$ since differentiation shows that $B(R)$ is monotone increasing in $R \geq 0$. QED.

For a linear subspace $S \subset L^\infty(m)$ we define $\hat{S}$ to consist of the functions $f \in L^\infty(m)$ such that there exists a sequence of functions $f_n \in S$ with $f_n \to f$ strictly. Of course $\hat{S} \subset L^\infty(m)$ is a linear subspace with $S \subset \hat{S}$, and $S = \hat{S}$ whenever S is weak*closed.

Let $S \subset L^\infty(m)$ be a linear subspace. A linear functional $\lambda : \hat{S} \to \mathbb{C}$ is defined to be strictly continuous iff $f_n \in S$ and $f_n \to f$ strictly implies that $\lambda(f_n) \to \lambda(f)$. We note two immediate properties. i) If $\lambda(f) = \int fG\,dm$ $\forall f \in \hat{S}$ for some $G \in L^1(m)$ then λ is strictly continuous. ii) If λ is strictly continuous then $\lambda|S$ is $L^\infty(m)$-norm continuous.

The decisive property of the strict convergence is formulated in the subsequent lemma. The proof rests upon the weak sequential completeness of the sequence space l^1 (see KÖTHE [1966] p.283).

<u>2.2 LEMMA</u>: Let $S \subset L^\infty(m)$ be a linear subspace. Assume that $\lambda_l, \lambda : \hat{S} \to \mathbb{C}$ are linear functionals $(l=1,2,\ldots)$ such that $\lambda_l(f) \to \lambda(f)$ $\forall f \in \hat{S}$ for $l \to \infty$. If all λ_l are strictly continuous then λ is strictly continuous as well.

Proof: Fix $f_n \in S$ with $f_n \to f \in \hat{S}$ strictly. If we put $g_1 := f_1$ and $g_n := f_n - f_{n-1}$ for $n \geq 2$ then $f = \sum_{n=1}^{\infty} g_n$ and $\sum_{n=1}^{\infty} |g_n| \leq$ const. We have to prove that the series $\sum_{n=1}^{\infty} \lambda(g_n)$ is convergent with sum $= \lambda(f)$. For $\alpha = (\alpha_n)_n \in l^\infty$ we define $f^\alpha := \sum_{n=1}^{\infty} \alpha_n g_n \in \hat{S}$. For each $l \geq 1$ then the series $\sum_{n=1}^{\infty} \alpha_n \lambda_l(g_n)$ is convergent with sum = $\lambda_l(f^\alpha)$. In particular $\tau^l := (\lambda_l(g_n))_n \in l^1$. Thus for each $\alpha \in l^\infty$ we have the convergence

$$<\alpha,\tau^l> = \sum_{n=1}^{\infty} \alpha_n \tau_n^l = \sum_{n=1}^{\infty} \alpha_n \lambda_l(g_n) = \lambda_l(f^\alpha) \to \lambda(f^\alpha) \quad \text{for } l\to\infty.$$

From the weak sequential completeness of l^1 we obtain an element $\tau = (\tau_n)_n \in l^1$ such that $<\alpha,\tau^l> \to <\alpha,\tau>$ $\forall \alpha \in l^\infty$. Thus $<\alpha,\tau> = \lambda(f^\alpha)$ $\forall \alpha \in l^\infty$. Now choose first $\alpha = e^n = (0,\ldots,0,1,0,\ldots)$ to obtain $\tau_n = \lambda(g_n)$, so that in particular $\sum_{n=1}^{\infty} |\lambda(g_n)| < \infty$. Then choose $\alpha = (1,1,\ldots)$ to obtain $\sum_{n=1}^{\infty} \lambda(g_n) = \lambda(f)$. QED.

3. Characterization Theorem and Main Result

We fix a reduced Hardy algebra situation (H,φ) on (X,Σ,m). After the preparations in Sections 1 and 2 we can now prove the following theorem which characterizes the situation that M is compact in $\sigma(\mathrm{ReL}^1(m),\mathrm{ReL}^\infty(m))$ by a list of equivalent properties. The basic idea is that weak compactness of M means that H is not too small. Recall that the situation has already been considered in IV.4.5 and VI.6.3.

3.1 THEOREM: The subsequent properties are equivalent.

A1) M is compact in $\sigma(\mathrm{ReL}^1(m),\mathrm{ReL}^\infty(m))$.

A2) If $0 \leq f_n \in \mathrm{ReL}^\infty(m)$ with $f_n \downarrow 0$ then $\theta(f_n) \to 0$.

A3) If $E(n) \in \Sigma$ with $X = E(0) \supset E(1) \supset \ldots \supset E(n) \supset E(n+1) \supset \ldots$ and $m(E(n)) \to 0$ then there exist $h \in H^+$ and $0 \leq c_n \uparrow \infty$ such that $\mathrm{Re}\, h \geq c_n$ on $E(n)$ $(n=0,1,2,\ldots)$.

B1) If $\tau \in (L^\infty(m))'$ is singular and $\tau|H$ is strictly continuous then $\tau|H=0$.

B2) If $\tau \in (L^\infty(m))'$ is singular and $\tau|H$ is weak* continuous then $\tau|H=0$.

B3) (of F.and M.Riesz type) Assume that $\psi \in (L^\infty(m))'$ is decomposed into $\psi = Gm + \tau$ with $G \in L^1(m)$ and $\tau \in (L^\infty(m))'$ singular. Then $\psi|H=0$ implies that $\tau|H=0$.

Proof: A1)⇒A2) is in VI.6.3. A2)⇒A3) Consider a sequence of subsets $E(n)\in\Sigma$ with $X=E(0)\supset E(1)\supset\ldots\supset E(n)\supset E(n+1)\supset\ldots$ and $m(E(n))\to 0$. Then $\theta(\chi_{E(n)})\to 0$ after A2). Choose numbers $\varepsilon_n>0$ with $\varepsilon_n\to 0$ and then numbers $\alpha_n\geqq 0$ with

$$\sum_{n=1}^{\infty}\alpha_n\Big(\theta(\chi_{E(n)})+\varepsilon_n\Big)<\infty \quad\text{and}\quad \sum_{n=1}^{\infty}\alpha_n=\infty.$$

From IV.3.10 and IV.2.5 we obtain functions $u_n\in H$ such that $\operatorname{Re}u_n\geqq\chi_{E(n)}$ and $\operatorname{Re}\varphi(u_n)\leqq\theta(\chi_{E(n)})+\varepsilon_n$. Let us put $v_n:=\alpha_o u_o+\ldots+\alpha_n u_n\in H$. For all $V\in M$ then

$$\int(\operatorname{Re}v_n)Vdm=\operatorname{Re}\varphi(v_n)=\sum_{l=0}^{n}\alpha_l\operatorname{Re}\varphi(u_l)\leqq\sum_{l=0}^{\infty}\alpha_l\Big(\theta(\chi_{E(l)})+\varepsilon_l\Big)<\infty.$$

Since there is at least one function $V\in M$ which is >0 on the whole of X the Beppo Levi theorem tells us that $\operatorname{Re}v_n\uparrow P$ to some $0\leq P\in L(m)$, and that $\int PVdm=\lim_{n\to\infty}\int(\operatorname{Re}v_n)Vdm=\lim_{n\to\infty}\operatorname{Re}\varphi(v_n)$ for all $V\in M$. Thus

$$\theta(|P-\operatorname{Re}v_n|)=\theta(P-\operatorname{Re}v_n)=\int(P-\operatorname{Re}v_n)Vdm\to 0\text{ for any }V\in M.$$

It follows from VI.3.8 that $P\in E$. In particular $h:=P+iP^*\in H^+$. And

$$\operatorname{Re}h=P\geqq\operatorname{Re}v_n=\sum_{l=0}^{n}\alpha_l\operatorname{Re}u_l\geqq\sum_{l=0}^{n}\alpha_l\chi_{E(l)},$$

$$\operatorname{Re}h=P\geqq\sum_{l=0}^{n}\alpha_l=:c_n\text{ on }E(n)\quad(n=0,1,2,\ldots).$$

A3)⇒B1) Assume that $\tau\in(L^\infty(m))'$ is singular with $\tau|H$ strictly continuous. Choose $E(n)\in\Sigma$ with $X=E(0)\supset E(1)\supset\ldots\supset E(n)\supset E(n+1)\supset\ldots$ and $m(E(n))\to 0$ such that $\tau(f)=\tau(f\chi_{E(n)})$ $\forall f\in L^\infty(m)$. If now $h\in H^+$ and $c_n\geqq 0$ are as in A3) then for each $t>0$ we have

$$\left|\tau\Big(\frac{f}{1+th}\Big)\right|=\left|\tau\Big(\frac{f}{1+th}\chi_{E(n)}\Big)\right|\leqq\|\tau\|\frac{\|f\|}{1+tc_n},$$

and hence $\tau(\frac{f}{1+th})=0$ $\forall f\in L^\infty(m)$. If in particular $f\in H$ then $\frac{f}{1+th}\in H$ $\forall t>0$ after V.4.1. Thus for $t\downarrow 0$ we deduce from 2.1 that $\tau(f)=0$.

B1)⇒B2) and B2)⇒B3) are obvious. B3)⇒A1) In view of IV.4.5 we have to prove that each linear functional $\psi\in(\operatorname{Re}L^\infty(m))^*$ with $\psi\leqq\alpha^o=\theta|\operatorname{Re}L^\infty(m)$ is of the form $\psi(f)=\int fVdm$ $\forall f\in\operatorname{Re}L^\infty(m)$ for some $V\in M$.

Let us fix such a functional $\psi \in (\mathrm{ReL}^{\infty}(m))^*$. From the remarks which follow the definition of α° in IV.2 we know that $\psi \leq$ Sup and $\psi(\mathrm{Re}\, u) = \mathrm{Re}\varphi(u)$ $\forall u \in H$. We extend ψ to $L^{\infty}(m)$ via $\psi(f) = \psi(\mathrm{Re}\, f) + i\psi(\mathrm{Im}\, f)$ $\forall f \in L^{\infty}(m)$. Then ψ is $L^{\infty}(m)$-norm continuous and positive and satisfies $\psi|H = \varphi$. From the decomposition theorem 1.1 we obtain $\psi = Fm + \tau$ with $0 \leq F \in L^1(m)$ and a positive singular functional $\tau \in (L^{\infty}(m))'$. We fix an arbitrary $V \in M$ and apply B3) to the functional $\psi - Vm = (F-V)m + \tau$. It follows that $\tau|H = 0$ and in particular $\tau(1) = 0$ which implies that $\tau = 0$ since τ is positive. Hence $\psi = Fm$ and therefore $F \in M$. QED.

For the remainder of the section we assume that M is compact in $\sigma(\mathrm{ReL}^1(m), \mathrm{ReL}^{\infty}(m))$.

3.2 COROLLARY: Assume that M is compact in $\sigma(\mathrm{ReL}^1(m), \mathrm{ReL}^{\infty}(m))$. A linear functional $\lambda: H \to \mathbb{C}$ is strictly continuous iff it is weak*continuous.

Proof: We have to show that a strictly continuous $\lambda: H \to \mathbb{C}$ must be weak* continuous. Since $H = \hat{H}$ we know from Section 2 that λ is $L^{\infty}(m)$-norm continuous. Now extend λ to $\lambda \in (L^{\infty}(m))'$ and then decompose it after 1.1 into $\lambda = Gm + \tau$ with $G \in L^1(m)$ and $\tau \in (L^{\infty}(m))'$ singular. Then $\tau|H$ is strictly continuous and hence $\tau|H = 0$ after 3.1.B1). It follows that $\lambda|H = Gm|H$. QED.

3.3 THEOREM: Assume that M is compact in $\sigma(\mathrm{ReL}^1(m), \mathrm{ReL}^{\infty}(m))$. Let the functions $G_n \in L^1(m)$ $(n=1,2,\ldots)$ be such that

$$\lim_{n\to\infty} \int f G_n dm \quad \text{exists for all } f \in H.$$

Then there exists a function $G \in L^1(m)$ such that

$$\lim_{n\to\infty} \int f G_n dm = \int f G dm \quad \text{for all } f \in H.$$

Proof: In view of 2.2 the functional $\lambda: \lambda(f) = \lim_{n\to\infty} \int f G_n dm$ $\forall f \in H$ is strictly continuous, so that 3.2 can be applied. QED.

Let K_o consist of the functions $h \in K$ with $\int h dm = 0$. Then $(L^1(m)/K_o)' = K_o^{\perp} \subset L^{\infty}(m)$ from the fundamentals of Banach space theory, and $K_o^{\perp} = H$ from IV.1.1.iii). Thus we can reformulate 3.3 as follows.

3.4 REFORMULATION: Assume that M is compact in $\sigma(\mathrm{ReL}^1(m), \mathrm{ReL}^{\infty}(m))$. Then $L^1(m)/K_o$ is weakly sequentially complete.

Notes

The main result 3.3 relative to the unit disk situation has been an open problem for quite some time. For its evolution form classical results we refer to PIRANIAN-SHIELDS-WELLS [1967]. Partial results in direction to 3.3 are in KAHANE [1967] for the unit disk situation and in HEARD [1967], GLICKSBERG [1970], BARBEY-KÖNIG [1972] for certain abstract function algebra situations. In BARBEY-KÖNIG [1972] the notion of strict convergence has been invented in order to replace a highly complicated construction due to KAHANE [1967]. The full theorem 3.3 in the unit disk situation is an independent result of MOONEY [1972] and HAVIN [1973]. The proof in MOONEY [1972] is rather non-transparent. It uses the Kahane construction as does the subsequent proof due to AMAR [1973]. Then 3.3 was proved via strict convergence in BARBEY [1975] in the unit disk and Szegö situation, by König under the assumption $\dim N < \infty$ (unpublished), and in BARBEY [1976] under the present weak compactness assumption for M. The latter proofs follow AMAR [1973] in the decisive use of the Gelfand transformation. Instead of it HAVIN [1973] uses the decomposition theorem due to HEWITT-YOSIDA [1952]. The present treatment adopts this idea. Our proof of the Hewitt-Yosida theorem 1.1 is close to the initial one. The final form of Sections 2 and 3 is from BARBEY-KÖNIG [1976]. An application of 3.4 is in PELCZYNSKI [1974].

Chapter IX

Logmodular Densities and Small Extensions

The principal aim of the present chapter is the maximality theorem 3.5: If (H,φ) is a reduced Hardy algebra situation with $\dim N<\infty$ and if $(\tilde{H},\tilde{\varphi})$ is a small extension of (H,φ) in the sense of Section 3 then $\tilde{H}=H$. Since it is not a priori clear that $(\tilde{H},\tilde{\varphi})$ is reduced it will be important to have certain intermediate results for non-reduced Hardy algebra situations as well. Thus we assume a fixed Hardy algebra situation (H,φ) which need not be reduced. The small extension property is in close relation to the logmodular representative functions (=densities) with which the chapter starts.

1. Logmodular Densities

Let (H,φ) be a Hardy algebra situation. We introduce

$$ML:=\{0\leqq V\in L^1(m):\log|\varphi(u)|=\int(\log|u|)Vdm\ \ \forall u\in H^\times\},$$

the class of logmodular functions (=densities), and

$$NL:=\text{real-linear span}(ML-ML)=\{c(U-V):U,V\in ML \text{ and } c>0\}.$$

It is clear that $MJ\subset ML\subset M$. Also ML is a convex and $\sigma(\mathrm{Re}L^1(m),\mathrm{Re}L^\infty(m))$ closed subset $\subset \mathrm{Re}L^1(m)$. And the equivalence remark VI.6.1 on M carries at once over to ML.

Next we introduce a close relative of the functional $\alpha:\mathrm{Re}L(m)\rightarrow[-\infty,\infty]$. The functional $\beta:\mathrm{Re}L^\infty(m)\rightarrow\mathbb{R}$ is defined to be

$$\beta(f)=\mathrm{Inf}\{-\log|\varphi(u)|:u\in H^\times \text{ with } -\log|u|\geqq f\}\qquad \forall f\in \mathrm{Re}L^\infty(m).$$

We list some immediate properties. i) $\mathrm{Inf} f\leqq\alpha(f)\leqq\beta(f)\leqq\alpha^o(f)=\theta(f)\leqq \mathrm{Sup} f$ $\forall f\in\mathrm{Re}L^\infty(m)$. Here $\beta\leqq\alpha^o$ results from IV.2.5 via exponentiation. ii) β is subadditive, but is not claimed to be sublinear. iii) β is isotone. iv) Lemma IV.2.3 permits to introduce the functionals $\beta^o,\beta^\infty:\mathrm{Re}L^\infty(m)\rightarrow\mathbb{R}$, defined to be

$$\beta^{o}(f) = \lim_{t\downarrow 0}\frac{1}{t}\beta(tf) = \operatorname{Sup}_{t>0}\frac{1}{t}\beta(tf),$$

$$\beta^{\infty}(f) = \lim_{t\uparrow\infty}\frac{1}{t}\beta(tf) = \operatorname{Inf}_{t>0}\frac{1}{t}\beta(tf) \quad \forall f\in ReL^{\infty}(m).$$

It follows that $\alpha^{\infty}\leq\beta^{\infty}\leq\beta\leq\beta^{o}=\alpha^{o}=\theta$. And β^{∞} is sublinear and isotone. v) $\beta(\log|u|) = \log|\varphi(u)|$ $\forall u\in H^{\times}$. It follows that $\beta^{\infty}(Reu) = \beta(Reu) = Re\varphi(u)$ $\forall u\in H$.

1.1 REMARK: Let $V\in ReL^{1}(m)$. Then $\int fVdm \leq \beta^{\infty}(f)$ $\forall f\in ReL^{\infty}(m) \Leftrightarrow \int fVdm \leq \beta(f)$ $\forall f\in ReL^{\infty}(m) \Leftrightarrow V\in ML$.

Proof: Direct verification as for IV.2.1.i) and IV.2.4. QED.

1.2 PROPOSITION: Assume that M is $\sigma(ReL^{1}(m),ReL^{\infty}(m))$ compact. Then $ML \neq \emptyset$ and $\beta^{\infty}(f) = \operatorname{Max}\{\int fVdm : V\in ML\}$ $\forall f\in ReL^{\infty}(m)$.

Proof: This follows from IV.4.5 via 1.1 as the corresponding result on MJ and α^{∞} in IV.4.5. QED.

The next result is a fundamental step.

1.3 AHERN-SARASON LEMMA: Assume that $ML \neq \emptyset$ and $\dim\frac{N}{NL} <\infty$. Then to each $\varepsilon>0$ there exists a constant $c(\varepsilon)>0$ such that

$$\beta(f) \leq \varepsilon + c(\varepsilon)\beta^{\infty}(f) \qquad \forall\ 0\leq f \in ReL^{\infty}(m).$$

We shall need the subsequent two lemmata. The first one is a well-known result from number theory which is reproduced for the sake of completeness, while the second one is a simple fact from linear algebra.

1.4 DIRICHLET SIMULTANEOUS APPROXIMATION LEMMA: To $a_1,\ldots,a_n\in\mathbb{R}$ and $r\in\mathbb{N}$ there exist numbers $q\in\mathbb{N}$ and $p_1,\ldots,p_n\in\mathbb{Z}$ such that $|qa_\ell-p_\ell|<\frac{1}{r}(\ell=1,\ldots,n)$ and $q\leq r^n$.

Proof of 1.4: Let us subdivide the cube $Q:=\{x=(x_1,\ldots,x_n):0\leq x_\ell<1$ $(\ell=1,\ldots,n)\}\subset\mathbb{R}^n$ into the r^n subcubes $Q(u)=\{x=(x_1,\ldots,x_n):\frac{1}{r}(u_\ell-1)\leq x_\ell<\frac{1}{r}u_\ell$ $(\ell=1,\ldots,n)\}$ via the r^n lattice points $u=(u_1,\ldots,u_n)\in\mathbb{Z}^n$ with $u_\ell=1,\ldots,r$ $(\ell=1,\ldots,n)$. For each $s\in\mathbb{Z}$ then $(sa_1,\ldots,sa_n)$modulo $\mathbb{Z}^n$ is in Q and hence in one of the $Q(u)$. It follows that in the set of the r^n+1 numbers $0,1,\ldots,r^n$ there is at least one pair of numbers $s<t$ such that $(sa_1, \ldots, sa_n) \bmod \mathbb{Z}^n$ and $(ta_1,\ldots,ta_n) \bmod \mathbb{Z}^n$ are in the same subcube $Q(u)$. Thus for $1 \leq q:=t-s \leq r^n$ we have numbers $p_1,\ldots,p_n \in \mathbb{Z}$ such that $|qa_\ell - p_\ell| < \frac{1}{r}$ $(\ell=1,\ldots,n)$ as required. QED.

1.5 LEMMA: Let W be a vector space and $T \subset W$ such that W = linear span of T. Assume that $\varphi_1,\ldots,\varphi_n \in W^*$ are linearly independent. Then there exist $\psi_1,\ldots,\psi_n \in W^*$ linear combinations of $\varphi_1,\ldots,\varphi_n$ and $u_1,\ldots,u_n \in T$ such that $\langle \psi_i, u_\ell \rangle = \delta_{i\ell}$ $(i,\ell=1,\ldots,n)$.

Proof of 1.5: It suffices to show that there are $u_1,\ldots,u_n \in T$ such that $\det(\langle \varphi_i, u_\ell\rangle)_{i,\ell=1,\ldots,n} \neq 0$. But this assertion has an obvious proof via induction. QED.

Proof of 1.3: i) We fix $F \in ML$ and $\varepsilon>0$ and prove the existence of a constant $c(\varepsilon)>0$ such that

$$\beta(f) \leq \varepsilon + c(\varepsilon)\int fF dm \quad \forall\, 0 \leq f \in L^\infty(m) \text{ with } f \perp N \cap (\log|H^\times|)^\perp .$$

In case that $N=N\cap(\log|H^\times|)^\perp$ this is clear with $c(\varepsilon)=1$ since for the $f \in \mathrm{Re}L^\infty(m)$ as above then $\int fF dm = \theta(f) \geq \beta(f)$. Thus we can assume that $N \supsetneq N \cap (\log|H^\times|)^\perp \supset NL$ with $\dim\left(N/N\cap(\log|H^\times|)^\perp\right) =: n \in \mathbb{N}$. We apply 1.5 to W:=real-linear span of $\log|H^\times| \subset \mathrm{Re}L^\infty(m)$ and $T:=\log|H^\times|$ and consider $N/N\cap(\log|H^\times|)^\perp \subset W^*$ via the canonical inclusion $N \subset \mathrm{Re}L^1(m) \subset (\mathrm{Re}L^\infty(m))^*$ and the restriction $(\mathrm{Re}L^\infty(m))^* \to W^*: \psi \mapsto \psi|W$. We start from functions $F_1,\ldots,F_n \in N$ which are linearly independent over $N\cap(\log|H^\times|)^\perp$ and obtain functions $G_1,\ldots,G_n \in N$ and $f_1,\ldots,f_n \in \log|H^\times|$ such that $\int G_i f_\ell dm = \delta_{i\ell}$ $(i,\ell=1,\ldots,n)$. It follows that

$$N = \left(N\cap(\log|H^\times|)^\perp\right) \oplus \text{real-linear span}(G_1,\ldots,G_n),$$

where the sum is direct and hence $L^1(m)$-norm bounded: there is a constant $c>0$ such that

$$|t_1|,\ldots,|t_n| \leq c\|G + \sum_{\ell=1}^n t_\ell G_\ell\|_{L^1(m)} \quad \forall\, G \in N\cap(\log|H^\times|)^\perp \text{ and } t_1,\ldots,t_n \in \mathbb{R}.$$

Consider now $0 \leq f \in L^\infty(m)$ with $f \perp N\cap(\log|H^\times|)^\perp$. For $q \in \mathbb{N}$ and $p_1,\ldots,p_n \in \mathbb{Z}$ we put $g:=p_1f_1+\ldots+p_nf_n \in \log|H^\times|$ and obtain

$$\beta(f) \leq \beta(qf) \leq \beta(qf-g) + \beta(g) \leq \theta(qf-g) + \int gFdm$$

$$= \sup_{V\in M} \int (qf-g)Vdm + \int gFdm = \sup_{V\in M} \int (qf-g)(V-F)dm + q\int fFdm.$$

But for $V\in M$ we have $V-F\in N$ with $L^1(m)$-norm ≤ 2 so that from the above direct sum decomposition

$$V-F = G + \sum_{\ell=1}^{n} t_\ell G_\ell \text{ with } G\in N\cap(\log|H^\times|)^\perp \text{ and } |t_1|,\dots,|t_n| \leq 2c.$$

It follows that

$$\int (qf-g)(V-F)dm = \sum_{\ell=1}^{n} t_\ell \int (qf-g)G_\ell dm$$

$$= \sum_{\ell=1}^{n} t_\ell \left(q\int fG_\ell dm - p_\ell\right) \leq 2cn \max_{1\leq \ell\leq n} |q\int fG_\ell dm - p_\ell| \ \forall\ V\in M,$$

$$\beta(f) \leq 2cn \max_{1\leq \ell\leq n} |q\int fG_\ell dm - p_\ell| + q\int fFdm,$$

valid for all $q\in\mathbb{N}$ and $p_1,\dots,p_n\in\mathbb{Z}$. If now $r\in\mathbb{N}$ is prescribed then after 1.4 the numbers $q\in\mathbb{N}$ and $p_1,\dots,p_n\in\mathbb{Z}$ can be chosen so that $|q\int fG_\ell dm - p_\ell| < \frac{1}{r}(\ell=1,\dots,n)$ and $q\leq r^n$. Then $\beta(f) \leq \frac{2cn}{r} + r^n\int fFdm$. Thus for $\varepsilon>0$ we have to take an $r=r(\varepsilon)\in\mathbb{N}$ with $\frac{2cn}{r}\leq\varepsilon$ and then obtain the desired extimation with $c(\varepsilon):=(r(\varepsilon))^n$.

ii) Now we can prove the assertion with the same $c(\varepsilon)>0$ as above. Let $0\leq f\in L^\infty(m)$. Fix $\delta>0$ and take $t>0$ such that $\frac{1}{t}\beta(tf)<\beta^\infty(f)+\delta$. Then there exists a function $u\in H^\times$ with $-\log|u| \geq tf$ and $-\frac{1}{t}\log|\varphi(u)| \leq \beta^\infty(f)+\delta$. For $v:=\frac{1}{u}\in H^\times$ thus $f\leq\frac{1}{t}\log|v|$ and $\frac{1}{t}\log|\varphi(v)|<\beta^\infty(f)+\delta$. Now $0\leq\frac{1}{t}\log|v|\in$ real-linear span of $\log|H^\times| \subset (\log|H^\times|)^{\perp\perp} \subset \left(N\cap(\log|H^\times|)^\perp\right)^\perp$ so that from i) we obtain

$$\beta(f) \leq \beta\left(\tfrac{1}{t}\log|v|\right) \leq \varepsilon + c(\varepsilon)\int \tfrac{1}{t}(\log|v|)Fdm = \varepsilon + \frac{c(\varepsilon)}{t}\log|\varphi(v)|$$

$$< \varepsilon + c(\varepsilon)\left(\beta^\infty(f)+\delta\right).$$

For $\delta\downarrow 0$ the assertion follows. QED.

One of the main consequences of 1.3 is theorem 1.7 below. In its proof we need the fact that a certain result from Section VI.6 remains true without the reducedness assumption.

1.6 REMARK: Let $F\in M$. Assume that if $0\leqq f_n\in ReL^\infty(m)$ then $\int f_n F dm\to 0$ implies that $\alpha(f_n)\to 0$ (condition VI.6.2.∞)). Then $N\subset\overline{N\cap F(ReL^\infty(m))}^{L^1(m)}$ (condition VI.6.4.vi)).

Proof: We know from VI.6.4 that the assertion is true when (H,φ) is reduced. Recall from IV.1.11 the associated reduced Hardy algebra situation (H_*,φ_*) on $(Y(X),\Sigma|Y(X),m|Y(X))$ which can be formed for each (H,φ). i) We claim that $\alpha_*(f|Y(X))=\alpha(f)$ for all $f\in ReL(m)$. In fact, we see from IV.1.10-11 that

$$\alpha_*(f|Y(X)) = \operatorname{Inf}\{-\log|\varphi_*(u)| : u\in H_* \text{ with } -\log|u| \geqq f|Y(X)\}$$

$$= \operatorname{Inf}\{-\log|\varphi(u)| : u\in H \text{ with } -\log|u| \geqq f \text{ on } Y(X)\}$$

$$= \operatorname{Inf}\{-\log|\varphi(u)| : u\in H \text{ with } -\log|u| \geqq f\} = \alpha(f).$$

ii) We conclude from i) that condition VI.6.2.∞) is fulfilled for (H,φ) iff it is fulfilled for (H_*,φ_*). Furthermore it is obvious from IV.1.11 that condition VI.6.4.vi) is true for (H,φ) iff it is true for (H_*,φ_*). Thus the desired implication carries over from (H_*,φ_*) to (H,φ). QED.

1.7 THEOREM: Assume that $\dim N<\infty$. Let $F\in M$ be such that each member of ML is $\leqq cF$ for some constant $c>0$. Then F is an internal point of M. In particular: If $F\in ML$ is an internal point of ML then it is an internal point of M as well.

Proof: i) From the proof of VI.6.1.i) ⇒ i') applied to ML and from 1.2 we conclude that there is a constant $c>0$ with $\beta^\infty(f)\leqq c\int fFdm\ \forall\ 0\leqq f\in L^\infty(m)$. ii) If $0\leqq f_n\in L^\infty(m)$ with $\int f_n Fdm\to 0$ then $\beta^\infty(f_n)\to 0$ from i) and hence $\beta(f_n)\to 0$ from 1.3. In particular $\alpha(f_n)\to 0$. iii) Now 1.6 implies that $N\subset\overline{N\cap F(ReL^\infty(m))}^{L^1(m)}=N\cap F(ReL^\infty(m))$. The assertion follows. QED.

In conclusion we list several properties of $\log|H^\times|$ and of NL which will be needed in the sequel.

1.8 REMARK: Assume that (H,φ) is reduced. i) $E^\infty\subset\log|H^\times|\subset ReL^\infty(m)$, and $\log|H^\times|$ is an additive subgroup. ii) $\log|H^\times|=\{f\in ReL^\infty(m):\alpha(f)+\alpha(-f)=0\}=$

$=\{f\in \mathrm{ReL}^\infty(m):\alpha(f)+\alpha(-f)\leqq 0\}$. iii) Assume that $f_n\in\log|H^\times|$ with $f_n\to f\in \mathrm{ReL}^\infty(m)$ and $|f_n|\leqq$ some G with $e^G\in L^\#$. Then $f\in\log|H^\times|$.

Proof: i) is obvious. ii) is a direct consequence of V.1.3. iii) From IV.3.12 we see that $\alpha(f)\leqq\liminf_{n\to\infty}\alpha(f_n)$ and $\alpha(-f)\leqq\liminf_{n\to\infty}\alpha(-f_n)$. Thus $\alpha(f)+$ $+\alpha(-f)\leqq 0$ and hence $f\in\log|H^\times|$ from ii). QED.

<u>1.9</u> <u>PROPOSITION</u>: Assume that (H,φ) is reduced. i) We have

$$NL\subset \overline{NL}^{\mathrm{ReL}^1(m)}=(NL)^{\perp\perp}\subset(\log|H^\times|)^\perp\subset N^{\perp\perp}=\overline{N}^{\mathrm{ReL}^1(m)}.$$

ii) If $\dim N<\infty$ then

$$NL=\overline{NL}^{\mathrm{ReL}^1(m)}=(\log|H^\times|)^\perp,$$

$$(NL)^\perp=(\log|H^\times|)^{\perp\perp}=\overline{\text{real-lin span}(\log|H^\times|)}^{\text{weak}*}=\text{real-lin span}(\log|H^\times|).$$

Proof: i) Is clear from $\log|H^\times|\subset(NL)^\perp$ and $N^\perp=E^\infty\subset\log|H^\times|$. ii) In order to prove $(\log|H^\times|)^\perp\subset NL$ let $F\in ML$ be an internal point of ML, so that after 1.7 it is an internal point of M as well. If now $f\in(\log|H^\times|)^\perp$ then $f\in\overline{N}^{\mathrm{ReL}^1(m)}=N$ from i). Thus $f=c(U-F)$ with $U\in M$ and $c>0$ from VI.6.1.iii). But then $f\perp\log|H^\times|$ and $F\in ML$ imply that $U\in ML$. Hence $f\in NL$ as claimed. In the last assertion it remains to prove that real-linear span$(\log|H^\times|)$ $\subset\mathrm{ReL}^\infty(m)$ is weak* closed. We have

$$N^\perp=E^\infty\subset\log|H^\times|\subset\text{real-linear span}(\log|H^\times|)\subset\mathrm{ReL}^\infty(m).$$

But with $'$:= the $L^1(m)$-norm dual we have

$$\dim N=\dim N'=\dim\left((\mathrm{ReL}^1(m))'/N^\perp\right)=\dim(\mathrm{ReL}^\infty(m)/N^\perp),$$

and this is assumed to be finite. It follows that real-linear span$(\log|H^\times|)$ is the direct sum of $N^\perp=E^\infty$ with some finite-dimensional linear subspace and hence is weak* closed. QED.

2. The Closed Subgroup Lemma

In the proof of the maximality theorem 3.5 we shall need a lemma from topological algebra which deserves some interest of its own. It will be established in the present section.

Let V be a fixed real Hausdorff topological vector space. For $S \subset V$ a closed additive subgroup we define

$$E(S) := \overline{\text{linear span}(S)} \quad \text{the closed linear span of } S,$$

$$D(S) := \bigcap_{t>0} tS \quad \text{the largest linear subspace of } S,$$

so that $D(S) \subset S \subset E(S)$ and $D(S)$ is closed as well.

2.1 <u>REMARK</u>: Assume that $\dim E(S) < \infty$. If $D(S) = \{0\}$ then S is discrete, that means it consists of isolated points (the converse is obvious).

Proof: Let us fix a norm $\|\cdot\|$ on $E(S)$. Assume that S is not discrete. Then there exists a sequence of nonzero points $u_\ell \in S$ with $u_\ell \to 0$ or $\|u_\ell\| \to 0$. We can assume that $\|u_\ell\| \leq 1$. Take the numbers $n(\ell) \in \mathbb{N}$ with $\frac{1}{n(\ell)+1} < \|u_\ell\| \leq \frac{1}{n(\ell)}$ or $1 - \|u_\ell\| < n(\ell)\|u_\ell\| \leq 1$. Then $n(\ell) \to \infty$ and $\|n(\ell)u_\ell\| \to 1$. We can pass to a subsequence and assume that $n(\ell)u_\ell \to$ some $u \in S$ with $\|u\| = 1$. We claim that $u \in D(S)$. In fact, for fixed $t>0$ take the numbers $p(\ell) \in \mathbb{N}$ with $p(\ell) - 1 < \frac{n(\ell)}{t} \leq p(\ell)$ or $\frac{p(\ell)}{n(\ell)} - \frac{1}{n(\ell)} < \frac{1}{t} \leq \frac{p(\ell)}{n(\ell)}$. Then $\frac{p(\ell)}{n(\ell)} \to \frac{1}{t}$. Hence the $p(\ell)u_\ell = \frac{p(\ell)}{n(\ell)} n(\ell) u_\ell \in S$ converge to $\frac{1}{t}u$ so that $\frac{1}{t}u \in S$ or $u \in tS$ for each $t>0$. It follows that $u \in D(S)$ which is a contradiction. QED.

2.2 <u>REMARK</u>: Assume that S is discrete and $\dim E(S) =: n \in \mathbb{N}$. Then there exist $u_1, \ldots, u_n \in S$ such that $S = \mathbb{Z}u_1 + \ldots + \mathbb{Z}u_n$. In this case $E(S) = \mathbb{R}u_1 + \ldots + \mathbb{R}u_n$. In particular $u_1, \ldots, u_n$ are linearly independent.

Proof: Let us fix a nonzero determinant function on $E(S)$. i) Let $a_1, \ldots, a_r \in S$ form a maximal linearly independent subset of S. Then $r \leq n$. We have $S \subset \mathbb{R}a_1 + \ldots + \mathbb{R}a_r$ and hence $E(S) \subset \mathbb{R}a_1 + \ldots + \mathbb{R}a_r$. Thus $r = n$, and we see that $E(S) = \mathbb{R}a_1 + \ldots + \mathbb{R}a_n$.

ii) We claim that there exists a maximal linearly independent subset $u_1, \ldots, u_n \in S$ such that $|\det(u_1, \ldots u_n)|$ is minimal. To see this start with a fixed maximal linearly independent subset $a_1, \ldots a_n \in S$ and form $A := \{t_1 a_1 + \ldots + t_n a_n : 0 \leq t_1, \ldots, t_n \leq 1\} \subset E(S)$. Then A is compact, so that $A \cap S$

is compact and discrete and hence finite. Consider now the coordinate functionals

$$\varphi_1,\dots,\varphi_n \in (E(S))^*: u = \sum_{\ell=1}^{n} \varphi_\ell(u) a_\ell \qquad \forall\, u\in E(S).$$

Then $\varphi_\ell(S)\subset\mathbb{R}$ is an additive subgroup with $1\in\varphi_\ell(S)$. Since each $u\in S$ is of the form $u=p_1a_1+\dots+p_na_n+v$ with $p_1,\dots,p_n\in\mathbb{Z}$ and $v\in A\cap S$ we have $\varphi_\ell(S)=\mathbb{Z}+$ $+\varphi_\ell(A\cap S)$. Thus the finiteness of $A\cap S$ implies that $\varphi_\ell(S)=\frac{1}{m(\ell)}\mathbb{Z}$ for some $m(\ell)\in\mathbb{N}$. It follows that each $u\in S$ is of the form $u=\frac{1}{m(1)}p_1a_1+\dots+\frac{1}{m(n)}p_na_n$ with $p_1,\dots,p_n\in\mathbb{Z}$. From this we see that for each collection $u_1,\dots,u_n\in S$ we have

$$\det(u_1,\dots,u_n) = \frac{p}{m(1)\dots m(n)}\det(a_1,\dots,a_n) \text{ for some } p\in\mathbb{Z}.$$

It is obvious that we can chose $u_1,\dots,u_n\in S$ with minimum $|\det(u_1,\dots,u_n)|>$ >0 as we have claimed.

iii) Let us fix $u_1,\dots,u_n\in S$ with minimum $|\det(u_1,\dots,u_n)|>0$ as obtained in ii). It remains to show that $S\subset\mathbb{Z}u_1+\dots+\mathbb{Z}u_n$ and hence $S=\mathbb{Z}u_1+\dots+\mathbb{Z}u_n$. To see this write $u\in S$ in the form $u=p_1u_1+\dots+p_nu_n+v$ with $p_1,\dots,p_n\in\mathbb{Z}$ and $v=t_1u_1+\dots+t_nu_n$ with $0\leq t_1,\dots,t_n<1$. Then $v\in S$ and hence the equation

$\det(u_1,\dots,u_{\ell-1},v,u_{\ell+1},\dots,u_n)=t_\ell\det(u_1,\dots,u_n)$ implies that $t_\ell=0\,(\ell=1,\dots,n)$. It follows that $v=0$ and hence the assertion. QED.

2.3 <u>REMARK</u>: Assume that $\dim(E(S)/D(S))=:n\in\mathbb{N}$. Then there exist $u_1,\dots,$ $u_n\in S$ such that $S=D(S)+\mathbb{Z}u_1+\dots+\mathbb{Z}u_n$. In this case $E(S)=D(S)+\mathbb{R}u_1+\dots+\mathbb{R}u_n$. In particular $u_1,\dots,u_n$ are linearly independent over $D(S)$.

Proof: Let $G\subset E(S)$ be an n-dimensional linear subspace such that $E(S)=$ $=D(S)\oplus G$. From $D(S)\subset S$ we have $S=D(S)+(G\cap S)$ with $G\cap S$ a closed subgroup. It follows that $D(G\cap S)\subset G\cap D(S)=\{0\}$ so that 2.1 tells us that $G\cap S$ is discrete. Furthermore $E(S)\subset D(S)+E(G\cap S)$ implies that $E(G\cap S)$ must be $=G$ and hence of dimension $=n$. Thus from 2.2 we obtain $u_1,\dots,u_n\in G\cap S$ such that $G\cap S=\mathbb{Z}u_1+\dots+\mathbb{Z}u_n$ and hence $S=D(S)+\mathbb{Z}u_1+\dots+\mathbb{Z}u_n$. The assertion follows. QED.

2.4 <u>CLOSED SUBGROUP LEMMA</u>: Let $S,T\subset V$ be closed additive subgroups with $S\subset T$ and $\dim(E(S)/D(S))<\infty$. Then $E(S\cap D(T))=E(S)\cap D(T)$.

Proof: i) We have $D(S)\subset E(S\cap D(T))\subset E(S)\cap D(T)\subset E(S)$. Let us put

$$\dim \frac{E(S\cap D(T))}{D(S)} =: p,\ \dim \frac{E(S)}{E(S)\cap D(T)} =: q \text{ and } \dim \frac{E(S)}{D(S)} =: n,$$

so that $0\leq p,q,n<\infty$ with $p+q\leq n$. The assertion is obvious if $q=0$ since then $S\subset E(S)\subset D(T)$. So we assume that $q\geq 1$ and hence $n\geq 1$. The assertion is $p+q=n$. We assume that $p+q<n$ or $(p+1)+q\leq n$ and shall obtain a contradiction.

ii) We introduce $B:=E(S)\cap T$ which is a closed subgroup with

$$\begin{array}{ccc} D(S)\subset E(S\cap D(T))\subset E(S)\cap D(T) & & E(S) \\ \| & & \| \\ D(B) & \subset B \subset & E(B). \end{array}$$

Then we apply 2.3 to obtain $v_1,\dots,v_q\in B$ linearly independent over $D(B)$ with

$$B = D(B) + \mathbb{Z}v_1+\dots+\mathbb{Z}v_q \text{ and } E(B) = D(B) + \mathbb{R}v_1+\dots,\mathbb{R}v_q,$$

and $x_1,\dots,x_n\in S$ linearly independent over $D(S)$ with

$$S = D(S) + \mathbb{Z}x_1+\dots+\mathbb{Z}x_n \text{ and } E(S) = D(S) + \mathbb{R}x_1+\dots+\mathbb{R}x_n.$$

Now $x_\ell\in S\subset B$ can be written

$$x_\ell = y_\ell + \sum_{j=1}^{q} \beta_\ell^j v_j \quad \text{with } y_\ell\in D(B) \text{ and } \beta_\ell^j\in\mathbb{Z}\ (\ell=1,\dots,n).$$

Consider the vectors $\beta^j:=(\beta_1^j,\dots,\beta_n^j)\in\mathbb{Z}^n\subset\mathbb{Q}^n$ $(j=1,\dots,q)$. In view of $(p+1)+$ $+q\leq n$ we can find nonzero vectors $\alpha^i:=(\alpha_1^i,\dots,\alpha_n^i)\in\mathbb{Q}^n$ $(i=0,1,\dots,p)$ which are orthogonal to each other and orthogonal to $\beta^1,\dots,\beta^q$ with respect to the usual scalar product in $\mathbb{Q}^n$. And we can of course assume that $\alpha^0,\alpha^1,\dots,\alpha^p\in\mathbb{Z}^n$. iii) Consider then

$$u_i := \sum_{\ell=1}^{n} \alpha_\ell^i x_\ell = \sum_{\ell=1}^{n} \alpha_\ell^i\Big(y_\ell + \sum_{j=1}^{q} \beta_\ell^j v_j\Big)$$

$$= \sum_{\ell=1}^{n} \alpha_\ell^i y_\ell + \sum_{j=1}^{q} \langle\alpha^i,\beta^j\rangle v_j = \sum_{\ell=1}^{n} \alpha_\ell^i y_\ell \quad (i=0,1,\dots,p).$$

We see that $u_i\in S\cap D(B)=S\cap D(T)\subset E(S\cap D(T))$. We have the desired contradiction when we can show that $u_0,u_1,\dots,u_p$ are linearly independent over $D(S)$. But for $\lambda_o,\lambda_1,\ldots,\lambda_p \in \mathbb{R}$ we have

$$\sum_{i=0}^{p} \lambda_i u_i = \sum_{\ell=1}^{n} \Big(\sum_{i=0}^{p} \lambda_i \alpha_\ell^i \Big) x_\ell .$$

If this is $\in D(S)$ then the choice of $x_1,\ldots,x_n$ implies that

$$\sum_{i=0}^{p} \lambda_i \alpha_\ell^i = 0 \ (\ell=1,\ldots,n) \quad \text{or} \quad \sum_{i=0}^{p} \lambda_i \alpha^i = 0.$$

But $\alpha^o,\alpha^1,\ldots,\alpha^p$ are nonzero and pairwise orthogonal and hence linearly independent as members of the vector space $\mathbb{R}^n$ over the scalar field $\mathbb{R}$. It follows that $\lambda_o=\lambda_1=\ldots=\lambda_p=0$ so that $u_o,u_1,\ldots,u_p$ are indeed linearly independent over $D(S)$. QED.

3. Small Extensions

Let (H,φ) be a Hardy algebra situation. An extension of (H,φ) is defined to be a Hardy algebra situation $(\tilde{H},\tilde{\varphi})$ on the same measure space such that $H\subset\tilde{H}$ and $\varphi=\tilde{\varphi}|H$. The entities which come from $(\tilde{H},\tilde{\varphi})$ will be written $\tilde{M},\tilde{N},\ldots$.

3.1 REMARK: Let $(\tilde{H},\tilde{\varphi})$ be an extension of (H,φ). Then i) $\tilde{M}\subset M$ and $\tilde{N}\subset N$. ii) $\tilde{M}=M \Leftrightarrow \tilde{N}=N$. iii) If $(\tilde{H},\tilde{\varphi})$ is reduced then (H,φ) is reduced (but the converse need not be true).

Proof: i) and iii) are obvious. ii) $\Rightarrow$ is obvious. To see $\Leftarrow$ fix $F\in\tilde{M}$. For $V\in M$ then $V-F\in N=\tilde{N}$. Thus $V=(V-F)+F\in\tilde{M}$. QED.

3.2 REMARK: Assume that (H,φ) is reduced. Let $(\tilde{H},\tilde{\varphi})$ be an extension of (H,φ). Then the subsequent properties are equivalent. i) $\tilde{M}=M$. ii) $\tilde{N}=N$. iii) $(\tilde{H},\tilde{\varphi})$ is reduced and $\tilde{E}^\infty=E^\infty$. iv) $\tilde{H}=H$.

Proof: The equivalence i) $\Leftrightarrow$ ii) is in 3.1. ii) $\Rightarrow$ iii) follows from $E^\infty=N^\perp$ and $\tilde{E}^\infty=\tilde{N}^\perp$. iii) $\Rightarrow$ iv) Take an $F\in\tilde{M}\subset M$ which is dominant over X. Then $H=(E^\infty+iE^\infty)\cap(H_\varphi F)^\perp$ and $\tilde{H}=(\tilde{E}^\infty+i\tilde{E}^\infty)\cap(\tilde{H}_{\tilde{\varphi}}F)^\perp$ from VI.4.5, so that $\tilde{H}\subset H$ and hence $\tilde{H}=H$. iv) $\Rightarrow$ i) is obvious. QED.

We add one further remark which shows that the requirements in the definition of an extension can sometimes be relaxed.

3.3 REMARK: Assume that M is $\sigma(\mathrm{ReL}^1(m),\mathrm{ReL}^\infty(m))$ compact. Let $\tilde{H}\subset L^\infty(m)$ be a subalgebra with $H\subset\tilde{H}$ and $\tilde{\varphi}:\tilde{H}\to\mathbb{C}$ a multiplicative linear functional with $\varphi=\tilde{\varphi}|H$. If $\tilde{\varphi}$ is $L^\infty(m)$-norm continuous then it is weak$*$ continuous. Thus $(\tilde{H},\tilde{\varphi})$ is a Hardy algebra situation whenever $\tilde{H}$ is weak$*$ closed.

Proof: If $\tilde{\varphi}$ is $L^\infty(m)$-norm continuous then $\|\tilde{\varphi}\|=1$ from the multiplicativity. We proceed as in the proof of VII.6.4: Let $\Lambda:L^\infty(m)\to\mathbb{C}$ be a bounded linear extension of $\tilde{\varphi}$ with $\|\Lambda\|=1$. Then $\|\Lambda\|=\Lambda(1)=1$ implies that Λ is positive and hence real. Thus for $f\in\mathrm{ReL}^\infty(m)$ we obtain

$$u\in H \text{ with } \mathrm{Re}\,u\geqq f \Rightarrow \mathrm{Re}\varphi(u)=\mathrm{Re}\Lambda(u)=\Lambda(\mathrm{Re}\,u)\geqq\Lambda(f),$$

and hence $\alpha^{\circ}(f)\geqq\Lambda(f)$. It follows from IV.4.5 that $\Lambda(f)=\int fV dm\ \forall f\in L^\infty(m)$ for some $V\in M$, and hence the assertion. QED.

An extension $(\tilde{H},\tilde{\varphi})$ of (H,φ) is defined to be a small extension iff

$$\mathrm{Re}\,\tilde{H}\subset(\log|H^\times|)^{\perp\perp}=\overline{\text{real-linear span}(\log|H^\times|)}^{\text{weak}*}.$$

In this connection observe that $\mathrm{Re}H=\log|e^H|\subset\log|H^\times|$.

3.4 REMARK: Assume that (H,φ) is reduced and dim $N<\infty$. Let $(\tilde{H},\tilde{\varphi})$ be an extension of (H,φ). Then the subsequent properties are equivalent. i) $(\tilde{H},\tilde{\varphi})$ is a small extension of (H,φ). ii) $NL\subset(\mathrm{Re}\tilde{H})^\perp$. iii) $ML\subset\tilde{M}$. iv) $NL\subset\tilde{N}$.

Proof: i)$\Leftrightarrow$ii) We have $(\log|H^\times|)^{\perp\perp}=(NL)^\perp$ from 1.9 so that smallness means that $\mathrm{Re}\tilde{H}\subset(NL)^\perp$ or $NL\subset(\mathrm{Re}\tilde{H})^\perp$. ii)$\Rightarrow$iii) We have $\widetilde{ML}\neq\emptyset$ from 1.2 since dim $\tilde{N}\leqq$dim $N<\infty$. Let us fix an $F\in\widetilde{ML}\subset\tilde{M}\cap ML$. For $V\in ML$ then $V-F\in NL\subset(\mathrm{Re}\tilde{H})^\perp$. Thus $F\in\tilde{M}$ implies that $V\in\tilde{M}$. iii)$\Rightarrow$iv) and iv)$\Rightarrow$ii) are obvious. QED.

We want to prove the subsequent maximality theorem.

3.5 MAXIMALITY THEOREM: Assume that (H,φ) is reduced and dim$N<\infty$. If $(\tilde{H},\tilde{\varphi})$ is a small extension of (H,φ) then $\tilde{H}=H$.

3.6 CONSEQUENCE: Assume that (H,φ) is reduced with dim$N<\infty$ and $NL=\{0\}$. If $(\tilde{H},\tilde{\varphi})$ is an extension of (H,φ) then $\tilde{H}=H$.

The implication 3.5$\Rightarrow$3.6 is clear from 3.4 which tells us that under the assumption $NL=\{0\}$ all extensions $(\tilde{H},\tilde{\varphi})$ of (H,φ) are small. And recall that 3.3 permits to reduce the assumptions in both 3.5 and 3.6. The proof of 3.5 will occupy the remainder of the section.

3.7 LEMMA: Assume that $\dim N<\infty$. Let $(\tilde{H},\tilde{\varphi})$ be a small extension of (H,φ). If $F\in\widetilde{ML}$ is an internal point of $\widetilde{ML}$ then it is an internal point of M as well. In particular: (H,φ) is reduced $\Rightarrow(\tilde{H},\tilde{\varphi})$ is reduced.

Proof: i) Let $F\in\widetilde{ML}$ be an internal point of $\widetilde{ML}$. From 1.7 we see that F is an internal point of $\tilde{M}$ as well. Also of course $F\in ML$. ii) In view of $\dim N<\infty$ it suffices to prove that $N\subset\overline{N\cap F(\mathrm{Re}L^{\infty}(m))}^{L^1(m)}$ since this is $=N\cap F(\mathrm{Re}L^{\infty}(m))$. After 1.6 we have to show that if $0\leqq f_n\in L^{\infty}(m)$ then $\int f_n F dm\to 0$ implies that $\alpha(f_n)\to 0$. And after 1.3 it suffices to show $\beta^{\infty}(f_n)\to 0$. iii) We have $\tilde{\alpha}^{o}(f_n)=\tilde{\theta}(f_n)\to 0$ since F is an internal point of $\tilde{M}$. Hence there exists a sequence of functions $v_n\in\tilde{H}$ with $\mathrm{Re}\,v_n\geqq f_n$ and $\mathrm{Re}\tilde{\varphi}(v_n)=$ $=\int(\mathrm{Re}\,v_n)Fdm\to 0$. Now $\mathrm{Re}\,v_n\in\mathrm{Re}\tilde{H}\subset(\log|H^{\times}|)^{\perp\perp}\subset(NL)^{\perp}$ and hence $\beta^{\infty}(\mathrm{Re}\,v_n)=$ $=\int(\mathrm{Re}\,v_n)Fdm$ after 1.2. It follows that $0\leqq\beta^{\infty}(f_n)\leqq\beta^{\infty}(\mathrm{Re}\,v_n)\to 0$. QED.

We next extablish - after a little remark which is to facilitate the subsequent proof - a technical lemma which prepares the reduction step 3.10 to small extensions of codimension one. After this the final step in the proof of 3.5 will be done.

3.8 REMARK: Assume that $\mathbb{Z}\subset G\subsetneqq\mathbb{R}$ and that G is closed and closed under addition. Then $G=\frac{1}{n}\mathbb{Z}$ for some $n\in\mathbb{N}$.

Proof: i) Let us put $a:=\mathrm{Inf}\{x\in G:x>0\}$ so that $a\in G$ and $0\leqq a\leqq 1$. We claim that $a>0$. To see this assume that $a=0$. Then there exist $0<a(n)\in G$ with $a(n)\to 0$. Fix any real $t>0$ and take the $p(n)\in\mathbb{N}$ with $p(n)-1<\frac{t}{a(n)}\leqq p(n)$. Then $p(n)a(n)-a(n)<t\leqq p(n)a(n)$ so that from $p(n)a(n)\in G$ and $p(n)a(n)\to t$ it follows that $t\in G$. But then $\mathbb{R}\subset G$ since $\mathbb{Z}\subset G$ and G is closed under addition. So we arrive at a contradiction which proves that indeed $a>0$. ii) The set $G^*:=-G$ fulfills the assumptions as well. Thus we can apply i) to G^* to obtain an $a^*>0$ such that $-a^*\in G$ and $x\leqq -a^*$ whenever $x\in G$ and $x<0$. iii) It follows that $a,-a^*\in G\Rightarrow a-a^*\in G$ and $a-a^*<a\Rightarrow a-a^*\leqq 0$ or $a\leqq a^*$ from the definition of a. Likewise $a-a^*\in G$ and $-a^*<a-a^*\Rightarrow 0\leqq a-a^*$ or $a^*\leqq a$ from the definition of a^*. Thus $a=a^*$. iv) We have $\pm a\in G$ and hence $a\mathbb{Z}\subset G$. We claim that $G=a\mathbb{Z}$. To see that $G\subset a\mathbb{Z}$ consider $x\in G$. Then there is an $m\in\mathbb{Z}$ such that $z:=$ $=ma+x$ satisfies $0\leqq z<a$. Since $z\in G$ it must be $z=0$ and hence $x=(-m)a\in a\mathbb{Z}$.

v) We have $1\in G$ so that $1=an$ for some $n\in\mathbb{N}$. It follows that $G=\frac{1}{n}\mathbb{Z}$. QED.

3.9 TECHNICAL LEMMA: Assume that (H,φ) is reduced and $\dim N<\infty$. Let $(\tilde{H},\tilde{\varphi})$ be an extension of (H,φ). And let $F\in\tilde{M}$ be an internal point of M (so that $F>0$ on X and hence $(\tilde{H},\tilde{\varphi})$ is reduced as well).

i) Let $P\in\tilde{E}^{\infty}$ and $P^{*}\in\operatorname{ReL}(m)$ be its conjugate function relative to $(\tilde{H},\tilde{\varphi})$. And assume that $P\in\log|H^{\times}|$ but $P\notin E^{\infty}$. Then there exists an $n\in\mathbb{N}$ such that $\forall t\in\mathbb{R}: e^{t(P+iP^{*})}\in H \Leftrightarrow t\in\frac{1}{n}\mathbb{Z}$.

ii) Let $P\in\tilde{E}^{\infty}$ and $P^{*}\in\operatorname{ReL}(m)$ be its conjugate function relative to $(\tilde{H},\tilde{\varphi})$. If $P\in\frac{N}{F}$ then $P^{*}\in\frac{N}{F}$, and hence in particular $P+iP^{*}\in\tilde{H}$.

iii) $\tilde{H}=H\oplus(\frac{N}{F}+i\frac{N}{F})\cap\tilde{H}$.

iv) If $(\tilde{H},\tilde{\varphi})$ is a small extension of (H,φ) then $\tilde{E}^{\infty}=E^{\infty}\oplus$ real-linear span$(\frac{N}{F}\cap(\log|H^{\times}|)\cap\tilde{E}^{\infty})$.

Proof: i) We have $e^{t(P+iP^{*})}\in\tilde{H}$ for all $t\in\mathbb{R}$. Also there exists a function $\in H^{\times}$ of modulus e^{P}. Thus the uniqueness assertion in V.1.3 implies that $e^{\pm(P+iP^{*})}\in H$ and hence $e^{n(P+iP^{*})}\in H$ $\forall n\in\mathbb{Z}$. Let now $G\subset\mathbb{R}$ consist of the $t\in\mathbb{R}$ such that $e^{t(P+iP^{*})}\in H$. Then G fulfills the assumptions in 3.8. It follows that $G=\frac{1}{n}\mathbb{Z}$ for some $n\in\mathbb{N}$ as claimed.

ii) From $F\in\tilde{M}$ we deduce that $\tilde{H}F\subset K$. Thus $e^{t(P+iP^{*})}\in\tilde{H}$ implies that $e^{t(P+iP^{*})}F\in K$ for all $t\in\mathbb{R}$. Now from VI.2.7.ii) we have

$$\left|\frac{1}{t}\left(e^{t(P+iP^{*})}-1\right)\right| \leqq \frac{1}{\varepsilon}\left|e^{(\tau+\varepsilon)(P+iP^{*})}\right|+\left|P+iP^{*}\right| \quad \forall 0<t\leqq\tau \text{ and } \varepsilon>0,$$

and the second member is $\in L^{p}(Fm)$ $\forall 1\leqq p<\infty$ after VI.5.5.i). Thus for $t\downarrow 0$ it follows that $(P+iP^{*})F\in K$. In view of VI.6.9.iii) this means that

$$P+iP^{*}=(A+iB)+(a+ib) \text{ with } A+iB\in H^{\#}\cap L^{1}(Fm) \text{ and } a,b\in\frac{N}{F}.$$

Now $P+iP^{*}\in\tilde{H}^{\#}$ and hence $\int(P+iP^{*})Fdm=\tilde{\varphi}(P+iP^{*})=\tilde{\alpha}(P)=\int PFdm=0$ since $PF\in N$. Also $\int(a+ib)Fdm=0$, so that $\varphi(A+iB)=\int(A+iB)Fdm=0$. Furthermore $L^{0}(Fm)\subset L^{\#}$ from VI.6.4 and the substitution theorem V.2.2 imply that $e^{t(A+iB)}\in H^{\#}$ and $\varphi(e^{t(A+iB)})=1$ $\forall t\in\mathbb{R}$. Hence $A\in E$ and $B=A^{*}$. Therefore $P=A+a$ implies that $A\in E^{\infty}\cap\frac{N}{F}$ and hence $A=0$ after VI.6.8.i). Thus $B=A^{*}=0$ so that $P^{*}=b\in\frac{N}{F}$ as claimed.

iii) We know from ii) that $\tilde{H}F\subset K$. Thus VI.6.9.iii) implies that $\tilde{H}\subset H^{\#}+(\frac{N}{F}+i\frac{N}{F})$ and hence $\tilde{H}\subset H+(\frac{N}{F}+i\frac{N}{F})$ so that $\tilde{H}=H+(\frac{N}{F}+i\frac{N}{F})\cap\tilde{H}$. The decomposition is direct after VI.6.8.i).

iv) We apply the closed subgroup lemma 2.4 to $V:=\mathrm{Re}L^{\infty}(m)$ with the $L^{\infty}(m)$-norm topology and to $S:=\log|H^{\times}|$ and $T:=\log|\tilde{H}^{\times}|$ which are closed subgroups after 1.8. We have

$$D(S)=\bigcap_{t>0} tS=\{f\in\mathrm{Re}L^{\infty}(m):\alpha(tf)+\alpha(-tf) = 0\ \forall t>0\}=E^{\infty},$$

and likewise $D(T)=\tilde{E}^{\infty}$ after 1.8.ii). Furthermore $E(S)=(\log|H^{\times}|)^{\perp\perp}=$ real-linear span $(\log|H^{\times}|)$ after 1.9.ii), and

$$\dim\frac{E(S)}{D(S)}\leqq\dim\frac{\mathrm{Re}L^{\infty}(m)}{E^{\infty}}=\dim\frac{(\mathrm{Re}L^{1}(m))'}{N^{\perp}}=\dim N<\infty,$$

as in the proof of 1.9.ii). Thus from 2.4 we obtain

$$E\big((\log|H^{\times}|)\cap\tilde{E}^{\infty}\big)=\big(\log|H^{\times}|\big)^{\perp\perp}\cap\tilde{E}^{\infty}.$$

The second member is $=\tilde{E}^{\infty}$ since $\tilde{E}^{\infty}=\overline{\mathrm{Re}\tilde{H}}^{\text{weak}*}\subset(\log|H^{\times}|)^{\perp\perp}$ from VI.4.3 and in view of the smallness assumption. To evaluate the first member recall from VI.6.8.ii) that $\mathrm{Re}L^{\infty}(m)=E^{\infty}\oplus\frac{N}{F}$ and hence $\tilde{E}^{\infty}=E^{\infty}\oplus(\frac{N}{F}\cap\tilde{E}^{\infty})$. Thus

$$(\log|H^{\times}|)\cap\tilde{E}^{\infty}=E^{\infty}+\big(\tfrac{N}{F}\cap(\log|H^{\times}|)\cap\tilde{E}^{\infty}\big),$$

$$E\big((\log|H^{\times}|)\cap\tilde{E}^{\infty}\big)=E^{\infty}+\text{real-lin span}\big(\tfrac{N}{F}\cap(\log|H^{\times}|)\cap\tilde{E}^{\infty}\big),$$

since the real-linear span in question is finite-dimensional. It is clear that the sum is direct. The assertion follows. QED

<u>3.10 REDUCTION STEP</u>: Assume that (H,φ) is reduced and $\dim N<\infty$. Let $(\tilde{H},\tilde{\varphi})$ be a small extension of (H,φ) with $\tilde{H}\neq H$. Then $\dim\frac{\tilde{H}}{H}<\infty$. And there exists an intermediate complex algebra $B:H\subset B\subset\tilde{H}$ such that $\dim\frac{B}{H}=1$.

Proof: i) Let us fix $F\in\widetilde{ML}$ an internal point of $\widetilde{ML}$. After 3.7 then F is an internal point of M as well so that we have the assumptions of 3.9. From 3.9.iii) we see that $\dim\frac{\tilde{H}}{H}<\infty$. And from 3.9.iv) combined with 3.2 we conclude that $\frac{N}{F}\cap(\log|H^{\times}|)\cap\tilde{E}^{\infty}\neq\{0\}$. Let us fix a nonzero function

$P\in\frac{N}{F}\cap(\log|H^{\times}|)\cap\tilde{E}^{\infty}$ and let $P^*\in \mathrm{ReL}(m)$ be its conjugate function relative to $(\tilde{H},\tilde{\varphi})$. Then $P\notin E^{\infty}$ after VI.6.8.i). Thus from 3.9.i) we obtain an $n\in\mathbb{N}$ such that $\forall t\in\mathbb{R}: e^{t(P+iP^*)}\in H \Leftrightarrow t\in\frac{1}{n}\mathbb{Z}$.

ii) Let us put $G:=\exp\left(\frac{1}{2n}(P+iP^*)\right)\in\tilde{H}$. Then $G\notin H$ but $G^2\in H$. It follows that $A:=\{u+vG: u,v\in H\}$ is an intermediate algebra with $H\subsetneqq A\subset\tilde{H}$. In particular $\dim\frac{A}{H}<\infty$.

iii) Now we apply the commutative operator algebra lemma VII.5.2 as in the proof of VII.5.1: Each $u\in H$ defines a linear operator $\langle u\rangle\in L(A/H)$ via $\langle u\rangle:[f]\mapsto[uf]\ \forall f\in A$. It is obvious that $\langle u+v\rangle=\langle u\rangle+\langle v\rangle$, $\langle cu\rangle=c\langle u\rangle$ and $\langle uv\rangle=\langle u\rangle\langle v\rangle$ for $u,v\in H$ and $c\in\mathbb{C}$, as well as $\langle 1\rangle$=identity. It follows that there exists a $W\in A$, $W\notin H$ and a function $\varepsilon: H\to\mathbb{C}$ such that $\langle u\rangle[W]=\varepsilon(u)[W]$ or $uW-\varepsilon(u)W\in H$ for all $u\in H$. It is clear that $\varepsilon: H\to\mathbb{C}$ is unique and is a multiplicative linear functional with $\varepsilon(1)=1$. Now we have $W=a+bG$ with $a,b\in H$ and hence

$$W^2-2\varepsilon(a)W=a^2+b^2G^2+2abG-2\varepsilon(a)W=b^2G^2-a^2+2\big(a-\varepsilon(a)\big)W\in H.$$

It follows that $B:=\{u+cW: u\in H \text{ and } c\in\mathbb{C}\}$ is an intermediate algebra with $H\subsetneqq B\subset A\subset\tilde{H}$ and $\dim\frac{B}{H}=1$. QED.

Proof of 3.5: Let (H,φ) be reduced with $\dim N<\infty$. We assume the existence of a small extension $(\tilde{H},\tilde{\varphi})$ of (H,φ) with $\tilde{H}\neq H$ and shall deduce a contradiction. In view of 3.10 we can assume that $\dim\frac{\tilde{H}}{H}=1$.

i) The initial step is as in the proof of 3.10. Let $F\in\widetilde{ML}$ be a fixed internal point of $\widetilde{ML}$. Then F is an internal point of M as well so that we have the assumptions of 3.9. From 3.9.iv) we conclude that $\frac{N}{F}\cap(\log|H^{\times}|)\cap\tilde{E}^{\infty}\neq\{0\}$. Let us fix a nonzero function $P\in\frac{N}{F}\cap(\log|H^{\times}|)\cap\tilde{E}^{\infty}$ and let $P^*\in \mathrm{ReL}(m)$ be its conjugate function relative to $(\tilde{H},\tilde{\varphi})$. Then $P\notin E^{\infty}$ since $E^{\infty}\cap\frac{N}{F}=\{0\}$. From 3.9.i) we have $e^{t(P+iP^*)}\in H$ at least for all $t\in\mathbb{Z}$. And from 3.9.ii) we see that $P^*\in\frac{N}{F}$ and $P+iP^*\in\tilde{H}$. Since $P+iP^*\notin H$ it follows that $\tilde{H}=\{u+c(P+iP^*): u\in H \text{ and } c\in\mathbb{C}\}$.

ii) We have $P,P^*\in \mathrm{Re}\tilde{H}\subset\tilde{E}^{\infty}$. Let us show that P and P^* are linearly independent over E^{∞}: From $aP+bP^*\in E^{\infty}$ with $a,b\in\mathbb{R}$ not both $=0$ it would follow that $aP+bP^*=0$ since $E^{\infty}\cap\frac{N}{F}=\{0\}$, and hence that $b(P+iP^*)=(b-ia)P\in\tilde{H}$

or $P\in\tilde{H}$ which is nonsense.

iii) From ii) we deduce that the dimension of $\tilde{E}^\infty/E^\infty$ must be ≥ 2. Thus 3.9.iv) implies that there exist functions $Q\in\frac{N}{F}\cap(\log|H^\times|)\cap\tilde{E}^\infty$ which are not real multiples of P (and hence in particular $\neq 0$). Let us fix such a function Q and let $Q^*\in\text{ReL}(m)$ be its conjugate function relative to $(\tilde{H},\tilde{\varphi})$. Then we see as in i) that $Q\notin E^\infty$ but $e^{t(Q+iQ^*)}\in H$ at least for all $t\in\mathbb{Z}$, and also $Q^*\in\frac{N}{F}$ and $Q+iQ^*\in\tilde{H}$.

iv) It follows that $Q+iQ^*=u+c(P+iP^*)$ with $u\in H$ and $c=a+ib\in\mathbb{C}$. From $\text{Re}\,u, \text{Im}\,u\in E^\infty\cap\frac{N}{F}$ we see that $u=0$ and hence $Q+iQ^*=c(P+iP^*)$. In particular $Q=aP-bP^*$ so that $b\neq 0$ after the choice of Q. So we arrive at $e^{t(P+iP^*)}\in H$ at least for $t=n$ and for $t=nc$ $\forall n\in\mathbb{Z}$ for some complex $c=a+ib$ with $b\neq 0$.

v) The proof has an unexpected finale. Consider the entire function

$$\phi:\phi(z) = \int e^{z(P+iP^*)}(P-iP^*)Fdm \qquad \forall\, z\in\mathbb{C}.$$

For $n\geq 1$ we have

$$D^n\phi(z) = \int e^{z(P+iP^*)}(P+iP^*)^n(P-iP^*)Fdm \qquad \forall\, z\in\mathbb{C}.$$

We list a series of properties. 1) $\phi'(0)=\int|P+iP^*|^2Fdm>0$ so that $\phi\neq 0$. 2) If $z\in\mathbb{C}$ is such that $e^{z(P+iP^*)}\in H$ then $\phi(z)=0$ in view of $PF, P^*F\in N$. Thus $\phi(n)=\phi(nc)=0$ for all $n\in\mathbb{Z}$. 3) There exist $\sigma,\tau\in\mathbb{C}$ such that $\phi''-\sigma\phi'-\tau\phi=0$. This is seen as follows: We have

$$(P+iP^*)^2 = u+\sigma(P+iP^*) \quad \text{with } u\in H \text{ and } \sigma\in\mathbb{C},$$

$$u(P+iP^*) = v+\tau(P+iP^*) \quad \text{with } v\in H \text{ and } \tau\in\mathbb{C}.$$

It follows that $(u-\tau)(P+iP^*)=\cdots\in H$ so that $(u-\tau)\tilde{H}\subset H$. But this means that $\left((P+iP^*)^2-\sigma(P+iP^*)-\tau\right)\tilde{H}\subset H$. Thus we obtain

$$\phi''(z)-\sigma\phi'(z)-\tau\phi(z) = \int e^{z(P+iP^*)}\left[(P+iP^*)^2-\sigma(P+iP^*)-\tau\right](P-iP^*)Fdm=0 \quad \forall\, z\in\mathbb{C},$$

as above in view of $PF, P^*F\in N$.

Now the elementary theory of ordinary differential equations tells us that ϕ has one of the two forms

$$\phi(z) = \alpha e^{sz} + \beta e^{tz} = \left(\alpha+\beta e^{(t-s)z}\right)e^{sz} \quad \text{with complex } s \neq t,$$

$$\phi(z) = (\alpha+\beta z)e^{sz} \quad \text{with complex } s,$$

where $\alpha,\beta \in \mathbb{C}$ are not both $= 0$. But in either case ϕ cannot possess as many zeros as found in 2) above. This is the desired contradiction. QED.

Notes

The Ahern-Sarason lemma 1.3 is an alienated extract from AHERN-SARASON [1967a]. The present form is from KÖNIG [1969b]. The idea to use the Dirichlet simultaneous approximation lemma 1.4 appears to be due to O'NEILL [1968]. For 1.4 we refer to HARDY-WRIGHT [1968] p.170. Theorem 1.7 is due to GAMELIN [1969], the special case of a unique logmodular density is a main result in AHERN-SARASON [1967a].

The restricted version 3.6 of the maximality theorem 3.5 appeared (up to the reducedness point settled in 3.7) in GAMELIN-LUMER [1968] and GAMELIN [1968], see also GAMELIN [1969] Section IV.7. The full theorem 3.5 is from KÖNIG [1969b]. The present proof is modelled after the elementary proof of the restricted version in GAMELIN-LUMER [1968]. Its proof in GAMELIN [1968][1969] is based on the Arens-Royden theorem and is much shorter.

Chapter X

Function Algebras on Compact Planar Sets

The present final chapter is devoted to the standard algebras of analytic functions on compact subsets of the complex plane. For nonvoid compact $K \subset \mathbb{C}$ these are

- $P(K)$ the supnorm closure in $C(K)$ of the polynomials,
- $R(K)$ the supnorm closure in $C(K)$ of the rational functions with poles off K,
- $A(K)$ the class of functions in $C(K)$ which are holomorphic in the interior of K.

All these are supnorm-closed complex algebras with $1 \in P(K) \subset R(K) \subset A(K) \subset$ $\subset C(K)$ and hence are function algebras on K in the sense of Chapter III. They are the most prominent concrete examples to the abstract theory of function algebras.

The inclusions $P(K) \subset R(K) \subset A(K) \subset C(K)$ can all be proper as simple examples reveal. Hence it is natural to ask for geometric conditions on K under which the individual inclusions become equalities. It is clear that $A(K)=C(K)$ iff the interior of K is $=\emptyset$. And it is simple to transform $P(K)=R(K)$ into a very intuitive geometric property of K: that K have no holes (= bounded components of the complement $\mathbb{C}-K$). Then even $P(K)=R(K)=A(K)$ after the Mergelyan polynomial approximation theorem. But it is an extremely difficult problem to characterize those K for which $R(K)=A(K)$, renowned as the rational approximation problem. A sufficient condition is that K have a finite number of holes. It is due to Mergelyan as well as the more general sufficient condition that the diameters of the holes of K be bounded away from zero. Another more general sufficient condition is that each boundary point of K be in the boundary of some component of $\mathbb{C}-K$. The latter one results from the quasi-geometric necessary and sufficient conditions due to Vitushkin which are in terms of the so-called continuous-analytic capacity.

The above conditions had been discovered via the constructive techniques of approximation theory. Thereafter it has been realized that an essential part of them can be based on the functional-analytic theory of function algebras with rather few concrete additives. But in spite of important contributions due to the abstract theory, the situation from its point of view is not quite favorable, since the abstract theory has not yet been able to develop conceptual methods in order to replace certain complicated concrete constructions.

In the present work we do not intend to enter into the depths of constructive rational approximation theory. We shall restrict ourselves to those results which are more or less direct outflows from the abstract theories developed hitherto - at least when combined with certain natural and basic concrete tools: these are the Cauchy transformation and logarithmic transformation of Baire measures in the complex plane.

The initial section remains within the abstract theory. Its purpose is to transfer certain main results from the abstract Hardy algebra theory of Chapters IV-IX into the function algebra theory of Chapters II-III. The road of transfer will be the direct image construction IV.1.3 while the vehicle will be the main F.and M.Riesz consequence II.4.5. The choice of material will of course be adapted to the aim of the present chapter: the main results 1.6 and 1.9 can be considered as abstract Mergelyan type theorems.

1. Consequences of the abstract Hardy Algebra Theory

We return to the bounded-measurable situation of Chapter II: We fix a measurable space (X,Σ) and a complex subalgebra $A\subset B(X,\Sigma)$ which contains the constants. Recall the weak topology $\omega=\sigma(B(X,\Sigma),ca(X,\Sigma))$. The closure $\overline{A}^{\omega}\subset B(X,\Sigma)$ of A relative to ω is a complex subalgebra as well. We have $\overline{A}^{\omega}=A^{\perp\perp}$ from the bipolar theorem A.1.7. Note that in the compact-continuous situation of Chapter III the F.Riesz representation theorem A.2.1 implies that $\omega|C(X)=\sigma(C(X),ca(X,\Sigma))=\sigma(C(X),C(X)')$ and hence that $\overline{A}^{\omega}\cap C(X) = \overline{A}^{\text{supnorm}} = A$.

For nonzero $m\in Pos(X,\Sigma)$ we consider the closure $A^m:= \overline{A \bmod m}^{\text{weak*}} \subset L^{\infty}(m)$ in the weak*topology $\sigma(L^{\infty}(m),L^1(m))$ which is a complex subalgebra as well. From the bipolar theorem we have for $f\in B(X,\Sigma)$: $f \bmod m \in A^m \Leftrightarrow$ $\Leftrightarrow \int f d\theta=0 \ \forall\theta\in A^{\perp}$ with $\theta\ll m$. Let us note a useful technical remark.

<u>1.1</u> <u>REMARK</u>: Fix $1\leq p<\infty$. For $f\in B(X,\Sigma)$ the subsequent properties are equivalent.

i) $f\in\overline{A}^{\omega}$.

ii) For each nonzero $m\in Pos(X,\Sigma)$ we have $f \bmod m \in A^m$.

iii) For each nonzero $m\in Pos(X,\Sigma)$ we have $f \bmod m \in \overline{A \bmod m}^{L^p(m)\text{-norm}}$.

Proof: i)⇒ii) is clear from the above. ii)⇒iii) If this were false then there were an $h\in L^q(m)$ with $\int uhdm=0$ $\forall u\in A$ and $\int fhdm\neq 0$, which were a contradiction to $f \bmod m \in A^m$. iii)⇒i) For $\theta\in A^{\perp}$ there are functions $u_n\in A$ such that $\int|u_n-f|^p d|\theta|\to 0$. Then $\int u_n d\theta=0$ implies that $\int fd\theta=0$. Thus $f\in\overline{A}^{\omega}$. QED.

We proceed to connect the basic notions for $\overline{A}^{\omega}$ with those for A. The assertions in the subsequent remark are all clear from the above. In iv) one has to proceed as in I.3.7. One could also rest on IV.1.3.

<u>1.2</u> <u>REMARK</u>: i) $(\overline{A}^{\omega})^{\perp}=A^{\perp}$.

ii) Each $\varphi\in\Sigma(A)$ admits a unique continuation $\in\Sigma(\overline{A}^{\omega})$ which in the sequel will be named φ as well. Thus $\Sigma(\overline{A}^{\omega})=\Sigma(A)$.

iii) For each $\varphi\in\Sigma(A)$ we have $M(\overline{A}^{\omega},\varphi) = M(A,\varphi)$. Thus $M(\overline{A}^{\omega}) = M(A)$.

iv) For each $\varphi\in\Sigma(A)$ we have $MJ(\overline{A}^{\omega},\varphi) = MJ(A,\varphi)$.

We turn to the serious characterizations of the functions in $\overline{A}^{\omega}$ which are our main concern. A fundamental step of reduction is the main F.and M.Riesz consequence II.4.5 which can be formulated as follows.

<u>1.3</u> <u>MAIN F.and M.RIESZ CONSEQUENCE</u>: The function $f\in B(X,\Sigma)$ is in $\overline{A}^{\omega}$ iff it satisfies

o) $\int fd\theta=0$ for all $\theta\in A^{\perp}\cap M(A)^{\wedge}$,

and furthermore satisfies the subsequent conditions which in view of the above are all equivalent.

i) For each $m\in M(A)$: $\int fd\theta=0$ $\forall\theta\in A^{\perp}$ with $\theta\ll m$.

ii) For each $m\in M(A)$: $f \bmod m \in A^m$.

i') For each $\sigma\in M(A)$ there is an $m\in Pos(X,\Sigma)$ with $\sigma\ll m$ such that $\int fd\theta=0$ $\forall\theta\in A^{\perp}$ with $\theta\ll m$.

ii′) For each $\sigma\in M(A)$ there is an $m\in Pos(X,\Sigma)$ with $\sigma\ll m$ such that $f \bmod m \in A^m$.

For the important concrete examples of the present chapter we shall see that $A^{\perp}\cap M(A)^{\wedge}=0$, so that we end up with the residual equivalent conditions. At this point it is most natural to invoke the direct image construction IV.1.3: For fixed $\varphi\in\Sigma(A)$ and $m\in Pos(X,\Sigma)$ such that $M(A,\varphi)$ contains measures $\ll m$ that construction produces the Hardy algebra situation (A^m,φ^m) with

$$M = \{0\leq V\in L^1(m): Vm\in M(A,\varphi)\},$$
$$N \subset \{f\in ReL^1(m): fm\in N(A,\varphi)\},$$
$$MJ = \{0\leq V\in L^1(m): Vm\in MJ(A,\varphi)\},$$
$$NJ \subset \{f\in ReL^1(m): fm\in NJ(A,\varphi)\}.$$

And if m itself is $\in M(A,\varphi)$ then the Hardy algebra situation (A^m,φ^m) is reduced with $1\in M$. Thus we can apply the abstract Hardy algebra theory of Chapters IV-IX. To start with we use central results from Section VI.4 to obtain the next characterization of the functions in $\overline{A}^{\omega}$.

<u>1.4 THEOREM</u>: For $f\in B(X,\Sigma)$ each of the subsequent conditions is equivalent to the conditions i)ii)i′)ii′) of 1.3.

iii) $\int f d\sigma = \int f d\tau$ $\quad\forall\ \varphi\in\Sigma(A)$ and $\sigma,\tau\in M(A,\varphi)$, and
$\int fu dm = \int f dm \int u dm$ $\quad\forall\ m\in M(A)$ and $u\in A$.

iv) $\int (f+u)^2 dm = (\int (f+u) dm)^2$ $\quad\forall\ m\in M(A)$ and $u\in A$.

v) $\int f^2 dm = (\int f dm)^2$ $\quad\forall\ m\in M(A)$, and
$\int fu dm = \int f dm \int u dm$ $\quad\forall\ m\in M(A)$ and $u\in A$.

Thus $f\in B(X,\Sigma)$ is in $\overline{A}^{\omega}$ iff it satisfies o) and some of the equivalent conditions iii)iv)v).

Proof: i)⇒iii) is clear: the first assertion since $\sigma-\tau\in A^{\perp}$ and $\ll \frac{1}{2}(\sigma+\tau)\in M(A)$, and the second assertion since $um - (\int u dm)m \in A^{\perp}$ and $\ll m$. ii)⇒iv) is obvious after the direct image construction. iv)⇒v) is trivial. So it remains to prove iii)⇒ii) and v)⇒ii). For this purpose fix $m\in M(A,\varphi)$ for some $\varphi\in\Sigma(A)$ and consider the reduced Hardy algebra situation (A^m,φ^m) with $1\in M$. We have to show that $\vec{f}:= f \bmod m \in L^{\infty}(m)$ is $\in A^m$. Under the assumption iii) we have 1) $U,V\in M \Rightarrow Um,Vm\in M(A,\varphi) \Rightarrow \int \vec{f}(U-V)dm=0$ and 2) $\int \vec{f}u dm = \int \vec{f} dm \int u dm$ $\forall u\in A$ mod m and hence $\forall u\in A^m$. Thus $\vec{f}\in A^m$ after

VI.4.5. Under the assumption v) we have 1) $V\in M \Rightarrow Vm\in M(A,\varphi) \Rightarrow \int \vec{f}^2 V dm = (\int \vec{f} V dm)^2$ and 2) as before. Thus $\vec{f}\in A^m$ after VI.4.7. QED.

The most prominent consequence is the subsequent maximality theorem.

<u>1.5</u> <u>THEOREM</u>: Assume that $B\subset B(X,\Sigma)$ is a complex subalgebra with $A\subset B$. Then $B\subset\overline{A}^\omega$ iff $A^\perp\cap M(A)^\wedge\subset B^\perp$ and $M(A)\subset M(B)$ (which means that $M(A)=M(B)$).

Proof: If $B\subset\overline{A}^\omega$ then the two assertions follow from 1.2. The converse follows upon application of 1.4 to the individual functions $f\in B$. QED.

<u>1.6</u> <u>THEOREM</u>: Assume that $B\subset B(X,\Sigma)$ is a complex subalgebra with $A\subset B$ such that

0) $A^\perp\cap M(A)^\wedge\subset B^\perp$,

1) each $\varphi\in\Sigma(A)$ has an extension $\psi\in\Sigma(B)$,

2) $N(A,\varphi)\subset B^\perp$ for each $\varphi\in\Sigma(A)$.

Then $B\subset\overline{A}^\omega$.

Proof: In view of 1.5 we have to show that $M(A)\subset M(B)$. Assume that $\sigma\in M(A,\varphi)$ for some $\varphi\in\Sigma(A)$. Choose an extension $\psi\in\Sigma(B)$ and some $\tau\in M(B,\psi)\subset\subset M(A,\varphi)$. From 2) then $\sigma-\tau\in N(A,\varphi)\subset B^\perp$ and hence $\sigma\in M(B,\psi)$. QED.

Theorem 1.6 is the first of the desired abstract Mergelyan theorems. Condition 2) is satisfied in particular if each $\varphi\in\Sigma(A)$ has a unique representative measure $\in M(A,\varphi)$. Also it is sufficient to assume the overall condition $\mathrm{Re}\, B\subset\overline{\mathrm{Re}\, A}^\omega$ which resembles the Dirichlet condition. Condition 2) is often difficult to be verified. A much more powerful theorem is obtained when the abstract Hardy algebra theory contributes the maximality theorem IX.3.5. This requires some preliminaries on logmodular measures.

For fixed $\varphi\in\Sigma(A)$ we introduce

$$ML(A,\varphi):=\{\sigma\in Pos(X,\Sigma):\ \log|\varphi(u)|=\int(\log|u|)d\sigma\ \ \forall u\in(\overline{A}^\omega)^\times\},$$

the class of logmodular measures for φ, and $NL(A,\varphi):=$ real-linear span$(ML(A,\varphi)-ML(A,\varphi))=\{c(\sigma-\tau):\ \sigma,\tau\in ML(A,\varphi)$ and $c>0\}$ in case that $ML(A,\varphi)$ is nonvoid. Thus $MJ(A,\varphi)\subset ML(A,\varphi)\subset M(A,\varphi)$ from 1.2 and via exponentiation, and $ML(\overline{A}^{\omega},\varphi) = ML(A,\varphi)$ from the definition itself. Note that in the compact-continuous situation of Chapter III we have $ML(A,\varphi)\neq\emptyset$ in view of III.1.1.

The direct image construction IV.1.3 connects the logmodular measures with the respective notion from Section IX.1. For nonzero $m\in Pos(X,\Sigma)$ we have $(\overline{A}^{\omega})^{\times} \bmod m \subset (A^m)^{\times}$ from the definitions involved. From this fact we conclude that for $\varphi\in\Sigma(A)$ and $m\in Pos(X,\Sigma)$ such that $M(A,\varphi)$ contains measures $\ll m$ the Hardy algebra situation (A^m,φ^m) satisfies

$$ML \subset \{0\leq V\in L^1(m): Vm\in ML(A,\varphi)\},$$
$$NL \subset \{f\in ReL^1(m): fm\in NL(A,\varphi)\} \text{ whenever } ML\neq\emptyset.$$

We insert some consequences from the abstract Hardy algebra theory.

<u>1.7</u> <u>REMARK</u>: Let $\varphi\in\Sigma(A)$ such that $\dim N(A,\varphi) < \infty$. Then $MJ(A,\varphi) \neq\emptyset$ and hence $ML(A,\varphi) \neq\emptyset$.

Proof: Fix any $m\in M(A,\varphi)$ and consider the Hardy algebra situation (A^m,φ^m). Then $\dim N < \infty$. Hence IV.4.5 shows that $\{0\leq V\in L^1(m): Vm\in MJ(A,\varphi)\}$ $= MJ \neq \emptyset$. QED.

<u>1.8</u> <u>THEOREM</u>: Let $\varphi\in\Sigma(A)$ such that $\dim N(A,\varphi) < \infty$. Assume that $\sigma\in M(A,\varphi)$ is such that each member of $ML(A,\varphi)$ is $\leq c\sigma$ for some constant $c>0$. Then σ is an internal point of $M(A,\varphi)$.

Proof: In view of $\dim N(A,\varphi) < \infty$ we can choose an internal point $m\in M(A,\varphi)$ of $M(A,\varphi)$. Then all $\tau\in M(A,\varphi)$ are $\ll m$. In particular $\sigma=Fm$ with $F\in M$. Consider the Hardy algebra situation (A^m,φ^m) with $\dim N < \infty$. We have $V\in ML \Rightarrow Vm\in ML(A,\varphi) \Rightarrow Vm\leq c\sigma=cFm$ or $V\leq cF$ for some $c>0$. Thus from IX.1.7 we obtain a constant $c>0$ such that $G\leq cF$ $\forall G\in M$. It follows that $\tau\in M(A,\varphi) \Rightarrow \tau=Gm$ with $G\in M \Rightarrow G\leq cF \Rightarrow \tau\leq c\sigma$. QED.

We conclude the section with the second abstract Mergelyan theorem.

<u>1.9</u> <u>THEOREM</u>: Assume that $B\subset B(X,\Sigma)$ is a complex subalgebra with $A\subset B$ such that

0) $A^{\perp}\cap M(A)^{\wedge} \subset B^{\perp}$,

1) each $\varphi\in\Sigma(A)$ has an extension $\psi\in\Sigma(B)$,

*) $\dim N(A,\varphi) < \infty$ for each $\varphi\in\Sigma(A)$,

2) $NL(A,\varphi)\subset B^{\perp}$ for each $\varphi\in\Sigma(A)$.

Then $B\subset\overline{A}^{\omega}$.

Theorem 1.9 has the obvious weak point of condition *) which is neither necessary nor adequate but which we are unable to eliminate. Condition 2) is satisfied in particular if each $\varphi\in\Sigma(A)$ has a unique logmodular measure $\in ML(A,\varphi)$. Also it is sufficient to assume the overall condition

$$\mathrm{Re}\, B \subset \overline{\text{real-linear span}(\log|(\overline{A}^{\omega})^{\times}|)}^{\omega},$$

which corresponds to the definition of smallness in Section IX.3.

Proof: In view of 1.3 it is sufficient to show that for each $\sigma\in M(A)$ there is an $m\in Pos(X,\Sigma)$ with $\sigma\ll m$ such that $B^m\subset A^m$. Assume that $\sigma\in M(A,\varphi)$ for some $\varphi\in\Sigma(A)$, choose an extension $\psi\in\Sigma(B)$ and some $\tau\in M(B,\psi)$, and put $m = \frac{1}{2}(\sigma+\tau) \in M(A,\varphi)$. Then (A^m,φ^m) is a reduced Hardy algebra situation with $\dim N < \infty$ and (B^m,ψ^m) is an extension of (A^m,φ^m) in the sense of Section IX.3. Now for $f\in NL\subset \mathrm{Re}L^1(m)$ we have $fm\in NL(A,\varphi)\subset B^{\perp}$ in view of 2), so that $\int ufdm = 0$ $\forall u\in B^m$ and hence $\forall u\in \mathrm{Re}B^m$. It follows that $NL\subset(\mathrm{Re}B^m)^{\perp}$. Thus after IX.3.4 (B^m,ψ^m) is a small extension of (A^m,φ^m) so that the maximality theorem IX.3.5 implies that $B^m=A^m$. QED.

2. The Cauchy Transformation of Measures

The Cauchy transformation is defined to transform the measure $\theta\in ca(\mathbb{C})$ into the function

$$\theta^C:\ \theta^C(z) = \int\frac{d\theta(u)}{u-z},$$

defined for those $z\in\mathbb{C}$ where $\theta^{AC}(z) := \int\frac{d|\theta|(u)}{|u-z|} < \infty$.

We shall see at once that this is true for Lebesgue-almost all $z\in\mathbb{C}$. In view of its kernel function the Cauchy transformation is a basic device for the study of the algebras $R(K)$.

In the present section we fix a measure $\theta\in ca(\mathbb{C})$. We need two simple lemmata. Let us introduce the notations

$$V(u,s) = \{z\in\mathbb{C}: |z-u|<s\},$$
$$\bar{V}(u,s) = \{z\in\mathbb{C}: |z-u|\leq s\} \quad \text{for } u\in\mathbb{C} \text{ and } s>0.$$

2.1 LEMMA: Let (X,Σ,σ) be a (finite or infinite) positive measure space. Then for each measurable function $f:X\to[0,\infty[$ we have

$$\int_X f d\sigma = \int_{]0,\infty[} \sigma([f\geq t])dt.$$

Proof: If $\sigma([f\geq t])=\infty$ for some $t>0$ then both integrals are $=\infty$. So we can assume that $\sigma([f\geq t])<\infty$ $\forall t>0$. Put $M = \{(x,t): 0<t\leq f(x)\} \subset X\times]0,\infty[$. We can apply the Fubini theorem to obtain

$$\int_X \Big(\int_{]0,\infty[} \chi_M(x,t)dt\Big)d\sigma(x) = \int_{]0,\infty[} \Big(\int_X \chi_M(x,t)d\sigma(x)\Big)dt.$$

But this is the assertion. QED.

2.2 LEMMA: For each Baire set $E\subset\mathbb{C}$ we have

$$\int_E \frac{dL(u)}{|u-z|} \leq 2\sqrt{\pi L(E)} \quad \forall z\in\mathbb{C}.$$

Proof: We can assume that $0<L(E)<\infty$. For $r>0$ we obtain from 2.1

$$\int_E \frac{dL(u)}{|u-z|} = \int_0^\infty L\Big(\Big\{u\in E: \frac{1}{|u-z|}\geq t\Big\}\Big)dt = \int_0^\infty L\Big(\Big\{u\in E: |u-z|\leq s\Big\}\Big)\frac{ds}{s^2}$$
$$= \int_0^r L\Big(E\cap\bar{V}(z,s)\Big)\frac{ds}{s^2} + \int_r^\infty L\Big(E\cap\bar{V}(z,s)\Big)\frac{ds}{s^2}$$
$$\leq \int_0^r L\Big(\bar{V}(z,s)\Big)\frac{ds}{s^2} + \int_r^\infty L(E)\frac{ds}{s^2} = \pi r + \frac{1}{r}L(E),$$

which is $= 2\sqrt{\pi L(E)}$ for $r = \left(\frac{1}{\pi}L(E)\right)^{1/2}$. QED.

2.3 PROPOSITION: i) The function $\theta^{AC}:\mathbb{C}\to[0,\infty]$ is Baire measurable.

ii) $N:= \{z\in\mathbb{C}: \theta^{AC}(z)=\infty\}$ has $L(N)=0$.

iii) The function $\theta^C:\mathbb{C}-N\to\mathbb{C}$ is Baire measurable.

iv) For each Baire set $E\subset\mathbb{C}$ we have

$$\int_E |\theta^C(z)|\,dL(z) \leq \int_E \theta^{AC}(z)\,dL(z) \leq 2\sqrt{\pi L(E)}\,\|\theta\|.$$

In particular $\theta^C \in L^1_{loc}(\mathbb{C},L)$.

Proof: The assertions follow from the Fubini theorem and from the above 2.2. QED.

2.4 PROPOSITION: $\theta^C(z)$ is defined $\forall z\in\mathbb{C}-\mathrm{Supp}(\theta)$, where $\mathrm{Supp}(\theta)$ denotes the support of θ. In other words: If $U\subset\mathbb{C}$ is open with $|\theta|(U)=0$ then $\theta^C(z)$ is defined $\forall z\in U$. Moreover $\theta^C|U$ is a holomorphic function.

2.5 PROPOSITION: If $\theta\in ca_*(\mathbb{C})$ then $\theta^C(z)$ is defined $\forall z\in\mathbb{C}$ with $|z|$ sufficiently large, and $\theta^C(z)\to 0$ for $z\to\infty$. If moreover $\theta=fL$ with $f\in L^\infty_*(\mathbb{C},L)$ then $\theta^C(z)$ is defined $\forall z\in\mathbb{C}$, and $\theta^C:\mathbb{C}\to\mathbb{C}$ is continuous and hence uniformly continuous and bounded.

Proof: $\theta^C(z)\to 0$ for $z\to\infty$ is an immediate estimation. Now assume that $\theta=fL$ with $f\in L^\infty_*(\mathbb{C},L)$ and put $E:=\mathrm{Supp}(f)\subset\subset\mathbb{C}$. Then $\theta^C(z)$ is defined $\forall z\in\mathbb{C}$ in view of

$$\theta^{AC}(z) = \int_E \frac{|f(u)|}{|u-z|}\,dL(u) \leq 2\ \mathrm{const}\sqrt{\pi L(E)}<\infty \quad \forall z\in\mathbb{C}.$$

It remains to show that θ^C is continuous. If $a,z\in\mathbb{C}$ with $a\neq z$ then

$$\left|\theta^C(z)-\theta^C(a)\right| = \left|\int f(u)\,\frac{z-a}{(u-z)(u-a)}\,dL(u)\right| \leq \mathrm{const}\int \frac{|z-a|}{|u-z|\,|u-a|}\,dL(u)$$

$$= \mathrm{const}\int_{E-\nabla(z,\frac{1}{2}|z-a|)} \frac{|z-a|}{|u-z|\,|u-a|}\,dL(u) + \mathrm{const}\int_{E\cap\nabla(z,\frac{1}{2}|z-a|)} \frac{|z-a|}{|u-z|\,|u-a|}\,dL(u)$$

The first integral tends $\to 0$ for $z\to a$ after the Lebesgue dominated convergence theorem. In order to estimate the second integral note that for $u\in\nabla(z,\frac{1}{2}|z-a|)$ we have $|z-a| \leq |u-z|+|u-a| \leq \frac{1}{2}|z-a|+|u-a|$ and hence $\frac{1}{2}|z-a|\leq|u-a|$. Thus 2.2 implies that the second integral is $\leq 2\pi|z-a|$. QED.

We turn to the question how to re-obtain θ from its Cauchy transform θ^C. It is convenient to use a bit of distribution theory. The distribution which corresponds to $\theta \in ca(¢)$ is

$$[\theta] : <[\theta],f> = \int f d\theta \quad \forall f \in C_*^\infty(¢).$$

The crucial point is the subsequent special case of the Cauchy formula A.3.5.

2.6 LEMMA: For $f \in C_*^1(¢)$ we have

$$f(z) = -\frac{1}{\pi}\int \frac{1}{u-z}\frac{\partial f}{\partial \bar{z}}(u)dL(u) \quad \forall z \in ¢.$$

In terms of the Cauchy transformation this means that $f = -\frac{1}{\pi}\left(\frac{\partial f}{\partial \bar{z}} L\right)^C$.

2.7 PROPOSITION: We have $[\theta] = -\frac{1}{\pi}\frac{\partial \theta^C}{\partial \bar{z}}$, with differentiation of $\theta^C \in L_{loc}^1(¢,L)$ in the distributional sense.

Proof: The distributional derivative is

$$\langle \frac{\partial \theta^C}{\partial \bar{z}}, f\rangle = -\langle \theta^C, \frac{\partial f}{\partial \bar{z}}\rangle = -\int \theta^C \frac{\partial f}{\partial \bar{z}} dL \quad \forall f \in C_*^\infty(¢).$$

Thus we have to show that $\int f d\theta = \frac{1}{\pi}\int \theta^C \frac{\partial f}{\partial \bar{z}} dL \ \forall f \in C_*^\infty(¢)$. But this is immediate after 2.6 and the Fubini theorem. QED.

From 2.7 a fortified converse to 2.4 can be derived.

2.8 PROPOSITION: If $U \subset ¢$ is open such that $\theta^C|U \in L_{loc}^1(U,L)$ is holomorphic (that means Lebesgue-almost everywhere equal to a holomorphic function) then $|\theta|(U)=0$. In particular $\theta^C|U=0$ implies that $|\theta|(U)=0$.

Proof: From 2.7 we see that $[\theta]|U=0$ or $\int f d\theta = 0 \ \forall f \in C_*^\infty(U)$. It follows that $|\theta|(U)=0$. QED.

2.9 COROLLARY: Define $CS(\theta) := \{z \in ¢: \theta^{AC}(z) < \infty \text{ and } \theta^C(z) \neq 0\}$ the Cauchy support of θ. Then $CS(\theta) \subset ¢$ is a Baire set with $L(CS(\theta)) > 0$ unless $\theta = 0$. Furthermore $Supp(\theta) \subset \overline{CS(\theta)}$.

The remainder of the section will be devoted to a new transformation of technical nature associated with $\theta \in ca(¢)$. Define

$$<\theta>: <\theta>f = (f\theta)^C - f\theta^C \quad \forall f \in B(¢, Baire).$$

It follows that

$$<\theta>f(z) \text{ exists and } = \int \frac{f(u)-f(z)}{u-z} d\theta(u) \quad \forall\ z\in\mathbb{C} \text{ with } \theta^{AC}(z)<\infty.$$

2.10 THEOREM: Assume that $f\in C^1_*(\mathbb{C})$. Then

$$<\theta>f = \frac{1}{\pi}\left(\frac{\partial f}{\partial \bar{z}}\,\theta^C L\right)^C \quad \text{Lebesgue-almost everywhere.}$$

Proof: The Fubini theorem shows that

$$\theta^{AC}(z)<\infty \quad \text{and} \quad \int_{\mathrm{Supp}(f)} \frac{\theta^{AC}(v)}{|v-z|} dL(v)<\infty \quad \text{in Lebesgue-almost all } z\in\mathbb{C}.$$

In these points $z\in\mathbb{C}$ both functions in question are defined, and we have

$$<\theta>f(z) - \frac{1}{\pi}\left(\frac{\partial f}{\partial \bar{z}}\,\theta^C L\right)^C(z) =$$

$$= \int \frac{f(u)-f(z)}{u-z} d\theta(u) - \frac{1}{\pi}\int \frac{\partial f}{\partial \bar{z}}(v)\,\theta^C(v)\frac{1}{v-z}dL(v)$$

$$= \int \frac{f(u)-f(z)}{u-z} d\theta(u) - \frac{1}{\pi}\int\left(\int \frac{\partial f}{\partial \bar{z}}(v)\frac{1}{(u-v)(v-z)}d\theta(u)\right)dL(v)$$

$$= \int\left(\frac{f(u)-f(z)}{u-z} - \frac{1}{\pi}\int \frac{\partial f}{\partial \bar{z}}(v)\frac{1}{(u-v)(v-z)}dL(v)\right)d\theta(u)$$

$$= \int \frac{1}{u-z}\left(f(u)-f(z) - \frac{1}{\pi}\int \frac{\partial f}{\partial \bar{z}}(v)\left(\frac{1}{v-z} - \frac{1}{v-u}\right)dL(v)\right)d\theta(u),$$

which is = 0 in view of 2.6. QED.

2.11 COROLLARY: If $U\subset\mathbb{C}$ is open such that $f\in B(\mathbb{C},\text{Baire})$ is holomorphic in U then $<\theta>f$ is holomorphic in U((that means Lebesgue-almost everywhere equal to a holomorphic function).

Proof: 1) If $f|U=0$ then $<\theta>f|U = (f\theta)^C|U$ and the assertion follows from 2.4. 2) If $f\in C^1_*(\mathbb{C})$ then the assertion follows from 2.4 combined with 2.10. 3) In the general case we fix an open V with $\bar{V}$ compact $\subset U$ and choose a function $h\in C^1_*(\mathbb{C})$ which is =1 on V and =0 outside of some compact subset of U. Then $f = (f-fh) + fh$ where $(f-fh)|V = 0$ and $fh\in C^1_*(\mathbb{C})$ with $fh|V$ holomorphic. Thus 1) and 2) show that $<\theta>f$ is Lebesgue-almost everywhere equal to a holomorphic function in V. QED.

In the next section we shall need the subsequent particular consequence.

2.12 COROLLARY: Assume that $\theta = -\frac{1}{\pi}\frac{\partial h}{\partial \bar{z}} L$ for some $h \in C^1_*(\mathbb{C})$ and $f \in CB(\mathbb{C})$. Then

i) $\langle\theta\rangle f$ exists everywhere and is continuous on $\mathbb{C}$.

ii) If Supp(h) is contained in some $\nabla(a,\varepsilon)$ then we have the supnorm estimation $\|\langle\theta\rangle f\| \leq 2\varepsilon \left\|\frac{\partial h}{\partial \bar{z}}\right\| \omega(f,\nabla(a,\varepsilon))$, where $\omega(f,M)$ denotes the oscillation $= \operatorname{Sup}\{|f(u)-f(v)| : u,v \in M\}$ of f on M.

iii) If $U \subset \mathbb{C}$ is open such that $f|U$ is holomorphic then $\langle\theta\rangle f|U$ is holomorphic.

iv) If $U \subset \mathbb{C}$ is open such that $h|U$ is holomorphic then $fh + \langle\theta\rangle f|U$ is holomorphic.

Proof: In view of 2.5 the functions θ^c and $(f\theta)^c$ and hence $\langle\theta\rangle f$ exist everywhere and are continuous on $\mathbb{C}$ and tend to 0 at infinity. From 2.6 we see that $\theta^c = h$ and hence $fh + \langle\theta\rangle f = (f\theta)^c$. Thus we have i), and iii) from 2.11 and iv) from 2.4. In order to prove ii) we estimate

$$|\langle\theta\rangle f| = \left|\int \frac{f(u)-f(z)}{u-z} d\theta(u)\right| \leq \omega(f,\nabla(a,\varepsilon)) \frac{1}{\pi}\left\|\frac{\partial h}{\partial \bar{z}}\right\| \int_{\nabla(a,\varepsilon)} \frac{dL(u)}{|u-z|}$$

$$\leq 2\varepsilon \left\|\frac{\partial h}{\partial \bar{z}}\right\| \omega(f,\nabla(a,\varepsilon)) \quad \forall z \in \nabla(a,\varepsilon).$$

But since $\langle\theta\rangle f$ is holomorphic outside of $\nabla(a,\varepsilon)$ in view of iv) and tends to 0 at infinity the same estimation is true all over $\mathbb{C}$. QED.

3. Basic Facts on $P(K) \subset R(K) \subset A(K)$

Let K be a fixed compact subset $\neq \emptyset$ of $\mathbb{C}$. For the remainder of the chapter we introduce the notations

- X the boundary ∂K of K,
- K^o the interior of K,
- Ω the complement $\mathbb{C}-K$ of K,
- Ω^∞ the unique unbounded component of Ω.

The bounded components of Ω (if there are any) are called the holes of K. In the present section we start to explore the algebras $P(K) \subset R(K) \subset$

$\subset A(K)$ defined in the Introduction. After some simple direct observations we present several basic applications of the Cauchy transformation.

3.1 EXAMPLE: In the unit disk situation $K = \{z\in\mathbb{C}:|z|\leq 1\} = D\cup S$ we have $A(K) = C\mathrm{Hol}(D)$ per definitionem. The Taylor series expansion shows that $P(K) = R(K) = A(K)$ (see I.3.3.i)).

3.2 PROPOSITION: We have $P(K)=R(K)$ iff $\Omega=\Omega^{\infty}$, that is iff K has no holes.

Proof: We put $F_u = \frac{1}{Z-u} \in R(K)$ for $u\in\Omega$. Define $M:=\{u\in\Omega:F_u\in P(K)\}$. Our aim is to show that $M=\Omega^{\infty}$ which proves the assertion. But this will result from the subsequent four remarks. 1) M is closed in Ω. In fact, an obvious estimation shows that the map $u \mapsto F_u$ is supnorm continuous $\Omega \to C(K)$. 2) M is open. To see this let $a\in M$ and $0<\delta<\mathrm{dist}(a,K)$. For $u\in\nabla(A,\delta)$ then $\left|\frac{u-a}{z-a}\right|\leq\delta/\mathrm{dist}(a,K)<1$ and hence

$$F_u(z) = \frac{1}{z-u} = \frac{1}{z-a}\,\frac{1}{1-\frac{u-a}{z-a}} = F_a(z)\sum_{k=0}^{\infty}(u-a)^k(F_a(z))^k \qquad \forall z\in K,$$

with the series uniformly convergent on K. Thus $\nabla(a,\delta)\subset M$.

3) $M\cap\Omega^{\infty}\neq\emptyset$ and hence $\Omega^{\infty}\subset M$ after 1) and 2). In fact, if $u\in\Omega^{\infty}$ with $|u|>\mathrm{Max}\{|z|:z\in K\}$ then

$$F_u(z) = \frac{1}{z-u} = -\frac{1}{u}\,\frac{1}{1-\frac{z}{u}} = -\frac{1}{u}\sum_{k=0}^{\infty}\frac{z^k}{u^k} \qquad \forall z\in K,$$

with the series uniformly convergent on K. Thus $u\in M$. 4) $M\cap G = \emptyset$ for each hole G of K. In fact, if $u\in G$ were in M and P_n polynomials with $P_n\to F_u$ uniformly on K, then $(Z-u)P_n\to 1$ uniformly on K and hence uniformly on $\overline{G}$ since $\partial G\subset X$. But this is nonsense since $(Z-u)P_n$ vanishes at u. QED.

We next determine the spectra of $P(K)\subset R(K)\subset A(K)$. To start with $P(K)$ define the hull $\hat{K}$ of K to consist of K and of the union of its holes. Thus $\hat{K} = \mathbb{C}-\Omega^{\infty}$ so that $\hat{K}$ is compact with $\partial\hat{K} = \partial\Omega^{\infty} \subset \partial\Omega = X$. It follows that the restriction map $C(\hat{K})\to C(K):f\mapsto f|K$ produces a supnorm isomorphism $P(\hat{K})\to P(\hat{K})|K = P(K)$. Therefore the spectrum $\Sigma(P(K)) = \Sigma(P(\hat{K}))$ contains the point evaluations φ_u in the points $u\in\hat{K}$. As usual we identify $u\equiv\varphi_u$ for $u\in\hat{K}$.

3.3 PROPOSITION: We have $\Sigma(P(K)) = \{\varphi_u : u\in\hat{K}\} = \hat{K}$.

Proof: We have to prove that each $\varphi\in\Sigma(P(K))$ is $=\varphi_u$ for some $u\in\hat{K}$. Fix $\varphi\in\Sigma(P(K))$ and put $u:=\varphi(Z)$. Then $u\in\hat{K}$ since otherwise $u\in\Omega^\infty$ and hence $F_u\in$ $\in P(K)$ after the proof of 3.2 which would lead to the contradiction $1 =$ $= \varphi(1) = \varphi((Z-u)F_u) = \varphi(Z-u)\varphi(F_u) = 0$. Now $\varphi(f) = f(u) = \varphi_u(f)$ for each polynomial f and hence $\forall f\in P(K)$. QED.

3.4 PROPOSITION: We have $\Sigma(R(K)) = \{\varphi_u : u\in K\} = K$.

Proof: As above we have to prove that each $\varphi\in\Sigma(R(K))$ is $=\varphi_u$ for some $u\in K$. Fix $\varphi\in\Sigma(R(K))$ and put $u:= \varphi(Z)$. Then $u\in K$ as above. Now $\varphi(P)=P(u)$ for each polynomial P. Thus if $f = P/Q$ is a rational function with poles off K then $P(u) = \varphi(P) = \varphi(f)\varphi(Q) = \varphi(f)Q(u)$ or $\varphi(f) = f(u)$. Hence $\varphi(f)=$ $= f(u) = \varphi_u(f)$ $\forall f\in R(K)$. QED.

The determination of the spectrum $\Sigma(A(K))$ is much more difficult. Of course it contains the point evaluations φ_u in the points $u\in K$. But the proof that each $\varphi\in\Sigma(A(K))$ is $=\varphi_u$ for some $u\in K$ requires the subsequent deeper fact which will be deduced from 2.12.

3.5 ARENS LEMMA: For each $a\in K$ the set $\{f|K: f\in C_*(\mathbb{C})$ holomorphic in K° and in some neighborhood of $a\}\subset A(K)$ is supnorm dense in $A(K)$ (observe that this is trivial for $a\in K^\circ$!).

Proof: Fix a function $f\in A(K)$ and extend it to some function $f\in C_*(\mathbb{C})$. Now choose for $\varepsilon>0$ some function $h_\varepsilon\in C^1_*(\mathbb{C})$ with $h_\varepsilon=1$ in $\nabla(a,\frac{\varepsilon}{2})$ and $h_\varepsilon=0$ off $\nabla(a,\varepsilon)$ such that

$$\left|\frac{\partial h_\varepsilon}{\partial \bar{Z}}\right| \leq \frac{c}{\varepsilon} \quad \text{with some constant } c>0 \text{ independent of } \varepsilon,$$

and form $\theta_\varepsilon\in ca_*(\mathbb{C})$ as in 2.12. It follows that $f_\varepsilon := f + \langle\theta_\varepsilon\rangle f \in CB(\mathbb{C})$ is holomorphic in K° and in $\nabla(a,\frac{\varepsilon}{2})$ and satisfies $\|f_\varepsilon - f\| \leq 2c\omega(f,\nabla(a,\varepsilon))$. Thus we obtain functions as required upon multiplication of the f_ε with a suitable function $\in C_*(\mathbb{C})$. QED.

3.6 PROPOSITION: We have $\Sigma(A(K)) = \{\varphi_u : u\in K\} = K$.

Proof: Fix $\varphi\in\Sigma(A(K))$ and conclude $u:= \varphi(Z)\in K$ as above. For $f\in A(K)$ the Arens lemma 3.5 provides us with a sequence of functions $f_n\in A(K)$ such that $f(u) + (Z-u)f_n \to f$ uniformly on K. It follows that $f(u) =$

$= \varphi(f(u)) = \varphi(f(u)+(Z-u)f_n) \to \varphi(f)$. Thus $\varphi(f) = f(u) = \varphi_u(f)$ $\forall f\in A(K)$. QED.

The next theorem expresses the decisive relation between R(K) and the Cauchy transformation.

3.7 THEOREM: For $\theta\in ca(K)$ the subsequent properties are equivalent.

i) $\int f d\theta=0$ $\forall f\in R(K)$, that is $\theta\in R(K)^{\perp}$.

ii) $\int f d\theta=0$ for each $f\in C(K)$ which has an extension $F\in C^1(U)$ to some open set $U\supset K$ such that $\frac{\partial F}{\partial \bar{Z}}|K = 0$.

iii) $\theta^C(x)=0$ for all $x\in\Omega$.

Proof: iii)⇒ii) Consider a function $f\in C(K)$ with an extension $F\in C^1(U)$ as above. We can assume that $F\in C^1_*(U)$ and hence $F\in C^1_*(\mathbb{C})$. Then 2.6 combined with the assumptions implies that

$$\int f d\theta = \int F(z)d\theta(z) = -\frac{1}{\pi}\int\left(\frac{\partial F}{\partial \bar{Z}}(u)\frac{1}{u-z}dL(u)\right)d\theta(z) = \frac{1}{\pi}\int\frac{\partial F}{\partial \bar{Z}}(u)\theta^C(u)dL(u) = 0.$$

ii)⇒i) and i)⇒iii) are obvious. QED.

3.8 COROLLARY: Assume that $f\in C(K)$ has an extension $F\in C^1(U)$ to some open set $U\supset K$ such that $\frac{\partial F}{\partial \bar{Z}}|K = 0$. Then $f\in R(K)$.

3.9 COROLLARY (Hartogs-Rosenthal): Assume that $L(K)=0$. Then $R(K) = A(K) = C(K)$.

Proof: For each $\theta\in R(K)^{\perp}$ we have $\theta^C=0$ Lebesgue-almost everywhere from 3.7 and hence $\theta=0$ from 2.8. Thus $R(K) = R(K)^{\perp\perp} = C(K)$. QED.

The next result is the important Bishop localization theorem for R(K). It will have a beautiful application in the proof of 8.4. Observe that the identical result is trivial for A(K).

3.10 THEOREM: Assume that $f\in C(K)$ is such that each point $x\in K$ has a closed neighborhood $U(x)$ with $f|K\cap U(x) \in R(K\cap U(x))$. Then $f\in R(K)$.

Proof: We choose points $x_1,\dots,x_r\in K$ with $K\subset U(x_1)^o\cup\dots\cup U(x_r)^o$ and functions $h_1,\dots,h_r\in C^1_*(\mathbb{C})$ with $h_k=0$ off $U(x_k)$ and $h_1+\dots+h_r=1$ in some neighborhood of K. Now fix $\theta\in R(K)^{\perp}$ so that $\theta\in ca(\mathbb{C})$ lives on K and $\theta^C|\Omega=0$. We have to show that $\int f d\theta=0$. Let us put

$$\theta_k := h_k\theta - \frac{1}{\pi}\frac{\partial h_k}{\partial \bar{z}}\,\theta^C L \in ca(\mathbb{C}), \text{ so that from 2.10}$$

$$\theta_k^C = (h_k\theta)^C - \frac{1}{\pi}\left(\frac{\partial h_k}{\partial \bar{z}}\,\theta^C L\right)^C = h_k\theta^C \quad \text{L-almost everywhere.}$$

It follows that $\theta_k^C|(\mathbb{C}-K\cap U(x_k))=0$ so that θ_k lives on $K\cap U(x_k)$ after 2.8 and $\theta_k \in R(K\cap U(x_k))^{\perp}$ after 3.7. Thus $\int f d\theta_k = 0$. Now $\theta_1+\ldots+\theta_r = \theta$ since θ and θ^C live on K. It follows that $\int f d\theta = 0$. QED.

For the measures $\theta \in A(K)^{\perp} \subset R(K)^{\perp}$ we have the subsequent addendum to 3.7.

<u>3.11</u> <u>PROPOSITION</u>: If $\theta \in A(K)^{\perp}$ then $\theta^C(x)=0$ for Lebesgue-almost all points $x \in X$.

Proof: Fix $f \in B(X,\text{Baire})$ and extend it to $f \in B(\mathbb{C},\text{Baire})$ via $f|(\mathbb{C}-X)=0$. Then $(fL)^C$ is defined and continuous on all of $\mathbb{C}$ after 2.5 and holomorphic on K^o after 2.4. Thus $(fL)^C|K \in A(K)$ and hence

$$\int_X f(z)\theta^C(z)\,dL(z) = \int_X\left(\int_K f(z)\frac{1}{u-z}\,d\theta(u)\right)dL(z) = -\int_K (fL)^C(u)\,d\theta(u) = 0.$$

It follows that $\theta^C=0$ Lebesgue-almost everywhere on X. QED.

We conclude the section with a famous example for $R(K) \neq A(K)$.

<u>3.12</u> <u>EXAMPLE</u> (The Swiss Cheese): We construct a compact subset $K \subset \mathbb{C}$ with $K^o=\emptyset$ such that $R(K) \neq A(K) = C(K)$. 1) The construction starts with the closed unit disk $D \cup S$. We choose points $a_n \in D$ and radii $0 < r_n < 1-|a_n|$ $(n=1,2,\ldots)$ such that

i) $\sum_{n=1}^{\infty} r_n < 1$, ii) $\nabla(a_n,r_n) \cap \nabla(a_m,r_m) = \emptyset$ whenever $n \neq m$,

iii) $K := (D\cup S) - \bigcup_{n=1}^{\infty} V(a_n,r_n)$ has no interior points.

To this end consider a dense sequence of points $z_n \in D$ $(n=1,2,\ldots)$. Put $a_1=z_1$ and fix $0<r_1<\text{Min}(\frac{1}{2},1-|a_1|)$. Then take the smallest index p with $z_p \notin \nabla(a_1,r_1)$, put $a_2=z_p$ and fix $0<r_2<\text{Min}(\frac{1}{4},1-|a_2|)$. Now proceed via induction.

2) In view of $K^{\circ}=\emptyset$ we have $A(K)=C(K)$. In order to prove $R(K)\neq A(K)$ we exhibit a nonzero measure $\theta\in R(K)^{\perp}$. Let σ and σ_n denote one-dimensional Lebesgue measure (= arc length) on $S=\partial V(0,1)$ and $S_n:=\partial V(a_n,r_n)$, and N and N_n the respective outer normal vector functions. Define

$$\theta = \sum_{n=1}^{\infty} N_n\sigma_n - N\sigma \quad \text{so that } 0\neq\theta\in ca(K).$$

For $z\in\Omega$ we have

$$\theta^C(z) = \sum_{n=1}^{\infty} \int_{S_n} \frac{1}{u-z} N_n(u)\,d\sigma_n(u) - \int_S \frac{1}{u-z} N(u)\,d\sigma(u).$$

If $|z|>1$ then all terms of the second member are $=0$ after the Cauchy theorem A.3.3. If $|z|\leq 1$ then $z\in V(a_m,r_m)$ for a certain $m\geq 1$: then the term of index m is $=2\pi$ after A.3.4 as is the last term of the second member, while the other terms $n\neq m$ are all $=0$. It follows that $\theta^C(z)=0$ for all $z\in\Omega$ and hence $\theta\in R(K)^{\perp}$. QED.

4. On the annihilating and the representing Measures for R(K) and A(K)

We retain the fixed compact subset $K\neq\emptyset$ of $\mathbb{C}$ and continue to explore the algebras $R(K)\subset A(K)$. We can expect to comprise $P(K)$ without explicit mention since under the restriction supnorm isomorphism combined with 3.2 we have $P(K) = P(\hat{K})|K \cong P(\hat{K}) = R(\hat{K})$. While until now we did not use our abstract theories the main results of the present section depend on the fundamental lemma II.1.4. Let us adopt the convention that R denotes either $R(K)$ or $A(K)$ in results which are true in both cases. As usual we write $M(R,u) = M(R,\varphi_u)$ for $u\in K$.

<u>4.1</u> <u>LEMMA</u>: Assume that $\theta\in R^{\perp}$ and $a\in K$ such that $\theta^{AC}(a)<\infty$.

i) If $\theta^C(a)=0$ then $\frac{1}{Z-a}\theta\in R^{\perp}$ as well.

ii) If $\theta^C(a)\neq 0$ then there exists an $m\in M(R,a)$ with $m\ll\theta$.

Proof: The functions $f\in R$ of the form $f = f(a) + (Z-a)F$ with $F\in R$ are dense in R. In the case $R=A(K)$ this follows from the Arens lemma 3.5 while in the case $R=R(K)$ one takes the rational functions $f = \frac{P}{Q}$ with poles off K in view of the equation

$$f = \frac{P(a)}{Q(a)} + (Z-a)\frac{1}{Q}(G - \frac{P(a)}{Q(a)}H) \text{ with } P=P(a)+(Z-a)G,\ Q=Q(a)+(Z-a)H.$$

Let us put $\sigma := \frac{1}{Z-a}\theta \in ca(K)$. For the functions $f \in R$ in question then

$$\int f d\sigma = f(a)\int d\sigma + \int F d\theta = f(a)\int d\sigma = f(a)\theta^C(a).$$

Thus the last equation is true for all $f \in R$. In the case $\theta^C(a)=0$ we are done. And in the case $\theta^C(a) \neq 0$ we obtain from II.1.6 a measure $m \in M(R,a)$ with $m << \sigma << \theta$. QED.

If $\theta \in R^{\perp}$ is nonzero then $CS(\theta)$ is $\neq \emptyset$ after 2.9 and contained in K after 3.7. Thus we obtain the subsequent fundamental consequence.

<u>4.2</u> <u>CONSEQUENCE</u>: If $\theta \in R^{\perp}$ is singular to all $m \in M(R)$ with $Supp(m) \subset Supp(\theta)$ then $\theta = 0$.

In the sequel we shall combine the above 4.1 with an elementary consideration on the supports of the measures in question.

<u>4.3</u> <u>REMARK</u>: i) Let $a \in X$. For each $m \in M(R,a)$ then $a \in Supp(m)$.

ii) Let $a \in K^o$ and G be the component of K^o which contains a. For each $m \in M(R,a)$ with $m \neq \delta_a$ and $G \cap Supp(m)$ discrete then $\partial G \subset Supp(m)$.

Proof: 1) Fix $a \in K$ and $m \in M(R,a)$ such that $a \notin Supp(m)$. Let S be the component of $\mathbb{C}-Supp(m)$ with $a \in S$. Put $\theta = \delta_a - m$ and consider the Cauchy transform

$$\theta^C:\ \theta^C(z) = \frac{1}{z-a} - \int \frac{dm(u)}{u-z} \quad \forall\ z \in (\mathbb{C}-Supp(m))-\{a\}.$$

Then θ^C is holomorphic on $(\mathbb{C}-Supp(m))-\{a\}$. Since $S-\{a\}$ is connected and $\theta^C(z) \to \infty$ for $z \to a$ we conclude that $\theta^C(z) \neq 0$ $\forall z \in S-\{a\}$ except on some discrete subset of $S-\{a\}$. On the other hand $\theta^C(z)=0$ $\forall\ z \in \Omega \subset (\mathbb{C}-Supp(m))-\{a\}$ after 3.7 since $\theta \in R^{\perp} \subset R(K)^{\perp}$. It follows that $S \cap \Omega = \emptyset$ and hence $S \cap X = \emptyset$. In particular $a \in X$ is impossible which proves i). 2) Let now G be the component of K^o which contains a and assume that $G \cap Supp(m)$ is discrete. Since $S \subset K^o$ is connected with $a \in S \cap G$ we conclude that $S \subset G$ and hence $S \subset G - G \cap Supp(m)$. Since $G - G \cap Supp(m)$ is connected in view of the discreteness assumption we see from $S \subset G - G \cap Supp(m) \subset \mathbb{C} - Supp(m)$ that $S = G - G \cap Supp(m)$. We conclude that $\partial G \in \partial S$. In fact, for $x \in \partial G$ there are $x_n \in G$ with $x_n \to x$, and we can assume that $x_n \notin Supp(m)$ or $x_n \in S$, so that $x \in \overline{S}$ and hence $x \in \partial S$ since $x \notin S$

in view of $x \notin G$. Thus we have shown that $\partial G \subset \partial S \subset \partial(\mathbb{C}-\mathrm{Supp}(m)) = \partial \mathrm{Supp}(m) \subset \mathrm{Supp}(m)$. 3) Let us now prove ii). In the case $a \notin \mathrm{Supp}(m)$ the assertion has been shown in 2). If $a \in \mathrm{Supp}(m)$ then in view of the discreteness assumption $0<m(\{a\})<1$ and $m = m(\{a\})\delta_a + (1-m(\{a\}))\sigma$ where $\sigma \in M(R,a)$ satisfies $\mathrm{Supp}(\sigma) = \mathrm{Supp}(m)-\{a\}$. Thus $\partial G \subset \mathrm{Supp}(\sigma)$ from 2) and hence $\partial G \subset \mathrm{Supp}(m)$ as well. QED.

We introduce one more notation. For a band $S \subset ca(K)$ we define the trace $\mathrm{Tr}(S) = \mathrm{Tr}(S,R) \subset K$ to consist of those $z \in K$ for which $M(R,z) \cap S \neq \emptyset$. Of course $\mathrm{Tr}(S)$ can be void. Note that 4.1.ii) can be formulated $CS(\theta) \subset \mathrm{Tr}(\{\theta\}^{\vee})$ for all $\theta \in R^{\perp}$.

<u>4.4</u> <u>PROPOSITION</u>: Let $S \subset ca(K)$ be a band such that $K^{o} \cap \mathrm{Supp}(\sigma)$ is discrete $\forall \sigma \in S$. Then there exists some $\sigma \in S$ with $\overline{\mathrm{Tr}(S)} \cap X \subset \mathrm{Supp}(\sigma)$.

Proof: 1) It suffices to prove that to each point $a \in \overline{\mathrm{Tr}(S)} \cap X$ there exists some $\sigma \in S$ with $a \in \mathrm{Supp}(\sigma)$. In fact, if $\{a_n : n=1,2,\ldots\}$ is dense in $\overline{\mathrm{Tr}(S)} \cap X$ and if $\sigma_n \in S$ with $a_n \in \mathrm{Supp}(\sigma_n)$ then $\sigma := \sum_{n=1}^{\infty} 2^{-n}|\sigma_n| \in S$ with $\overline{\mathrm{Tr}(S)} \cap X \subset \mathrm{Supp}(\sigma)$. 2) Let us fix $a \in \overline{\mathrm{Tr}(S)} \cap X$ and choose $a(n) \in \mathrm{Tr}(S)$ with $a(n) \to a$ and $m_n \in M(R,a(n)) \cap S$. Then $\sigma := \sum_{n=1}^{\infty} 2^{-n} m_n \in S$. For each $n \geq 1$ we have one of the three situations

i) $a(n) \in X$,

ii) $a(n) \in$ some component G_n of K^{o} and $m_n = \delta_{a(n)}$,

iii) $a(n) \in$ some component G_n of K^{o} and $m_n \neq \delta_{a(n)}$.

In case i) we have $a(n) \in \mathrm{Supp}(m_n) \subset \mathrm{Supp}(\sigma)$ after 4.3.i). In case ii) $a(n) \in \mathrm{Supp}(m_n) \subset \mathrm{Supp}(\sigma)$ is trivial. In case iii) we conclude $\partial G_n \subset \mathrm{Supp}(m_n)$ from 4.3.ii). Since $a(n) \in G_n$ but $a \notin G_n$ there exists a point $b(n) = ta+(1-t)a(n) \in \partial G_n \subset \mathrm{Supp}(m_n) \subset \mathrm{Supp}(\sigma)$ for some $0<t\leq 1$. In this case take $b(n)$ instead of $a(n)$. Then the modified sequence $\to a$ as well. Hence $a \in \mathrm{Supp}(\sigma)$. QED.

We combine the above 4.4 with the fundamental lemma consequence 4.1 to obtain the subsequent rather precise information.

<u>4.5</u> <u>THEOREM</u>: Assume that $\theta \in R^{\perp}$ such that $K^{o} \cap \mathrm{Supp}(\theta)$ is discrete. Then $\mathrm{Supp}(\theta) \cap X = \overline{CS(\theta)} \cap X = \overline{\mathrm{Tr}(\{\theta\}^{\vee})} \cap X$.

Proof: 1) We have $\mathrm{Supp}(\theta) \subset \overline{CS(\theta)}$ after 2.9. 2) We have $CS(\theta) \subset \subset \mathrm{Tr}(\{\theta\}^{\vee})$ after 4.1.ii) as remarked above. 3) From 4.4 applied to $S=\{\theta\}^{\vee}$ we see that $\overline{\mathrm{Tr}(\{\theta\}^{\vee})}\cap X$ is $\subset \mathrm{Supp}(\sigma)$ for some $\sigma \in S$ and hence $\subset \mathrm{Supp}(\theta)$. Combine all these facts to obtain the assertion. QED.

The above considerations lead to the fundamental idea to form the restriction to compact sets Y with $X\subset Y\subset K$ and in particular to X itself. In view of the maximum modulus principle the restriction map $C(K)\rightarrow C(Y)$: $f \mapsto f|Y$ is supnorm isometric on $A(K)$. Thus we have the supnorm isomorphisms

$$\begin{array}{ccccccc} 1 \in P(K) & \subset & R(K) & \subset & A(K) & \subset & C(K) \\ \wr\wr & & \wr\wr & & \wr\wr & & \downarrow \\ 1 \in P(K)|Y & \subset & R(K)|Y & \subset & A(K)|Y & \subset & C(Y). \end{array}$$

It follows that the three restriction algebras are supnorm closed and hence are function algebras on Y in the sense of Chapter III (which should not be confused with $P(Y)\subset R(Y)\subset A(Y)\subset C(Y)$ defined on Y itself: one verifies that $P(K)|Y=P(Y)$ but $R(K)|Y=R(Y)$ and $A(K)|Y=A(Y)$ iff $Y=K$). Let us adopt the convention that P denotes either $P(K)$ or $R(K)$ or $A(K)$. The restriction of course preserves the dual Banach space $P'=(P|Y)'$ and the spectrum $\Sigma(P)=\Sigma(P|Y)$. But the classes of measures associated with P become drastically smaller: Instead of $P^{\perp} \subset ca(K)$ we have $(P|Y)^{\perp} = P^{\perp}\cap ca(Y) \subset ca(Y)$, and for $\varphi\in\Sigma(P)=\Sigma(P|Y)$ we have $M(P,\varphi)\subset \mathrm{Prob}(K)$ reduced to $M(P|Y,\varphi) = M(P,\varphi)\cap \mathrm{Prob}(Y) \subset \mathrm{Prob}(Y)$. This reduction will be decisive in order to obtain the final results of the present chapter.

An immediate application is the subsequent specialization of 4.2 the importance of which is evident in view of Section 1.

<u>4.6 THEOREM</u> (Wilken): $(R|Y)^{\perp}\cap M(R|Y)^{\wedge}=0$ for each compact Y with $X\subset Y\subset K$.

We conclude with an important remark on the Gleason part theme which will be dealt with in the next section. The restriction under consideration preserves the Gleason part decomposition of $\Sigma(P)=\Sigma(P|Y)$: the Gleason part of $\varphi\in\Sigma(P)=\Sigma(P|Y)$ each time consists of the $\psi\in\Sigma(P)=\Sigma(P|Y)$ with $\|\varphi-\psi\|<2$ so that indeed $Gl(P,\varphi) = Gl(P|Y,\varphi)$. But after III.4.4 we have the equivalent characterizations $M(P,\varphi)^{\vee} = M(P,\psi)^{\vee}$ and $M(P|Y,\varphi)^{\vee} = M(P|Y,\psi)^{\vee}$. These two characterizations appear to be different but are in fact identical in view of the above.

5. On the Gleason Parts for R(K) and A(K)

We retain the fixed compact subset $K \neq \emptyset$ of $\mathbb{C}$ and the notation R for either R(K) or A(K). We consider the Gleason part decomposition of $K = \Sigma(R)$ for R. As before let $Gl(R,u) = Gl(R,\varphi_u)$ denote the Gleason part of $u \in K$. Note that the identification map $K \to \Sigma(R): u \mapsto \varphi_u$ is continuous in the Gelfand topology of $\Sigma(R)$ which therefore coincides with the metric topology of $K \subset \mathbb{C}$. Thus after III.4.5 each Gleason part of K for R is a countable union of compact sets and hence a Baire set.

5.1 REMARK: For $u,v \in K$ we have $\|\varphi_u - \varphi_v\|_{R(K)'} \leq \|\varphi_u - \varphi_v\|_{A(K)'}$. Therefore each Gleason part for R(K) is the union of one or several Gleason parts for A(K).

5.2 REMARK: Each component of K^o is contained in some Gleason part for R. Thus each Gleason part $P \subset K$ for R satisfies $\partial P \subset X$.

Note that there are examples which show that one Gleason part for R can contain several components of K^o (see GAMELIN [1969] p.145).

Proof of 5.2: Let G be a component of K^o and $a \in G$. If $\delta > 0$ is such that $V(a,\delta) \subset G$ then for $f \in R$ with $\|f\| \leq 1$ the H.A.Schwarz lemma (see CARATHÉODORY [1950] Vol.I p.137) implies that

$$|f(u)-f(a)| = \left|f\left(a+\delta\frac{u-a}{\delta}\right)-f(a)\right| \leq \frac{2}{\delta}|u-a| \quad \forall u \in V(a,\delta).$$

Thus we have $\|\varphi_u - \varphi_a\| \leq \frac{2}{\delta}|u-a| < 2 \ \forall u \in V(a,\delta)$ so that $V(a,\delta) \subset Gl(R,a)$. It follows that $G \subset Gl(R,a)$ since G is connected and the Gleason relation is transitive. QED.

The next results depend on the fundamental lemma 4.1.

5.3 REMARK: For $a \in K$ and $m \in M(R,a)$ we have $CS(\delta_a - m) \subset Gl(R,a)$.

Proof: If $z \in CS(\delta_a - m)$ then after 4.1.ii) there exists some $\sigma \in M(R,z)$ with $\sigma \ll \delta_a - m$ and hence with $\sigma \ll \frac{1}{2}(\delta_a + m) \in M(R,a)$. It follows that $z \in Gl(R,a)$. QED.

5.4 PROPOSITION: Let $a \in K$ and $P := Gl(R,a)$. For each $m \in M(R,a)$ then $Supp(m) \subset \overline{P}$. Furthermore $m \in M(R(\overline{P}),a)$.

Proof: The assertions are clear if $m=\delta_a$ so that we assume $m\neq\delta_a$. Then $Supp(m) \subset Supp(\delta_a-m) \subset \overline{CS(\delta_a-m)} \subset \overline{P}$ after 2.9 and 5.3. To prove the second assertion note that $u\in ¢-\overline{P}$ implies that $(\delta_a-m)^{AC}(u)<\infty$ and hence $(\delta_a-m)^C(u)=0$ since otherwise $u\in CS(\delta_a-m)\subset P$ after 5.3. Thus 3.7 implies that $\delta_a-m\in R(\overline{P})^{\perp}$ and hence $m\in M(R(\overline{P}),a)$. QED.

The first assertion in 5.4 cannot be improved to the assertion that $m(P)=1$, and it cannot even be claimed that $m(P)>0$. As an example take the unit disk $K = D\cup S$. For $a\in D$ we have $P:=Gl(R,a)=D$ from II.4.6 while $m:=P(a,.)\lambda$ is $\in M(R,a)$ with support $=S$. Thus $m(P)=0$.

Let us insert a weakened converse to the first assertion in 5.4 which depends on the elementary fact 4.4.

<u>5.5</u> <u>PROPOSITION</u>: Let $a\in K$ and $P:=Gl(R,a)$. There exists an $m\in M(R,a)$ such that $Supp(m) = \overline{P}\cap X$.

Proof: From 4.4 applied to $S:=M(R|X,a)^{\vee}\subset ca(X)\subset ca(K)$ we obtain some $\sigma\in S$ with $\overline{Tr(S)}\cap X \subset Supp(\sigma)$. And this is $\subset Supp(m)$ for each $m\in M(R|X,a)$ such that $\sigma<<m$. Now $Tr(S) = Gl(R|X,a) = Gl(R,a) = P$ per definitionem and after the final remark in Section 4. It follows that $\overline{P}\cap X\subset Supp(m)\subset X$. Combine this with 5.4 to obtain the result. QED.

<u>5.6</u> <u>PROPOSITION</u>: For each Gleason part $P\subset K$ for R the closure $\overline{P}$ is connected.

Proof: Assume that $\overline{P}$ is not connected. Then $\overline{P} = S\cup T$ with nonvoid closed S,T with $S\cap T = \emptyset$. We have $P\cap S, P\cap T \neq \emptyset$ since $P\cap S=\emptyset \Rightarrow P\subset T \Rightarrow \overline{P}\subset T \Rightarrow S=\emptyset$. Choose $a\in P\cap S, b\in P\cap T$ and measures $\sigma\in M(R,a), \tau\in M(R,b)$ such that $\sigma<<\tau$. Then $Supp(\sigma), Supp(\tau)\subset\overline{P}$ and $\sigma\in M(R(\overline{P}),a), \tau\in M(R(\overline{P}),b)$ after 5.4. Now the characteristic function χ_S is $\in R(\overline{P})$ in view of 3.8. It follows that

$$1 = \chi_S(a) = \int\chi_S d\sigma = \sigma(S) \quad \text{and} \quad 0 = \chi_S(b) = \int\chi_S d\tau = \tau(S),$$

which is a contradiction to $\sigma<<\tau$. QED.

Another consequence of 5.3 is of particular importance: If $M(R,a)$ is $\neq\{\delta_a\}$ then $Gl(R,a)$ has positive Lebesgue measure after 2.9. We thus obtain the subsequent theorem.

5.7 THEOREM: For $a \in X$ the subsequent properties are equivalent.

i) $Gl(R,a) \neq \{a\}$.   iii) $M(R,a) \neq \{\delta_a\}$.

ii) $L(Gl(R,a)) > 0$.   iv) $M(R|X,a) \neq \{\delta_a\}$.

Note that for $a \in K^o$ properties i)ii) are fulfilled as obvious consequences of 5.2.

Proof: iv)⇒iii) and ii)⇒i) are obvious and iii)⇒ii) follows from 5.3 as explained above. i)⇒iv) Choose $b \in Gl(R,a) = Gl(R|X,a)$ different from a and then $\tau \in M(R|X,b)$ and $\sigma \in M(R|X,a)$ such that $\tau << \sigma$. Then of course $\sigma \neq \delta_a$. QED.

Thus we see that the Gleason parts $P \subset K$ for R can only be of two types: The so-called nontrivial parts which have $L(P) > 0$, and the so-called trivial parts which are $P = \{a\}$ with $a \in X$ and $M(R,a) = \{\delta_a\}$. And there can be at most a countable number of nontrivial parts.

In the remainder of the section we consider the collection $\Gamma(R)$ of all Gleason parts for R. The basis is the final part of Section II.4. Recall for $P \in \Gamma(R)$ the associated band $b(P) := M(R,u)^\vee$ for any $u \in P$. Thus for trivial $P = \{a\}$ with $a \in X$ we have $b(P) = ¢\delta_a$. Therefore and in view of 4.6 the series expansion II.4.4 of the measures $\theta \in R^\perp$ simplifies as follows: Each $\theta \in R^\perp$ admits the representation

$$\theta = \sum_{\substack{P \in \Gamma(R) \\ \text{nontrivial}}} \theta_P \quad \text{with} \quad \sum_{\substack{P \in \Gamma(R) \\ \text{nontrivial}}} \|\theta_P\| < \infty,$$

where $\theta_P := \langle b(P) \rangle \theta \in R^\perp$.

5.8 CONSEQUENCE: The subsequent assertions are equivalent.
i) $R = C(K)$. ii) All Gleason parts $P \in \Gamma(R)$ are trivial. iii) $K^o = \emptyset$ and $M(R,a) = \{\delta_a\}$ $\forall a \in X = K$.

5.9 THEOREM: There exists a correspondence which associates with each nontrivial Gleason part $P \in \Gamma(R)$ a Baire set $E(P) \subset K$ such that

i) $P \subset E(P) \subset \overline{P}$,

ii) $E(P) \cap E(Q) = \emptyset$ whenever $P \neq Q$,

iii) if $a \in P$ then $m(E(P)) = 1$ for all $m \in M(R,a)$,

iv) for each $\theta \in R^{\perp}$ we have $\theta_P = \chi_{E(P)}\theta$ and

$$(\theta_P)^C(x) = \left\{ \begin{array}{ll} 0 & \text{if } x \notin P \\ \theta^C(x) & \text{if } x \in P \end{array} \right\} \; \forall \; x \in K \text{ with } \theta^{AC}(x) < \infty.$$

Proof: We can assume that there are at least two nontrivial Gleason parts. 1) For each ordered pair of different nontrivial $P,Q \in \Gamma(R)$ we form disjoint Baire sets $E_P^Q, E_Q^P \subset K$ such that

$$\sigma(E_P^Q) = 1 \quad \forall \; u \in P \text{ and } \sigma \in M(R,u),$$
$$\tau(E_Q^P) = 1 \quad \forall \; v \in Q \text{ and } \tau \in M(R,v).$$

To achieve this choose $a \in P$, $b \in Q$ and apply III.2.1 to $M(R,a), M(R,b) \subset \subset Pos(K)$ to obtain disjoint Baire sets $U,V \subset K$ such that $\sigma(U)=1$ $\forall \sigma \in M(R,a)$ and $\tau(V)=1$ $\forall \tau \in M(R,b)$. Thus we can take $U=:E_P^Q$ and $V=:E_Q^P$. 2) We have $P \subset E_P^Q$ since $u \in P \Rightarrow \delta_u \in M(R,u) \Rightarrow \delta_u(E_P^Q)=1$ or $u \in E_P^Q$. 3) For each nontrivial $P \in \Gamma(R)$ define $E(P)$ to be the intersection of the E_P^Q for all nontrivial $Q \neq P$ in $\Gamma(R)$ (the number of which is at most countable), intersected with $\overline{P}$. Then the $E(P) \subset K$ are Baire sets which fulfill i)ii)iii). 4) Let $\theta \in R^{\perp}$. For nontrivial $P \in \Gamma(R)$ then $\theta_P \in b(P) = M(R,u)^{\vee}$ for any $u \in P$ so that $\theta_P <<$ some $\sigma \in M(R,u)$. Thus $|\theta_P|(K-E(P))=0$. Then the above series expansion of θ shows that $\theta(B)=\theta_P(B)$ $\forall$ Baire sets $B \subset E(P)$. It follows that $\theta_P = \chi_{E(P)}\theta$ as claimed in iv). To prove the last assertion in iv) fix $x \in K$ with $\theta^{AC}(x)<\infty$. Then $(\theta_P)^{AC}(x)<\infty$ as well. If $(\theta_P)^C(x) \neq 0$ then after 4.1 there exists some $\sigma \in M(R,x)$ with $\sigma << \theta_P \in b(P)$ so that $x \in P$. Thus $(\theta_P)^C(x)=0$ if $x \notin P$. Then the above series expansion of θ shows that $\theta^C(x)=(\theta_P)^C(x)$ for $x \in P$. QED.

6. The Logarithmic Transformation of Measures and the Logarithmic Capacity of Planar Sets

The present section is to prepare the proofs of the Walsh-type theorems of Section 7. The first half is devoted to the logarithmic transformation of Baire measures in the complex plane. Its properties correspond to those of the Cauchy transformation except that its definition has to be restricted to measures of compact support. It is thus convenient to define

$$ca_*(E) = \{\theta \in ca(\mathbb{C}) : \mathrm{Supp}(\theta) \text{ is compact and } \subset E\} \quad \text{for nonvoid } E \subset \mathbb{C}.$$

Of course $ca_*(E) = ca(E)$ if E is compact. The second half of the section is devoted to a short ab-ovo discussion of the logarithmic potential of planar sets to the extent which will be needed in the next section.

The logarithmic transformation is defined to transform the measure $\theta \in ca_*(\mathbb{C})$ into the function

$$\theta^L : \theta^L(z) = \int \log|u-z| d\theta(u),$$

defined for those $z \in \mathbb{C}$ where $\theta^{AL}(z) := \int |\log|u-z|| d|\theta|(u) < \infty$.

In the case $\theta \in Pos_*(\mathbb{C})$ we define $\theta^L(z) = \int \log|u-z| d\theta(u)$ $\forall z \in \mathbb{C}$ in the extended sense $-\infty \leq \theta^L(z) < \infty$. For the first half of the section we fix a measure $\theta \in ca_*(\mathbb{C})$.

6.1 LEMMA: For each bounded Baire set $E \subset \mathbb{C}$ we have

$$\int_E |\log|u-z|| dL(u) \leq \sqrt{\pi L(E)} + L(E)(\frac{1}{2} + \sup_{u \in E}|u-z|) \quad \forall z \in \mathbb{C}.$$

Proof: Take an $R>0$ with $E \subset \nabla(z,R)$. Then from 2.1 we have

$$\begin{aligned} I := \int_E |\log|u-z|| dL(u) &= \int_{]0,\infty[} L(\{u \in E : |\log|u-z|| \geq t\}) dt \\ &= \int_{]0,\infty[} L(\{u \in E : \log|u-z \geq t\}) dt + \int_{]0,\infty[} L(\{u \in E : \log|u-z| \leq -t\}) dt \\ &= \int_1^\infty L(\{u \in E : |u-z| \geq s) \frac{ds}{s} + \int_0^1 L(\{u \in E : |u-z| \leq s\}) \frac{ds}{s} . \end{aligned}$$

The first integral is $\leq L(E) \log \mathrm{Max}(1,R) \leq L(E) R$. The second integral can for each $0<c\leq 1$ be estimated

$$\leq \int_0^c \pi s^2 \frac{ds}{s} + \int_c^1 L(E) \frac{ds}{s} = \frac{\pi}{2} c^2 + L(E) \log\frac{1}{c} .$$

In the case $L(E) \geq \pi$ put $c=1$ to obtain $\leq \frac{\pi}{2} \leq \frac{1}{2} L(E)$. In the case $0<L(E) \leq \pi$ put $c = (\frac{1}{\pi} L(E))^{1/2}$ to obtain $\leq \frac{1}{2} L(E) + \sqrt{\pi L(E)}$. QED.

6.2 PROPOSITION: i) The function $\theta^{AL} : \mathbb{C} \to [0,\infty]$ is Baire measurable.

ii) $N := \{z \in \mathbb{C} : \theta^{AL}(z) = \infty\}$ has $L(N) = 0$.

iii) The function $\theta^L:\mathbb{C}-N\to\mathbb{C}$ is Baire measurable.

iv) For each bounded Baire set $E\subset\mathbb{C}$ we have

$$\int_E |\theta^L(z)|dL(z) \leq \int_E \theta^{AL}(z)dL(z) \leq \Big(\sqrt{\pi L(E)} + L(E)(\tfrac{1}{2} + \sup_{\substack{u\in E\\ z\in \mathrm{Supp}(\theta)}} |u-z|)\Big)\|\theta\|.$$

In particular $\theta^L \in L^1_{loc}(\mathbb{C},L)$.

Proof: The assertions follow from the Fubini theorem and from the above 6.1. QED.

6.3 PROPOSITION: $\theta^L(z)$ is defined $\forall z\in\mathbb{C}-\mathrm{Supp}(\theta)$. In other words: If $U\subset\mathbb{C}$ is open with $|\theta|(U)=0$ then $\theta^L(z)$ is defined $\forall z\in U$. Moreover $\theta^L|U$ is a harmonic function.

6.4 PROPOSITION: $\theta^L(z)$ is defined $\forall z\in\mathbb{C}$ with $|z|$ sufficiently large. Moreover $\theta^L(z)-\theta(\mathbb{C})\log|z|\to 0$ for $z\to\infty$. And if $\theta=fL$ with $f\in L^\infty_*(\mathbb{C},L)$ then $\theta^L(z)$ is defined $\forall z\in\mathbb{C}$ and $\theta^L:\mathbb{C}\to\mathbb{C}$ is uniformly continuous.

Proof: The first assertion is obvious. 1) Fix $R>0$ with $\mathrm{Supp}(\theta)\subset\nabla(0,R)$ and take $z\in\mathbb{C}$ with $|z|>R$. The calculus inequality $\frac{x-1}{x} \leq \log x \leq x-1$ $\forall x>0$ implies that

$$\Big|\log\Big|1-\frac{u}{z}\Big|\Big| \leq \frac{R}{|z|-R} \qquad \forall u\in\mathrm{Supp}(\theta),$$

$$|\theta^L(z)-\theta(\mathbb{C})\log|z|| = \Big|\int\log\Big|1-\frac{u}{z}\Big|d\theta(u)\Big| \leq \frac{R}{|z|-R}\|\theta\|.$$

This proves the second assertion. 2) Suppose now that $\theta=fL$ with $f\in L^\infty_*(\mathbb{C},L)$ and put $E:=\mathrm{Supp}(f)$. After 6.1 then

$$\theta^{AL}(z) = \int|\log|u-z||\,|f(u)|dL(u) \leq c\int_E|\log|u-z||dL(u)<\infty \quad \forall z\in\mathbb{C}.$$

3) Next we show that θ^L is continuous. Fix $a\in\mathbb{C}$ and take $z\neq a$. Then

$$\theta^L(z)-\theta^L(a) = \int_E(\log|u-z|-\log|u-a|)f(u)dL(u),$$

$$|\theta^L(z)-\theta^L(a)| \leq c\int_{\nabla(a,2|z-a|)}(|\log|u-z||+|\log|u-a||)dL(u)$$

$$+\, c\int_{E-\nabla(a,2|z-a|)}\Big|\log\Big|1-\frac{z-a}{u-a}\Big|\Big|dL(u).$$

After 6.1 the first integral is $\leq 4\pi|z-a| + 4\pi|z-a|^2 + 20\pi|z-a|^3$ which tends $\rightarrow 0$ for $z\rightarrow a$. And the second integral is

$$= \int_E F_z(u)dL(u) \quad \text{with} \quad F_z(u) = \begin{cases} 0 & \text{for } |u-a|\leq 2|z-a| \\ \left|\log\left|1-\frac{z-a}{u-a}\right|\right| & \text{for } |u-a|>2|z-a| \end{cases}.$$

Observe that $0\leq F_z(u)\leq 1$ and $F_z(u)\rightarrow 0$ for $z\rightarrow a$ for all complex $u\neq a$. Hence the second integral tends $\rightarrow 0$ for $z\rightarrow a$ as well. The assertion follows. 4) It remains to show that θ^L is uniformly continuous. But θ^L is uniformly continuous on compact sets $\subset ¢$ after 3) and $z\mapsto \log|z|$ is uniformly continuous on $\{z\in ¢: |z|\geq r\}$ $\forall r>0$. Hence the assertion follows from 1). QED.

In order to re-obtain θ from its logarithmic transform θ^L we invoke the fundamental representation formula VI.5.9 which corresponds to the Cauchy transformation formula 2.6.

6.5 LEMMA: For $f\in C_*^2(¢)$ we have

$$f(z) = \frac{1}{2\pi}\int(\log|u-z|)\Delta f(u)dL(u) \quad \forall z\in ¢.$$

In terms of the logarithmic transformation this means $f = \frac{1}{2\pi}((\Delta f)L)^L$.

6.6 PROPOSITION: We have $[\theta] = \frac{1}{2\pi}\Delta(\theta^L)$, as earlier with $[\theta]$:= the distribution which corresponds to $\theta\in ca_*(¢)$ and with differentiation of $\theta^L \in L^1_{loc}(¢,L)$ in the distributional sense.

Proof: The distributional derivative is

$$\langle\Delta(\theta^L),f\rangle = \langle\theta^L,\Delta f\rangle = \int\theta^L\Delta f dL \quad \forall f\in C_*^\infty(¢).$$

Thus we have to show that $\int f d\theta = \frac{1}{2\pi}\int\theta^L\Delta f dL$ $\forall f\in C_*^\infty(¢)$. This is clear once we know that the Fubini theorem can be applied. But for $S:=\mathrm{Supp}(\theta)$ and $E:=\mathrm{Supp}(f)$ we see from 6.1 that

$$\int_S\left(\int_E |\log|u-z||\,|\Delta f(u)|dL(u)\right)d|\theta|(z)$$

$$\leq c\|\theta\|\left(\sqrt{\pi L(E)} + L(E)\left(\tfrac{1}{2} + \sup_{\substack{u\in E\\ z\in S}}|u-z|\right)\right) < \infty. \quad \text{QED.}$$

From 6.6 we obtain a fortified converse to 6.3 as in Section 2.

6.7 PROPOSITION: If $U \subset \mathbb{C}$ is open such that $\theta^L|U \in L^1_{loc}(U,L)$ is harmonic (that means Lebesgue-almost everywhere equal to a harmonic function) then $|\theta|(U) = 0$. In particular $\theta^L|U = 0$ implies that $|\theta|(U) = 0$.

6.8 COROLLARY: Define $LS(\theta) := \{z \in \mathbb{C} : \theta^{AL}(z) < \infty \text{ and } \theta^L(z) \neq 0\}$ the logarithmic support of θ. Then $LS(\theta) \subset \mathbb{C}$ is a Baire set with $L(LS(\theta)) > 0$ unless $\theta = 0$. Furthermore $Supp(\theta) \subset \overline{LS(\theta)}$.

The remainder of the section is devoted to the logarithmic capacity. It is defined to be

$$cap(E) = \exp\left(\underset{\theta \in Prob_*(E)}{Sup} \ \underset{z \in \mathbb{C}}{Inf} \ \theta^L(z) \right) \quad \forall \text{ nonvoid } E \subset \mathbb{C}.$$

Thus $0 \leq cap(E) \leq \infty$, and $A \subset B$ implies that $cap(A) \leq cap(B)$. For compact sets there is an important equivalent definition. The equivalence proof below is an impressive application of our Hahn-Banach version A.2.3.

6.9 THEOREM: For nonvoid compact $E \subset \mathbb{C}$ we have

$$cap(E) = Inf\left\{ \|P\|_E^{1/n} : P \text{ polynomial with } P(z) = z^n + \ldots \ (n = 1,2,\ldots) \right\},$$

where $\|.\|_E$ denotes the supnorm on E.

Proof: Write I for the Inf in question. 1) For $\theta \in Prob(E)$ and P: $P(z) = z^n + \ldots = (z-u_1)\ldots(z-u_n)$ we have

$$\underset{z \in \mathbb{C}}{Inf} \ \theta^L(z) \leq \frac{1}{n} \sum_{k=1}^n \theta^L(u_k) = \frac{1}{n} \int \left(\sum_{k=1}^n \log|z-u_k| \right) d\theta(z)$$
$$= \int \log|P(z)|^{1/n} d\theta(z) \leq \log \|P\|_E^{1/n} .$$

Hence $\log cap(E) \leq \log I$. 2) The set $T := \left\{ \frac{1}{n} \log|P| : P \text{ polynomial with } P(z) = z^n + \ldots \ (n=1,2,\ldots) \right\} \subset USC(E)$ fulfills $u,v \in T \Rightarrow \frac{1}{2}(u+v) \in T$. Hence after A.2.3 there exists a measure $\sigma \in Prob(E)$ with

$$\log I = \underset{P}{Inf} \log \|P\|_E^{1/n} = \underset{P}{Inf} \underset{z \in E}{Max} \frac{1}{n} \log|P(z)| = \underset{f \in T}{Inf} \underset{z \in E}{Max} f(z)$$
$$= \underset{f \in T}{Inf} \int f d\sigma = \underset{P}{Inf} \frac{1}{n} \int \log|P(z)| d\sigma(z)$$
$$\leq \underset{u \in \mathbb{C}}{Inf} \int \log|z-u| d\sigma(z) = \underset{u \in \mathbb{C}}{Inf} \sigma^L(u) \leq \log cap(E). \qquad \text{QED.}$$

6.10 COROLLARY: For nonvoid compact $E\subset \mathbb{C}$ we have $\operatorname{cap}(E) = \operatorname{cap}(\partial E)$.

6.11 EXAMPLE: $\operatorname{cap}(\nabla(a,R)) = \operatorname{cap}(\partial\nabla(a,R)) = R$ for all $a\in\mathbb{C}$ and $R\geq 0$.

Proof: Write $E:=\nabla(a,R)$. 1) For $P:P(z)=z-a$ we have $\operatorname{cap}(E)\leq \|P\|_E = R$. 2) If $P:P(z)=z^n+\dots$ then $Q:Q(z)=P(a+\frac{R}{z})z^n$ is a polynomial with $Q(0)=R^n$. It follows that

$$\|P\|_E = \sup_{|z|\leq 1} |P(a+Rz)| = \sup_{|z|=1} |P(a+Rz)| = \sup_{|z|=1} |P(a+\frac{R}{z})z^n|$$

$$= \sup_{|z|=1} |Q(z)| = \sup_{|z|\leq 1} |Q(z)| \geq Q(0) = R^n.$$

Hence $\operatorname{cap}(E)\geq R$. QED.

6.12 LEMMA: Assume that $\varphi:\mathbb{C}\to\mathbb{C}$ satisfies $|\varphi(u)-\varphi(v)|\leq c|u-v|$ $\forall u,v\in\mathbb{C}$ for some $c>0$. Then $\operatorname{cap}(\varphi(E))\leq c\operatorname{cap}(E)$ $\forall$ nonvoid $E\subset\mathbb{C}$.

Proof: For $P:P(z)=z^n+\dots=(z-u_1)\dots(z-u_n)$ put $Q:Q(z)=(z-\varphi(u_1))\dots(z-\varphi(u_n))$. Then

$$|Q(\varphi(z))|^{1/n} = \Big(\prod_{k=1}^{n}|\varphi(z)-\varphi(u_k)|\Big)^{1/n} \leq c\Big(\prod_{k=1}^{n}|z-u_k|\Big)^{1/n} = c|P(z)|^{1/n} \quad \forall z\in\mathbb{C},$$

and hence $\|Q\|_{\varphi(E)}^{1/n} \leq \|P\|_E^{1/n}$. The assertion follows. QED.

The above lemma will lead us to an important estimation for the logarithmic capacity.

6.13 LEMMA: There exists an $\varepsilon>0$ such that $\operatorname{cap}(E)\geq\varepsilon\ell(E)$ for all nonvoid compact $E\subset\mathbb{R}$, where ℓ denotes one-dimensional Lebesgue measure on $\mathbb{R}$.

Proof: 1) Let $E=[a,b]$ with $a<b$. Define $\varphi:\mathbb{C}\to\mathbb{C}$ by $\varphi(t)=\exp(2\pi i\frac{t-a}{b-a})$ for $t\in\mathbb{R}$ and $\varphi(z)=\varphi(\operatorname{Re} z)$ for $z\in\mathbb{C}$. Then $\varphi(E)$ is the unit circle S. And $|\varphi(u)-\varphi(v)| = |\exp(2\pi i\frac{u-v}{b-a})-1| \leq \frac{2\pi}{b-a}|u-v|$ $\forall u,v\in\mathbb{R}$ and hence $\forall u,v\in\mathbb{C}$. Thus 6.11 and 6.12 show that $1=\operatorname{cap}(\varphi(E)) \leq \frac{2\pi}{b-a}\operatorname{cap}(E)$ or $\operatorname{cap}(E)\geq\varepsilon\ell(E)$ with $\varepsilon=1/2\pi$. 2) Fix a compact set $E\subset\mathbb{R}$ with $\ell(E)>0$. Define $\varphi:\mathbb{C}\to\mathbb{C}$ by $\varphi(t) = \ell(E\cap]-\infty,t])$ for $t\in\mathbb{R}$ and $\varphi(z)=\varphi(\operatorname{Re} z)$ for $z\in\mathbb{C}$. Then φ is monotone increasing on $\mathbb{R}$ and fulfills $0 \leq \varphi(v)-\varphi(u) = \ell(E\cap]u,v]) \leq v-u$ $\forall u,v\in\mathbb{R}$ with $u<v$. It follows that $|\varphi(u)-\varphi(v)|\leq|u-v|$ $\forall u,v\in\mathbb{C}$. In particular φ is continuous. Therefore $\varphi(\mathbb{R})$ is an interval $\subset\mathbb{R}$ and hence $=[0,\ell(E)]$. Now φ is constant on each interval $\subset\mathbb{R}$ the interior of which does not meet E. Thus

$\varphi(E)=\varphi(\mathbb{R})=[0,\ell(E)]$. Therefore from 6.12 and 1) we obtain $\mathrm{cap}(E)\geq$ $\geq\mathrm{cap}(\varphi(E))=\mathrm{cap}([0,\ell(E)])\geq\varepsilon\ell(E)$. QED.

6.14 THEOREM: There exists an $\varepsilon>0$ such that $\mathrm{cap}(E)\geq\varepsilon\ell(\{|z-a|:z\in E\})$ for all $a\in\mathbb{C}$ and all nonvoid $E\subset\mathbb{C}$ which are countable unions of compact sets.

Proof: For fixed $a\in\mathbb{C}$ define $\varphi:\mathbb{C}\to\mathbb{C}$ by $\varphi(z)=|z-a|$ $\forall z\in\mathbb{C}$. It is clear that $|\varphi(u)-\varphi(v)|\leq|u-v|$ $\forall u,v\in\mathbb{C}$. From 6.12 and 6.13 thus $\mathrm{cap}(E)\geq\mathrm{cap}(\varphi(E))=$ $=\mathrm{cap}(\{|z-a|:z\in E\})\geq\varepsilon\ell(\{|z-a|:z\in E\})$ for all nonvoid compact sets $E\subset\mathbb{C}$. The extension to countable unions is then immediate. QED.

7. The Walsh Theorem

The Walsh theorem asserts that the Dirichlet problem is solvable for certain compact planar sets. Moreover it furnishes the concrete basis for the action of the abstract theory of Section 1.

We retain the fixed compact subset $K\neq\emptyset$ of $\mathbb{C}$ and introduce

$$D(K)=\{f\in\mathrm{ReC}(X):\exists F\in\mathrm{ReC}(K)\text{ with }F|X=f\text{ and }F|K^{\circ}\text{ harmonic}\}.$$

For $f\in D(K)$ the function $F\in\mathrm{ReC}(K)$ in question is unique and satisfies $\underset{K}{\mathrm{Max}}\,F=\underset{X}{\mathrm{Max}}\,f$ and $\underset{K}{\mathrm{Min}}\,F=\underset{X}{\mathrm{Min}}\,f$. It follows that $D(K)$ is a supnorm closed linear subspace $\subset\mathrm{ReC}(X)$. We define K to be Dirichlet iff $D(K)=$ $=\mathrm{ReC}(X)$, that is iff the Dirichlet problem is solvable for K.

7.1 REMARK: For each $a\in K$ there exists some $\eta\in\mathrm{Prob}(X)$ such that $F(a)=$ $=\int f d\eta$ $\forall f\in D(K)$. If K is Dirichlet then η is unique $=:\eta_a$ called the harmonic measure for $a\in K$. Of course $\eta_a=\delta_a$ if $a\in X$. This is immediate after A.2.5.

For the points $u\in\Omega$ we define $L_u:L_u(z)=\log|z-u|$ $\forall z\in K$. Thus $L_u|X\in D(K)$ with harmonic extension L_u. We define K to be Walsh iff the real-linear span of $\{L_u|X:u\in\Omega\}$ is supnorm dense in $\mathrm{ReC}(X)$. Therefore Walsh $\Rightarrow$ Dirichlet. Now we can formulate the Walsh theorem.

7.2 THEOREM: Assume that

$$\liminf_{\tau\downarrow 0} \frac{\log \operatorname{cap}(\Omega\cap V(a,\tau))}{\log \tau} < \infty \qquad \forall a\in X.$$

Then K is Walsh.

The proof to be presented below does not require the above results on the logarithmic capacity but merely its definition. Rather the final result 6.14 serves to deduce from 7.2 the much more tractable version of the Walsh theorem which follows.

7.3 THEOREM: Assume that

$$\liminf_{\tau\downarrow 0} \frac{\log \ell(\{|z-a|:z\in\Omega\cap V(a,\tau)\})}{\log \tau} < \infty \qquad \forall a\in X.$$

Then K is Walsh.

7.4 COROLLARY: Assume that each $a\in X$ is in the boundary of some component of Ω (which in particular is true when Ω has a finite number of components). Then K is Walsh.

Proof of 7.3 ⇒ 7.4: If $a\in X$ is in the boundary of the component G of Ω then $\{|z-a|:z\in G\}$ is an open intervall $\subset\mathbb{R}$ with left endpoint $=0$. For small $\tau>0$ therefore $]0,\tau[=\{|z-a|:z\in G\cap V(a,\tau)\}=\{|z-a|:z\in\Omega\cap V(a,\tau)\}$ so that the lim inf in 7.3 is $=1$. QED.

The assumptions in 7.2-7.4 are to express the requirement that the open set Ω be not too thin near the boundary point $a\in X$. The same is true for the so-called Lebesgue condition: the open set $S = \{|z-a|:z\in\Omega$ with $|z-a|<1\} \subset\,]0,1[$ is to fulfill $\int_S \frac{dx}{x} = \infty$. We claim that the Lebesgue condition implies the assumption in 7.3.

7.5 PROPOSITION: If $a\in X$ satisfies the Lebesgue condition then

$$\liminf_{\tau\downarrow 0} \frac{\log \ell(\{|z-a|:z\in\Omega\cap V(a,\tau)\})}{\log \tau} \leq 1.$$

Thus if each $a\in X$ satisfies the Lebesgue condition then K is Walsh.

Proof: Lemma 2.1 shows that

$$\int_S \frac{dx}{x} = \int_{]0,\infty[} \ell(\{x\in S:\tfrac{1}{x}\geq t\})\,dt = \int_{]0,1[} \ell(\{x\in S:x\leq\tfrac{1}{t}\})\,dt + \int_{[1,\infty[} \ell(\{x\in S:x\leq\tfrac{1}{t}\})\,dt \leq$$

$$\leq \ell(S) + \int_0^1 \ell(\{x\in S:x<t\})\frac{dt}{t^2} = \ell(S) + \int_0^1 \ell(\{|z-a|:z\in\Omega\cap V(a,t)\})\frac{dt}{t^2}.$$

Assume now that the lim inf in 7.3 be $>c>1$. Then $\ell(\{|z-a|:z\in\Omega\cap V(a,\tau)\}) <\tau^c$ for all sufficiently small $0<\tau<1$ so that $\int_S \frac{dx}{x} < \infty$ from the above. QED.

The remainder of the section is devoted to the proof of theorem 7.2. We first reformulate the Walsh property: From the bipolar theorem we conclude that K is Walsh iff $\{L_u|X:u\in\Omega\}^\perp = \{\theta\in ca(X):\theta^L|\Omega=0\}$ is $=0$. Thus 6.7 tells us that K is Walsh iff

(*) each $\theta\in ca(X)$ with $\theta^L(u)=0$ $\forall u\in\Omega$ satisfies $\theta^L(u)=0$ $\forall u\in X$ with $\theta^{AL}(u)<\infty$ and $\forall u\in K^o$.

The decisive step in the proof of (*) is anticipated in the subsequent lemma.

7.6 LEMMA: Let $G\subset\mathbb{C}$ be open and $a\in\partial G$ with

$$\liminf_{\tau\downarrow 0} \frac{\log \operatorname{cap}(G\cap V(a,\tau))}{\log \tau} < \infty.$$

Assume that $\theta\in ca_*(\mathbb{C}-G)$ satisfies $\theta^{AL}(a)<\infty$ and $\theta^L(z) \to$ some $\lambda\in\mathbb{C}$ for $z\to a$ on G. Then $\lambda=\theta^L(a)$.

Proof: There exist some $\alpha\geq 1$ and a sequence of numbers $0<\tau_n<1$ with $\tau_n\downarrow 0$ and $\log \operatorname{cap}(G\cap V(a,\tau_n))>\alpha \log \tau_n$. Hence after the definition of the logarithmic capacity there exist $\theta_n\in \mathrm{Prob}_*(G\cap V(a,\tau_n))$ with $\theta_n^L(z)>\alpha \log \tau_n$ $\forall z\in\mathbb{C}$. We put $S(n):=\mathrm{Supp}(\theta_n)\subset G\cap V(a,\tau_n)$ and $S:=\mathrm{Supp}(\theta)\subset\mathbb{C}-G$ and choose $c>0$ such that $|u-a|\leq c$ $\forall u\in S$. Then

$$\log|z-u| \leq \log(|z-a|+|u-a|) \leq \log(\tau_n+c) \leq c \quad \forall\ z\in S(n) \text{ and } u\in S,$$

so that the Fubini theorem can be applied to obtain

$$\int_{S(n)} \theta^L(z)\,d\theta_n(z) = \int_S \theta_n^L(u)\,d\theta(u).$$

Here the first member tends $\to\lambda$ for $n\to\infty$. Thus we have to show that the second member tends $\to\theta^L(a)$. In view of $|\theta|(\{a\})=0$ this follows from the Lebesgue dominated convergence theorem once we know that

i) $\theta_n^L(u) \to \log|u-a|$ for $n\to\infty$ $\qquad \forall\ u\in\mathbb{C}$ with $u\neq a$,

ii) $\theta_n^L(u) \geqq -\alpha|\log|u-a|-\log 2|$ $\qquad \forall\ u\in\mathbb{C}$ with $u\neq a$,

iii) $\theta_n^L(u) \leqq c$ $\qquad \forall\ u\in S$.

Here iii) is clear from the above. Consider now $u\in\mathbb{C}$ with $u\neq a$. Then

$$\theta_n^L(u)-\log|u-a| = \int_{S(n)} \log\left|\frac{u-z}{u-a}\right| d\theta_n(z) = \int_{S(n)} \log\left|1-\frac{z-a}{u-a}\right| d\theta_n(z).$$

In the case $\tau_n<|u-a|$ we have for $z\in S(n)$

$$\log\left|1-\frac{z-a}{u-a}\right| \leqq \log\left(1+\left|\frac{z-a}{u-a}\right|\right) \leqq \log\left(1+\frac{\tau_n}{|u-a|}\right),$$

$$\log\left|1-\frac{z-a}{u-a}\right| \geqq \log\left(1-\left|\frac{z-a}{u-a}\right|\right) \geqq \log\left(1-\frac{\tau_n}{|u-a|}\right), \text{ and hence}$$

$$\log\left(1-\frac{\tau_n}{|u-a|}\right) \leqq \theta_n^L(u)-\log|u-a| \leqq \log\left(1+\frac{\tau_n}{|u-a|}\right).$$

This proves i). Furthermore in the case $\tau_n \leqq \frac{1}{2}|u-a|$ we have

$$\theta_n^L(u) \geqq \log|u-a|+\log\left(1-\frac{\tau_n}{|u-a|}\right) \geqq \log|u-a|-\log 2,$$

while in the case $\tau_n \geqq \frac{1}{2}|u-a|$ after the definition of θ_n we have

$$\theta_n^L(u) > \alpha \log \tau_n \geqq \alpha(\log|u-a|-\log 2).$$

This proves ii). QED.

Proof of 7.2: Assume that each point $a\in X$ satisfies

$$\liminf_{\tau\downarrow 0} \frac{\log \operatorname{cap}(\Omega\cap V(a,\tau))}{\log \tau} < \infty.$$

We fix $\theta\in ca(X)$ with $\theta^L(u)=0$ $\forall u\in\Omega$. After 7.6 then $\theta^L(u)=0$ $\forall u\in X$ with $\theta^{AL}(u)<\infty$. Thus in view of the reformulation (*) we have to show that θ^L vanishes on K^o. Let us fix $b\in K^o$. 1) From 7.1 we obtain a measure $\sigma\in \operatorname{Prob}(X)$ with

$$L_u(b) = \log|b-u| = \int_X \log|z-u|\,d\sigma(z) = \sigma^L(u) \qquad \forall u\in\Omega.$$

We write this equation in the form

$$\int_X \log^-|u-z|\,d\sigma(z) = \int_X \log^+|u-z|\,d\sigma(z) - \log|u-b| \quad \forall u\in\Omega.$$

Thus for fixed $a\in X$ the Fatou lemma applied to $u\to a$ on Ω implies that

$$\int_X \log^-|a-z|\,d\sigma(z) \leqq \int_X \log^+|a-z|\,d\sigma(z) - \log|a-b|$$

$$\leqq \log^+\mathrm{diam}(X) - \log\,\mathrm{dist}(b,X) =: \alpha,$$

$$\int_X \big|\log|a-z|\big|\,d\sigma(z) \leqq \alpha + \log^+\mathrm{diam}(X) =: \beta<\infty.$$

2) The above 1) combined with 7.6 shows that $\log|u-b|=\sigma^L(u)$ $\forall u\in X$. Furthermore

$$\int_X\Big(\int_X \big|\log|u-z|\big|\,d|\theta|(u)\Big)d\sigma(z) = \int_X\Big(\int_X \big|\log|u-z|\big|\,d\sigma(z)\Big)d|\theta|(u) < \infty.$$

For σ-almost all $z\in X$ thus $\theta^{AL}(z)<\infty$ and hence $\theta^L(z)=0$ after 7.6. And the Fubini theorem can be applied to obtain

$$\theta^L(b) = \int_X \log|u-b|\,d\theta(u) = \int_X \sigma^L(u)\,d\theta(u) = \int_X\Big(\int_X \log|u-z|\,d\sigma(z)\Big)d\theta(u)$$

$$= \int_X\Big(\int_X \log|u-z|\,d\theta(u)\Big)d\sigma(z) = \int_X \theta^L(z)\,d\sigma(z) = 0. \quad \text{QED.}$$

8. Application to the Problem of Rational Approximation

As before we fix a nonvoid compact subset $K\subset\mathbb{C}$. We combine the Walsh theorem with the abstract theory of Section 1.

8.1 REMARK: o) For each $u\in\Omega$ we have $L_u := \log|Z-u| \in \log|R(K)^\times|$.

i) If $u,v \in$ the same component G of Ω then $L_u - L_v \in \mathrm{Re}R(K)$.

ii) If $u\in\Omega^\infty$ then $L_u\in\mathrm{Re}R(K)$.

Proof: o) is obvious. i) Fix $a\in G$ and $\delta>0$ with $V(a,\delta)\subset G$. For each $u\in V(a,\delta)$ then $\left|\frac{u-a}{z-a}\right| \leqq \frac{1}{\delta}|u-a| < 1$ $\forall z\in K$ and hence

$$\mathrm{Log}\,\frac{z-u}{z-a} = \mathrm{Log}\Big(1-\frac{u-a}{z-a}\Big) = -\sum_{k=1}^{\infty}\frac{1}{k}\Big(\frac{u-a}{z-a}\Big)^k \quad \forall z\in K,$$

with Log the main branch of the logarithm in the open halfplane $\Delta :=$ $\{s\in\mathbb{C}: \mathrm{Re}\, s > 0\}$, where the series is uniformly convergent on K. Thus $\mathrm{Log}\,\frac{Z-u}{Z-a} \in R(K)$ and hence $L_u - L_a = \log\left|\frac{Z-u}{Z-a}\right| = \mathrm{Re}\mathrm{Log}\,\frac{Z-u}{Z-a} \in \mathrm{Re}R(K)$. We obtain i) via connectedness. ii) If $a\in\mathbb{C}$ with $|a|>\mathrm{Max}\{|z|:z\in K\}$ then

$$\mathrm{Log}\left(1-\frac{z}{a}\right) = -\sum_{k=1}^{\infty}\frac{1}{k}\left(\frac{z}{a}\right)^k \quad \forall z\in K,$$

where the series is uniformly convergent on K. Thus $\mathrm{Log}(1-\frac{Z}{a})\in R(K)$ and hence $L_a = \log|Z-a| = \log|a|+\log|1-\frac{Z}{a}|\in \mathrm{Re}R(K)$. So we obtain ii) from i). QED.

8.2 PROPOSITION: If K is Walsh then

$$ML(R(K)|X,a) = ML(A(K)|X,a) = \{\eta_a\} \quad \forall a\in K,$$

with $\eta_a\in\mathrm{Prob}(X)$ the harmonic measure for $a\in K$.

Proof: We know from Section 1 that $\emptyset \neq ML(A(K)|X,a) \subset ML(R(K)|X,a)$. Hence we have to show that each $m\in ML(R(K)|X,a)$ must be $=\eta_a$. But for $u\in\Omega$ we have $L_u|X\in\log|(R(K)|X)^\times|$ and hence $L_u(a) = \int_X L_u dm$. Thus $F(a) = \int_X f dm$ $\forall$ $f\in$real-linear span$\{L_u|X:u\in\Omega\}$ and hence $\forall$ $f\in\mathrm{Re}C(X)$, with $F\in\mathrm{Re}C(K)$ the harmonic extension of f. It follows that $m=\eta_a$. QED.

We turn to the main theorems. We start with the case that K has no holes, that means $\Omega=\Omega^\infty$, which after 3.2 is equivalent to $P(K)=R(K)$. This case is particularly transparent: After 7.4 K is Walsh. Then 8.1. ii) implies that $\overline{\mathrm{Re}R(K)|X} = \mathrm{Re}C(X)$, that is the restriction algebra $R(K)|X \subset C(X)$ has the Dirichlet property in the sense of Section III.1. Hence 8.2 shows that $M(R(K)|X,a) = \{\eta_a\}$ $\forall a\in K$. We put this fact and the Wilken theorem 4.6 into the abstract Mergelyan theorem 1.6 to obtain the Mergelyan polynomial approximation theorem in the subsequent form.

8.3 THEOREM: Assume that K has no holes. Then

i) If $B\subset C(K)$ is a complex subalgebra with $P(K)=R(K)\subset B$ and $\|f\|=\|f|X\|$ $\forall f\in B$ then $B=R(K)$. In particular $P(K)=R(K)=A(K)$.

ii) $P(K)|X = R(K)|X \subset C(X)$ has the Dirichlet property and

$$M(P(K)|X,a) = M(R(K)|X,a) = \{\eta_a\} \quad \forall a\in K.$$

Proof: For $B\subset C(K)$ as in i) we want to apply 1.6 to the restrictions $R(K)|X \subset B|X \subset C(X) \subset B(X,\text{Baire})$. It suffices to remark that for each $a\in K = \Sigma(R(K)|X)$ we have a well-defined multiplicative linear functional $\psi: f|X \mapsto f(a)\ \forall f\in B$ on $B|X$ with $\|\psi\|=1$ and hence an extension $\psi\in\Sigma(B|X)$. From 1.6 we conclude that $B|X = R(K)|X$ and hence that $B = R(K)$. QED.

Theorem 8.3 combined with the Bishop localization theorem 3.10 forms a fast road to the Mergelyan rational approximation theorem.

8.4 THEOREM: Assume that there exists an $\alpha>0$ with $\operatorname{diam}(G)\geqq\alpha$ for all holes G of K (which in particular is true when K has a finite number of holes). Then $R(K)=A(K)$.

Proof: 1) Put $K(a):=K\cap\nabla(a,\delta)\ \forall a\in K$ with some $0<\delta<\frac{\alpha}{2}$. We claim that $\mathbb{C}-K(a)$ is connected. In fact, the points $z\in\mathbb{C}$ with $|z-a|>\delta$ are all in $\mathbb{C}-K(a)$. And if $u\in\mathbb{C}-K(a)$ with $|u-a|\leqq\delta$ then $u\notin K$ and hence $u\in$some component G of Ω. In view of $\operatorname{diam}(G)\geqq\alpha>2\delta$ there exists some $v\in G$ with $|v-a|>\delta$ and then a continuous curve on $G\subset\Omega=\mathbb{C}-K\subset\mathbb{C}-K(a)$ which connects u and v. Thus u is in the unbounded component of $\mathbb{C}-K(a)$ so that $\mathbb{C}-K(a)$ is indeed connected $\forall a\in K$. 2) If now $f\in A(K)$ then $f|K(a)\in A(K)|K(a)\subset$ $\subset A(K(a))$ which is $= R(K(a))$ after 8.3. Hence $f\in R(K)$ in view of the Bishop theorem 3.10. QED.

In the case that K has a finite number of holes we can prove more that the mere $R(K)=A(K)$. But the subsequent final theorem rests upon the abstract Mergelyan theorem 1.9 and hence upon the maximality theorem IX.3.5 so that the expenditure is much higher.

8.5 THEOREM: Assume that K has n holes ($n=0,1,2,\ldots$). Then

i) If $B\subset C(K)$ is a complex subalgebra with $R(K)\subset B$ and $\|f\|=\|f|X\|\ \forall f\in B$ then $B=R(K)$. In particular $A(K)=R(K)$.

ii) $\dim N(R(K)|X,a) \leqq \dim \dfrac{\operatorname{Re}C(X)}{\operatorname{Re}R(K)|X} \leqq n \quad \forall a\in K$.

iii) $ML(R(K)|X,a) = \{\eta_a\}\ \forall a\in K$. And η_a is an interior point of $M(R(K)|X,a)$.

Proof: ii) Choose points $a(1),\ldots,a(n)$ from the holes $G(1),\ldots,G(n)$ of K. Then 8.1 implies that

$$\{L_u : u \in \Omega\} \subset \mathrm{ReR}(K) + \text{real-linear span}\{L_{a(1)},\dots,L_{a(n)}\}.$$

Since K is Walsh after 7.4 we obtain

$$\mathrm{ReC}(X) = \overline{\mathrm{ReR}(K)|X} + \text{real-linear span}\{L_{a(1)}|X,\dots,L_{a(n)}|X\},$$

and hence the second inequality (the case n=0 is comprised as well). The first inequality is then clear from $N(R(K)|X,a) \subset \overline{\mathrm{ReR}(K)|X}^{\perp}$ combined with the usual norm isomorphism $\overline{\mathrm{ReR}(K)|X}^{\perp} \cong \left(\frac{\mathrm{ReC}(X)}{\overline{\mathrm{ReR}(K)|X}}\right)^{\hat{}}$. iii) then follows from 8.2 and 1.8. i) results from 1.9 applied to the restrictions $R(K)|X \subset B|X \subset C(X) \subset B(X,\mathrm{Baire})$ as we deduced above 8.3.i) from 1.6. QED.

Notes

The constructive theory of polynomial and rational approximation up to the decisive work of Mergelyan and Vitushkin is presented in ZALCMAN [1968] and GAMELIN [1969] Chapter VIII. See also VITUSHKIN [1975]. The action of the functional-analytic theory of function algebras is a main theme in each of the treatises of BROWDER [1969], GAMELIN [1969], LEIBOWITZ [1970] and STOUT [1971]. For the subsequent development see GLICKSBERG [1972] and the literature cited therein. The access to the Mergelyan polynomial approximation theorem 8.3 via abstract methods is due to BISHOP [1960] and GLICKSBERG-WERMER [1963]. See the beautiful presentations in WERMER [1964] and CARLESON [1964]. As to the Mergelyan rational approximation theorem 8.4-8.5 the access is due to AHERN-SARASON [1967a], GLICKSBERG [1968] and GARNETT [1968].

The functional-analytic theory led to an abstract version of the polynomial approximation theorem 8.3: it is theorem 1.6 in the compact-continuous situation under the overall assumption that A be Dirichlet. Theorem 1.6 as it stands and its close relative 1.5 depend on more elaborate versions of the abstract F.and M.Riesz theorem and of the abstract Hardy algebra theory. In the compact-continuous situation 1.5 is in GARNETT-GLICKSBERG [1967] after GLICKSBERG [1967] while 1.4-1.5 are in KÖNIG-SEEVER [1969]. A similar abstract version of the rational approximation theorem 8.5 is theorem 1.9 in the compact-continuous situation under the overall assumption that A be hypo-Dirichlet. It is due to GAMELIN-LUMER [1968], see also GAMELIN [1969] Chapter IV. The essence of theorem 1.9 as it stands is in KÖNIG [1969b].

The essence of theorem 1.8 is in GAMELIN [1969] Theorem IV.7.4. The present formulation is somewhat more comprehensive which is adequate since it permits to consider certain obvious variants of the definition of the logmodular measures.

Sections 2-5 contain much standard material for which we refer to the treatises cited above. In these Notes we restrict ourselves to some particular points. The introduction of the Cauchy transformation of Baire measures in the complex plane and its systematic use in the functional-analytic approximation theory is due to BISHOP [1959][1960]. The identification 3.6 of the spectrum $\Sigma(A(K))=K$ based on 3.5 is due to ARENS [1958]. For the Bishop localization theorem 3.10 we refer to GARNETT [1968] and ZALCMAN [1968] Section 15. Lemma 4.1 has the fundamental consequences 4.2,4.6 due to WILKEN [1968a] and 5.3-5.4,5.6-5.7 due to WILKEN [1967][1968b]. We have added the remarks 4.3-4.5 and 5.5 on the supports of the measures in question which are perhaps new. The selection theorem 5.9 is from GAMELIN [1969] Section VI.3. Fundamental further results on the Gleason parts for R(K) when K has a finite number of holes are due to AHERN-SARASON [1967b] and represented in GAMELIN [1969] Chapter VI and STOUT [1971] Section 26.

The WALSH approximation theorem [1928][1929] which is also attributed to LEBESGUE [1907] has found a new ab-ovo proof in CARLESON [1964]. The present Sections 6-7 are from KÖNIG [1975] where the Carleson idea is further developed to obtain a soft analysis proof of certain extended versions of the Walsh theorem. For a detailed treatment of the logarithmic capacity of planar sets we refer to TSUJI [1959] Chapter III.

Appendix

The Appendix is to introduce basic notations and to recall certain definitions and fundamental theorems from functional analysis, measure theory and advanced calculus in an unsystematic manner. In particular we quote unconventional versions of standard theorems which we found useful in applications. There are no proofs, except when we cannot name a convenient reference.

1. Linear Functionals and the Hahn-Banach Theorem

Let E be a real vector space and E* its dual space, defined to consist of all linear functionals $E \to \mathbb{R}$. A functional $\theta: E \to \mathbb{R}$ is defined to be sublinear iff $\theta(u+v) \leq \theta(u)+\theta(v)$ and $\theta(tu)=t\theta(u)$ for all $u,v \in E$ and real $t \geq 0$. We start with a powerful version of the Hahn-Banach theorem.

1.1 HAHN-BANACH THEOREM (CONVEX VERSION): Let $\theta: E \to \mathbb{R}$ be sublinear. Assume that $T \subset E$ is nonvoid and such that to $u,v \in T$ there exists $f \in T$ with $\theta(f-\frac{1}{2}(u+v)) \leq 0$. Then there exists $\sigma \in E^*$ with $\sigma \leq \theta$ such that

$$\operatorname*{Inf}_{f \in T} \sigma(f) = \operatorname*{Inf}_{f \in T} \theta(f).$$

If we specialize to T={f} and ={-f} for fixed $f \in E$ then we obtain the subsequent conventional form of the Hahn-Banach theorem: Let $\theta: E \to \mathbb{R}$ be sublinear. Then there exist linear functionals $\sigma \in E^*$ with $\sigma \leq \theta$. Furthermore

$$\{\sigma(f): \sigma \in E^* \text{ with } \sigma \leq \theta\} = [-\theta(-f), \theta(f)] \quad \forall f \in E.$$

We quote another important consequence of the convex version 1.1.

1.2 HAHN-BANACH THEOREM (CONE VERSION): Let $\theta: E \to \mathbb{R}$ be sublinear. Assume that $T \subset E$ is nonvoid and such that to $u,v \in T$ there exists $f \in T$ with $\theta(f-(u+v)) \leq 0$. If $\theta(f) \geq 0$ $\forall f \in T$ then there exists $\sigma \in E^*$ with $\sigma \leq \theta$ such that $\sigma(f) \geq 0$ $\forall f \in T$.

The most familiar form of the Hahn-Banach theorem is the extension theorem: Let $\theta:E\to\mathbb{R}$ be sublinear. Then each $\varphi\in S^*$ on a linear subspace $S\subset E$ with $\varphi\leq\theta|S$ can be extended to some $\sigma\in E^*$ with $\sigma\leq\theta$. This is a consequence of the above conventional form applied to an appropriate modified sublinear functional.

There is an important special case which is valid in both the real and the complex situation: Let E be a (real or complex) vector space with E^* its (real or complex) dual space, and let $\|.\|:E\to[0,\infty[$ be a seminorm (relative to real or complex coefficients). Then each $\varphi\in S^*$ on a (real or complex) linear subspace $S\subset E$ with $|\varphi|\leq\|.\|$ can be extended to some $\sigma\in E^*$ with $|\sigma|\leq\|.\|$. It is of course natural to formulate this version in terms of the norm dual space E' of E relative to $\|.\|$. The vehicle which carries over to the complex situation is the subsequent simple but important remark.

<u>1.3</u> <u>REMARK</u>: Let E be a complex vector space. Then the complex-linear functionals $\phi:E\to\mathbb{C}$ and the real-linear functionals $\varphi:E\to\mathbb{R}$ are in one-to-one correspondence with each other via

$$\phi\mapsto\varphi:\varphi(x) = \mathrm{Re}\phi(x)\ \forall x\in E,$$

$$\phi(x) = \varphi(x)-i\varphi(ix)\ \forall x\in E:\phi\leftarrow\varphi.$$

An important aspect for linear functionals is positivity. We do not intend to consider vector spaces with order structures. Instead we shall restrict ourselves to $B(X)$, defined to consist of the bounded complex-valued functions on the nonvoid set X, with its natural pointwise structure and with $\|.\|$ the supremum norm (= supnorm). The basic equivalence theorem will be formulated in both a real and a complex version. For $V\subset B(X)$ define $\mathrm{Re}V:=\{\mathrm{Re}\ f:f\in V\}$.

<u>1.4</u> <u>THEOREM</u>: Let $V\subset \mathrm{Re}B(X)$ be a real-linear subspace with $1\in V$ and $\varphi:V\to\mathbb{R}$ be a real-linear functional. Then the subsequent properties are equivalent.

i) If $f\in V$ and $f\geq 0$ then $\varphi(f)\geq 0$, and $\varphi(1)=1$.

ii) $\varphi(f)\leq \mathrm{Sup}\ f$ for all $f\in V$.

iii) $|\varphi(f)|\leq\|f\|$ for all $f\in V$, and $\varphi(1)=1$.

<u>1.5</u> <u>THEOREM</u>: Let $V\subset B(X)$ be a complex-linear subspace with $1\in V$ and $\phi:V\to\mathbb{C}$ be a complex-linear functional. Then the subsequent properties are equivalent.

i) If $f \in V$ and $\mathrm{Re}\, f \geq 0$ then $\mathrm{Re}\phi(f) \geq 0$, and $\phi(1)=1$.

ii) $\mathrm{Re}\phi(f) \leq \mathrm{Sup}\, \mathrm{Re}\, f$ for all $f \in V$.

iii) $|\phi(f)| \leq \|f\|$ for all $f \in V$, and $\phi(1)=1$.

iv) $\phi(f) \in \overline{\mathrm{Konv}\, f(X)}$ (= the closed convex hull) for all $f \in V$.

If V is closed under complex conjugation then the equivalence extends to

i*) If $f \in V$ and $f \geq 0$ then $\phi(f) \geq 0$, and $\phi(1)=1$.

The steps i)⇒ii)⇒iii)⇒i) in the proof of 1.4 and i)⇒ii)⇒iii)⇒iv)⇒i) and i)⇔i*) in the proof of 1.5 are all conventional, except iii)⇒iv) in 1.5 which is an immediate consequence of the subsequent lemma.

<u>1.6</u> <u>LEMMA</u>: Let $K \subset ¢$ be convex bounded $\neq \emptyset$ and $a \in ¢$. Then

$$|a-z| \leq \mathop{\mathrm{Sup}}_{u \in K} |u-z| \quad \forall z \in ¢ \quad \text{implies that } a \in \overline{K}.$$

Proof: Assume that $a \notin \overline{K}$. Then from 1.1 we obtain a complex number c of modulus $|c|=1$ such that

$$0 < \varepsilon := \mathop{\mathrm{Inf}}_{u \in K} |u-a| = \mathop{\mathrm{Inf}}_{u \in K} \mathrm{Re}\, \overline{c}(u-a) \quad \text{and hence } \mathrm{Re}\, \overline{c}(u-a) \geq \varepsilon \ \forall u \in K.$$

Take an $R>0$ such that $|u-a| \leq R \ \forall u \in K$. For $t>0$ then

$$|u-(a+tc)|^2 = |(u-a)-tc|^2 = |u-a|^2 - 2t\mathrm{Re}\overline{c}(u-a) + t^2 \leq R^2 - 2t\varepsilon + t^2,$$

$$t = |a-(a+tc)| \leq \mathop{\mathrm{Sup}}_{u \in K} |u-(a+tc)| \leq (R^2-2t\varepsilon+t^2)^{1/2}.$$

It follows that $2t\varepsilon \leq R^2 \ \forall t>0$ which is a contradiction. QED.

We conclude with an unconventional but useful version of the bipolar theorem.

<u>1.7</u> <u>BIPOLAR THEOREM</u>: Let E be a real vector space and $F \subset E^*$ be a linear subspace. Then the closed convex hull in the weak topology $\sigma(E,F)$ of $M \subset E$ is

$$\overline{\mathrm{Konv}\, M} = \{u \in E : \varphi(u) \leq \mathop{\mathrm{Sup}}_{x \in M} \varphi(x) \text{ for all } \varphi \in F\}.$$

2. Measure Theory

Let (X,Σ) be a measurable space, that is a nonvoid set X with a σ-algebra Σ of subsets. Let $B(X,\Sigma)\subset B(X)$ consist of the measurable bounded complex-valued functions on X. Define $ca(X,\Sigma)$ to consist of the complex-valued measures on Σ, in particular $Pos(X,\Sigma)\subset ca(X,\Sigma)$ to consist of the measures with values $\geqq 0$, and $Prob(X,\Sigma)$ of the so-called probability measures $\sigma\in Pos(X,\Sigma)$ with $\sigma(X)=1$. $B(X,\Sigma)$ and $ca(X,\Sigma)$ form a dual system via the bilinear functional $(f,\theta)\mapsto \int f d\theta$. For $\theta\in ca(X,\Sigma)$ the variation $|\theta|\in Pos(X,\Sigma)$ is defined to be

$$|\theta|(A) = \operatorname{Sup}\{|\int f d\theta| : f\in B(X,\Sigma) \text{ with } |f|\leqq\chi_A\} \quad \forall A\in\Sigma,$$

with χ_A the characteristic function of $A\subset X$, whence in particular the norm

$$\|\theta\| = |\theta|(X) = \operatorname{Sup}\{|\int f d\theta| : f\in B(X,\Sigma) \text{ with } |f|\leqq 1\}.$$

In this norm $ca(X,\Sigma)$ is complete.

For $m\in Pos(X,\Sigma)$ define $L(m) = L(X,\Sigma,m)$ to consist of the equivalence classes modulo m of the measurable complex-valued functions on X. With the obvious precautions we can use the same notations and expressions for the members of $L(m)$ as for the functions themselves. For example, for a sequence of members $f_n\in L(m)$ the symbol $[f_n\to 0]:=\{x\in X: f_n(x)\to 0\}$ defines a measurable set $\in\Sigma$ modulo m-null sets. In $L(m)$ we have the usual Banach spaces $L^p(m)$ for $1\leqq p\leqq\infty$. Furthermore define $L^{\circ}(m)$ to consist of the $f\in L(m)$ with $(\log|f|)^+\in L^1(m)$, that is with $|f|\leqq e^F$ for some $F\in L^1(m)$.

For $\theta,m\in ca(X,\Sigma)$ the notation $\theta<<m$ is to mean that θ is m-continuous (= absolutely continuous with respect to m), that is $|m|$-continuous in the usual sense. After the Radon-Nikodym theorem this means that $\theta=fm$ for some $f\in L^1(|m|)$. For $\theta,m\in ca(X,\Sigma)$ we have the Lebesgue decomposition $\theta=\theta_m+\theta_m^{\wedge}$ of θ into the m-continuous part $\theta_m = \frac{d\theta}{dm}m$ with $\frac{d\theta}{dm}\in L^1(|m|)$ and the m-singular part $\theta_m^{\wedge}$. The Lebesgue decomposition is a particular case of the notion of preband decomposition to be dealt with in Section II.2.

In the remainder of the section we consider the important particular measurable space (X,Σ) where X is a compact Hausdorff space and Σ is the σ-algebra of its Baire subsets. We write $B(X,\text{Baire})$ and observe that $C(X)\subset B(X,\text{Baire})$. Also we abbreviate $ca(X):=ca(X,\text{Baire})$ and likewise $Pos(X)$ and $Prob(X)$. In the present situation the fundamental fact is the F.Riesz representation theorem which states that the above bilinear functional $(f,\varphi)\mapsto\int f d\varphi$ in fact produces a norm isometric isomorphism between the supnorm dual space $(C(X))'$ and $ca(X)$.

2.1 F.RIESZ REPRESENTATION THEOREM: The linear functionals $\phi\in(C(X))'$ and the measures $\varphi\in ca(X)$ are in one-to-one correspondence with each other via $\phi(f) = \int f d\varphi \ \forall f\in C(X)$. Furthermore $\|\phi\|=\|\varphi\|$.

In particular $\varphi\in Pos(X)$ iff ϕ is a positive functional in the sense of 1.5. As usual the measures $\varphi\in ca(X)$ and their functionals $\phi\in(C(X))'$ will be identified, so that $\varphi(f) = \int f d\varphi$ for all $f\in C(X)$.

We do not intend to discuss the details of the standard extension procedure for Baire measures. Let us merely recall the extension of a positive measure $\sigma\in Pos(X)$ to the class $USC(X)$ of upper semicontinuous functions $X\rightarrow[-\infty,\infty[$ which is defined to be

$$\int f d\sigma := \operatorname{Inf}\{\sigma(F): F\in ReC(X) \text{ with } F\geqq f\} \geqq -\infty \quad \forall f\in USC(X).$$

For a subset $K\subset X$ we have $\chi_K\in USC(X)$ iff K is closed. We then write $\int f d\sigma=:\sigma(K)$. At this point we can define the support $Supp(\sigma)$ of $\sigma\in ca(X)$ to be the smallest closed subset $K\subset X$ with full measure $|\sigma|(K)=|\sigma|(X)$, the existence of which is not hard to see.

Now we combine the F.Riesz representation theorem with the Hahn-Banach versions of Section 1 to obtain some efficient representation theorems. Note that $Max: ReC(X)\rightarrow\mathbb{R}$ is a sublinear functional. We see from 1.4 and 1.5 that the linear functionals $\sigma: ReC(X)\rightarrow\mathbb{R}$ with $\sigma\leqq Max$ are precisely the probability measures $\sigma\in Prob(X)$. Let us prove a certain refinement.

2.2 REMARK: Let $K\subset X$ be closed $\neq\emptyset$. Then the linear functionals σ : $ReC(X)\rightarrow\mathbb{R}$ with $\sigma\leqq Max(.|K)$ are precisely the probability measures $\sigma\in Prob(X)$ with $\sigma(K)=1$.

Proof: Let $\sigma: ReC(X)\rightarrow\mathbb{R}$ be a linear functional. i) If $\sigma\leqq Max(.|K)\leqq Max$ then $\sigma\in Prob(X)$. And for $\chi_K \leqq f\in ReC(X)$ we have $1-\sigma(f) = \sigma(1-f) \leqq Max(1-f|K) \leqq 0$, so that $\sigma(K)\geqq 1$ and hence $\sigma(K)=1$. ii) For the converse let $\sigma\in Prob(X)$ with $\sigma(K)=1$. For $f\in ReC(X)$ then $Maxf - f\geqq(Maxf-Max(f|K))\chi_K$ implies that $\sigma(f)\leqq Max(f|K)$. QED.

2.3 THEOREM: Let $K\subset X$ be closed $\neq\emptyset$. Assume that $T\subset USC(X)$ is nonvoid and such that to $u,v\in T$ there exists $f\in T$ with $f\leq\frac{1}{2}(u+v)$ on K. Then there exists $\sigma\in Prob(X)$ with $\sigma(K)=1$ such that

$$\operatorname*{Inf}_{f\in T} \int f d\sigma = \operatorname*{Inf}_{f\in T} Max(f|K).$$

Proof: Follows upon application of 1.1 to the subset $\{F\in ReC(X):F \geq \text{some } f\in T\}\subset ReC(X)$. QED.

2.4 THEOREM: Let $K\subset X$ be closed $\neq\emptyset$. Assume that $T\subset USC(X)$ is nonvoid and such that to $u,v\in T$ there exists $f\in T$ with $f\leq u+v$ on K. If $Max(f|K)\geq 0$ $\forall f\in T$ then there exists $\sigma\in Prob(X)$ with $\sigma(K)=1$ such that $\int fd\sigma\geq 0$ $\forall f\in T$.

Proof: Follows upon application of 1.2 to the subset $\{F\in ReC(X):F \geq \text{some } f\in T\}\subset ReC(X)$. QED.

In conclusion we use 1.4 and 1.5 and the Hahn-Banach extension theorem to obtain another important representation theorem, as before in both a real and a complex version.

2.5 THEOREM: Let $V\subset ReC(X)$ be a real-linear subspace with $1\in V$ and $\varphi:V\rightarrow\mathbb{R}$ a real-linear functional such that $\varphi(f)\geq 0$ for all $0\leq f\in V$. Then there exists $\sigma\in Pos(X)$ such that $\varphi(f) = \int fd\sigma$ $\forall f\in V$.

2.6 THEOREM: Let $V\subset C(X)$ be a complex-linear subspace with $1\in V$ and $\phi:V\rightarrow\mathbb{C}$ a complex-linear functional such that $Re\phi(f)\geq 0$ for all $f\in V$ with $Re\ f \geq 0$. Then there exists $\sigma\in Pos(X)$ such that $\phi(f)=\int fd\sigma$ $\forall f\in V$.

Proofs: We restrict ourselves to 2.5. We can assume that $\varphi(1)=1$. Then φ fulfills the equivalent condition iii) in 1.4. Hence it admits an extension $\sigma:ReC(X)\rightarrow\mathbb{R}$ which preserves the equivalent properties i)-iii) in 1.4. Then 2.1 can be applied. QED.

3. The Cauchy Formula via the Divergence Theorem

It is well-known that the Cauchy formula can be deduced from an appropriate version of the divergence theorem. It then appears in what can be considered to be its natural form. We take the opportunity to present the details. Let us fix a nonvoid bounded open subset $G\subset\mathbb{C}=\mathbb{R}^2$.

A boundary point $u\in\partial G$ is called regular iff there exists a neighbourhood $U\subset\mathbb{C}$ of u such that $U\cap\partial G$ is a smooth curve (= one-dimensional C^1-manifold). Otherwise $u\in\partial G$ is called a singular boundary point. We introduce

$$R(G) = \{u\in\partial G : u \text{ is regular}\},$$
$$S(G) = \{u\in\partial G : u \text{ is singular}\}.$$

Then R(G) is a smooth curve and S(G) is compact with $\partial G=R(G)\cup S(G)$. A regular boundary point $u\in R(G)$ is called outer iff one of the two unit normal vectors N to R(G) at u is such that $u-tN\in G$ and $u+tN\notin\overline{G}$ for small $t>0$. Then this unique $N=:N(u)$ is called the outer normal of G at u. Otherwise $u\in R(G)$ is called inner, which means that u is an interior point of $\overline{G}$. We introduce

$$X(G) = \{u\in\partial G : u \text{ is outer}\},$$
$$Y(G) = \{u\in\partial G : u \text{ is inner}\}.$$

Then X(G) and Y(G) are smooth curves with $X(G)\cup Y(G)=R(G)$. The function $u\mapsto N(u)$ is continuous on X(G).

The divergence theorem holds true whenever S(G) is small. This statement is made precise in terms of the one-dimensional Minkowski content, defined for a compact set $S\subset\mathbb{C}$ to be

$$\tau(S) = \limsup_{\delta\downarrow 0} \frac{L(\{z\in\mathbb{C}:\operatorname{dist}(z,S)\leq\delta\})}{2\delta},$$

where L denotes two-dimensional Lebesgue masure on $\mathbb{C}$. Clearly $\tau(S)=0$ if S consists of finitely many points.

3.1 THEOREM: Assume that G satisfies $\tau(S(G))=0$. Let $A:G\cup R(G)\to\mathbb{R}^2$ be a continuous bounded vector function, differentiable on G (in the sense of real analysis) such that $\operatorname{div}A:G\to\mathbb{R}$ is continuous. Then we have

$$\int_{X(G)} \langle A(x),N(x)\rangle d\sigma(x) = \int_G \operatorname{div} A(x)dL(x),$$

whenever both integrals exists. Here σ denotes one-dimensional Lebesgue measure (= arc length) on X(G).

Observe that in the case $\tau(S(G))=0$ and $\sigma(X(G))<\infty$ all assumptions of the above theorem are fulfilled if A is defined and C^1 on some open set

$U\subset\mathbb{C}$ with $G\subset\overline{G}\subset U$. From 3.1 we obtain the Green formula via a well known argument.

3.2 COROLLARY: Assume that G satisfies $\tau(S(G))=0$ and $\sigma(X(G))<\infty$. Let $f,g\in C^2(U)$ on some open set $U\subset\mathbb{C}$ with $G\subset\overline{G}\subset U$. Let $\frac{\partial f}{\partial n}(x)$ denote the directional derivative of f at the point $x\in X(G)$ in the direction $N(x)$. Then we have

$$\int\limits_{X(G)}\left(f(x)\frac{\partial g}{\partial n}(x)-g(x)\frac{\partial f}{\partial n}(x)\right)d\sigma(x) = \int\limits_G (f(x)\Delta g(x)-g(x)\Delta f(x))\,dL(x).$$

We derive from 3.1 the Cauchy theorem and the Cauchy formula. We introduce the differential operators

$$\frac{\partial}{\partial Z} = \frac{1}{2}\left(\frac{\partial}{\partial X} + \frac{1}{i}\frac{\partial}{\partial Y}\right) \quad\text{and}\quad \frac{\partial}{\partial \overline{Z}} = \frac{1}{2}\left(\frac{\partial}{\partial X} - \frac{1}{i}\frac{\partial}{\partial Y}\right).$$

A function $f:G\to\mathbb{C}$ is holomorphic iff f is differentiable (in the sense of real analysis) with $\frac{\partial f}{\partial \overline{Z}} = 0$; in this case $f'(z) = \frac{\partial f}{\partial Z}(z) = \frac{\partial f}{\partial X}(z) = \frac{1}{i}\frac{\partial f}{\partial Y}(z)$ $\forall z\in G$.

3.3 THEOREM: Assume that G satisfies $\tau(S(G))=0$. Let $f:G\cup R(G)\to\mathbb{C}$ be a continuous bounded function such that f is differentiable on G with $\frac{\partial f}{\partial \overline{Z}}$ continuous. Then we have

$$\int\limits_{X(G)} f(x)N(x)\,d\sigma(x) = 2\int\limits_G \frac{\partial f}{\partial \overline{Z}}(x)\,dL(x),$$

whenever both integrals exist.

Proof: Apply 3.1 to the vector functions $A=(\operatorname{Re} f, -\operatorname{Im} f)$ and $B=(\operatorname{Im} f, \operatorname{Re} f)$ and sum the two equations. QED.

3.4 CONSEQUENCE: Assume that G satisfies $\tau(S(G))=0$ and $\sigma(X(G))<\infty$. Let $f:G\cup R(G)\to\mathbb{C}$ be a continuous bounded function such that f is differentiable on G with $\frac{\partial f}{\partial \overline{Z}}$ continuous. Suppose further that

$$\int\limits_G \left|\frac{\partial f}{\partial \overline{Z}}(x)\right| dL(x)<\infty.$$

Then we have

$$f(u) = \frac{1}{2\pi}\int\limits_{X(G)}\frac{f(x)}{x-u}N(x)\,d\sigma(x) - \frac{1}{\pi}\int\limits_G \frac{1}{x-u}\frac{\partial f}{\partial \overline{Z}}(x)\,dL(x) \quad \forall u\in G.$$

The subsequent special case will be sufficient for our purposes.

3.5 SPECIAL CASE: Assume that G satisfies $\tau(S(G))=0$ and $\sigma(X(G))<\infty$. Let $f\in C^1(U)$ on some open set $U\subset\mathbb{C}$ with $G\subset\overline{G}\subset U$. Then we have

$$f(u) = \frac{1}{2\pi}\int_{X(G)} \frac{f(x)}{x-u} N(x)d\sigma(x) - \frac{1}{\pi}\int_G \frac{1}{x-u}\frac{\partial f}{\partial \overline{z}}(x)dL(x) \quad \forall u\in G.$$

Notes

There is an extensive literature on fortified Hahn-Banach type theorems. The ancestor of numerous versions is a well-known theorem due to MAZUR-ORLICZ [1953], see also PTÁK [1956]. The above presentation of 1.1-1.2 and 2.3-2.4 follows KÖNIG [1968][1970b]. The version 1.7 of the bipolar theorem is from KÖNIG [1972]. The presentation of the divergence theorem is adapted from KÖNIG [1964].

References

P.R.AHERN

[1965] On the generalized F.and M.Riesz theorem.
Pacific J.Math.15(1965) 373-376.

P.R.AHERN and Donald SARASON

[1967a] The H^p spaces of a class of function algebras.
Acta Math.117(1967) 123-163.

[1967b] On some hypo-Dirichlet algebras of analytic functions.
Amer.J.Math.89(1967) 932-941.

A.C.ALLEN and E.R.KERR

[1953] The converse of Fatou's theorem.
J.London Math.Soc.28(1953) 80-89.

Eric AMAR

[1973] Sur un théorème de Mooney relatif aux fonctions analytiques bornées.
Pacific J.Math.49(1973) 311-314.

Richard ARENS

[1958] The maximal ideals of certain function algebras.
Pacific J.Math.8(1958) 641-648.

Klaus BARBEY

[1975] Ein Satz über abstrakte analytische Funktionen.
Arch.Math.26(1975) 521-527.

[1976] Zum Satz von Mooney für abstrakte analytische Funktionen.
Arch.Math.27(1976) 622-626.

Klaus BARBEY and Heinz KÖNIG

[1972] Ein Grenzwertsatz für annullierende Maße.
Arch.Math.23(1972) 509-512.

[1976] Zum Satz von Mooney für abstrakte analytische Funktionen II.
Preprint.

Herbert S.BEAR

[1965] A geometric characterization of Gleason parts.
Proc.Amer.Math.Soc.16(1965) 407-412.

[1970] Lectures on Gleason parts.
Lecture Notes Vol.121,Berlin 1970.

Herbert S.BEAR and Max L.WEISS

[1967] An intrinsic metric for parts.
Proc.Amer.Math.Soc.18(1967) 812-817.

Arne BEURLING

[1949] On two problems concerning linear transformations in Hilbert space.
Acta Math.81(1949) 239-255.

Errett BISHOP

[1959] A minimal boundary for function algebras.
Pacific J.Math.9(1959) 629-642.

[1960] Boundary measures of analytic differentials.
Duke Math.J.27(1960) 331-340.

[1963] Holomorphic completions, analytic continuation and the interpolation of semi-norms.
Ann.of Math.78(1963) 468-500.

[1964] Representing measures for points in a uniform algebra.
Bull.Amer.Math.Soc.70(1964) 121-122.

S.BOCHNER

[1959] Generalized conjugate and analytic functions without expansions.
Proc.Nat.Acad.Sci.USA 45(1959) 855-857.

Andrew BROWDER

[1969] Introduction to function algebras.
New York 1969.

Constantin CARATHÉODORY

[1950] Funktionentheorie.Vol.I.
Basel 1950.

Lennart CARLESON

[1964] Mergelyan's theorem on uniform polynomial approximation.
Math.Scand.15(1964) 167-175.

Allen DEVINATZ

[1966] Conjugate function theorems for Dirichlet algebras.
Rev.Un.Mat.Argentina 23(1966) 3-30.

Nelson DUNFORD and Jacob T.SCHWARTZ

[1958] Linear Operators.Vol.I: General theory.
New York 1958.

Peter L.DUREN

[1970] Theory of H^p spaces.
New York 1970.

Frank FORELLI

[1963] Analytic measures.
Pacific J.Math.13(1963) 571-578.

T.W.GAMELIN

[1968] Embedding Riemann surfaces in maximal ideal spaces.
J.Functional Analysis 2(1968) 123-146.

[1969] Uniform algebras.
Englewood Cliffs 1969.

T.W.GAMELIN and G.LUMER

[1968] Theory of abstract Hardy spaces and the universal Hardy class.
Adv.in Math.2(1968) 118-174.

John GARNETT

[1967] A topological characterization of Gleason parts.
Pacific J.Math.20(1967) 59-63.

[1968] On a theorem of Mergelyan.
Pacific J.Math.26(1968) 461-467.

John GARNETT and Irving GLICKSBERG

[1967] Algebras with the same multiplicative measures.
J.Functional Analysis 1(1967) 331-341.

F.W.GEHRING

[1957] The Fatou theorem and its converse.
Trans.Amer.Math.Soc.85(1957) 106-121.

Andrew M.GLEASON

[1957] Function algebras. Seminar on analytic functions.Vol.II, 213-226.
Institute for Advanced Study, Princeton 1957.

I.GLICKSBERG (see also GARNETT)

[1967] The abstract F.and M.Riesz theorem.
J.Functional Analysis 1(1967) 109-122.

[1968] Dominant representing measures and rational approximation.
Trans.Amer.Math.Soc.130(1968) 425-462.

[1970] Extensions of the F.and M.Riesz theorem.
J.Functional Analysis 5(1970) 125-136.

[1972] Recent results on function algebras.
Conference Board of the Mathematical Sciences Regional Conference Series in Mathematics,No.11.Amer.Math.Soc.
Providence 1972.

I.GLICKSBERG and J.WERMER

[1963] Measures orthogonal to Dirichlet algebras.
Duke Math.J.30(1963) 661-660.Errata,ibid.31(1964) 717.

A.GROTHENDIECK

[1954] Sur certain sous-espaces vectoriels de L^p.
Canad.J.Math.6(1954) 158-160.

G.H.HARDY and E.M.WRIGHT

[1968] An introduction to the theory of numbers.
Oxford 1968.

V.P.HAVIN

[1973] Weak completeness of the space L^1/H^1_o (Russian,English summary).
Vestnik Leningrad.Univ.Mat.Meh.Astronom.Vyp 3.13(1973) 77-81.

Elizabeth Ann HEARD

[1967] A sequential F.and M.Riesz theorem.
Proc.Amer.Mat.Soc.18(1967) 832-835.

Henry HELSON

[1964] Lectures on invariant subspaces.
New York 1964.

Henry HELSON and David LOWDENSLAGER

[1958] Prediction theory and Fourier series in several variables.
Acta Math.99(1958) 165-202.

Edwin HEWITT and Kosaku YOSIDA

[1952] Finitely additive measures.
Trans.Amer.Math.Soc.72(1952) 46-66.

I.J.HIRSCHMANN,Jr. and Richard ROCHBERG

[1974] Conjugate function theory in weak* Dirichlet algebras.
J.Functional Analysis 16(1974) 359-371.

Kenneth HOFFMAN

[1962a] Banach spaces of analytic functions.
Englewood Cliffs 1962.

[1962b] Analytic functions and logmodular Banach algebras.
Acta Math.108(1962) 271-317.

Kenneth HOFFMAN and Hugo ROSSI

[1965] Function theory and multiplicative linear functionals. Trans.Amer.Math.Soc.116(1965) 536-543.

[1967] Extensions of positive weak*continuous functionals. Duke Math.J.34(1967) 453-466.

Jean-Pierre KAHANE

[1967] Another theorem on bounded analytic functions. Proc.Amer.Math.Soc.18(1967) 827-831.

Yitzhak KATZNELSON

[1968] An introduction to harmonic analysis. New York 1968.

E.R.KERR (see ALLEN)

A.KOLMOGOROV

[1925] Sur les fonctions harmoniques conjugées et les séries de Fourier. Fund.Math.7(1925) 24-29.

Heinz KÖNIG (see also BARBEY)

[1960] Einige Eigenschaften der Fourier-Stieltjes-Transformation. Arch.Math.11(1960) 352-365.

[1964] Ein einfacher Beweis des Gaußschen Integralsatzes. Jber.Deutsch.Math.-Verein.66(1963/64) 119-138.

[1965] Zur abstrakten Theorie der analytischen Funktionen. Math.Z.88(1965) 136-165.

[1966a] Zur abstrakten Theorie der analytischen Funktionen II. Math.Ann.163(1966) 9-17.

[1966b] Lectures on abstract H^p theory. Summer School on topological algebra theory. Bruges, September 6-16,1966.

[1967a] Holomorphe Funktionen von H^p-Funktionen. Arch.Math.18(1967) 160-166.

[1967b] Zur abstrakten Theorie der analytischen Funktionen III. Arch.Math.18(1967) 273-284.

[1967c] Theory of abstract Hardy spaces. Lectures Notes, California Institute of Technology. Pasadena 1967.

[1968] Über das von Neumannsche Minimax Theorem. Arch.Math.19(1968) 482-487.

[1969a] Abstract Hardy space theory. Papers from the Summer Gathering on functions algebras at Aarhus,July 1969. Aarhus Universitet,Matematisk Institut, Various Publication Series No 9,56-62.

[1969b] Ein abstrakter Mergelyan Satz.
Arch.Math.20(1969) 405-412.

[1969c] On the Gleason and Harnack metrics for uniform algebras.
Proc.Amer.Math.Soc.22(1969) 100-101.

[1970a] Generalized conjugate functions in abstract Hardy algebra theory.
Journées de la Société Mathématique de France, Algèbres de Fonctions,Grenoble, 11-16 Septembre 1970, 36-44.

[1970b] On certain applications of the Hahn-Banach and minimax theorems.
Arch.Math.21(1970) 583-591.

[1972] Sublineare Funktionale.
Arch.Math.23(1972) 500-508.

[1975] Ein funktionalanalytischer Beweis des Approximationssatzes von Walsh.
J.Reine Angew.Math.(Crelle) 274/275(1975) 158-163.

[1978] On the Marcel Riesz estimation for conjugate functions in the abstract Hardy algebra theory.
Commentationes Math.1978.

Heinz KÖNIG and G.L. SEEVER

[1969] The abstract F.and M.Riesz theorem.
Duke Math.J.36(1969) 791-797.

Gottfried KÖTHE

[1966] Topologische lineare Räume I.
2.Auflage,Berlin 1966.

Nand LAL (see MERRILL)

Henri LEBESGUE

[1907] Sur le problème de Dirichlet.
Rend.Circ.Mat.Palermo 29(1907) 371-402.

Gerald M.LEIBOWITZ

[1970] Lectures on complex function algebras.
Glenview 1970.

Lynn H.LOOMIS

[1943] The converse of the Fatou theorem for positive harmonic functions.
Trans.Amer.Math.Soc.53(1943) 239-250.

David LOWDENSLAGER (see HELSON)

G.LUMER (see also GAMELIN)

[1964] Analytic functions and Dirichlet problem.
Bull.Amer.Math.Soc.70(1964) 98-104.

[1965] Herglotz-transformation and H^p-theory.
Bull.Amer.Math.Soc.71(1965) 725-730.

[1966a] Intégrabilité uniforme dans les algèbres de fonctions, classes H^Φ, et classe de Hardy universelle.
C.R.Acad.Sci.Paris Sér.A 262(1966) 1046-1049.

[1966b] H^∞ and the imbedding of the classical H^p spaces in arbitrary ones.
Function algebras. Proceedings of an international symposium on function algebras held at Tulane University, 1965, edited by Frank T.Birtel,Glenview 1966, 285-286.

[1968] Algèbres de fonctions et espaces de Hardy.
Lecture Notes Vol.75,Berlin 1968.

S.MAZUR and W.ORLICZ

[1953] Sur les espaces metriques linéaires (II).
Studia Math.13(1953) 137-179.

Samuel MERRILL

[1968] Maximality of certain algebras H^∞ (dm).
Math.Z.106(1968) 261-266.

Samuel MERRILL,III,and Nand LAL

[1969] Characterization of certain invariant subspaces of H^p and L^p spaces derived from logmodular algebras.
Pacific J.Math.30(1969) 463-474.

Michael C.MOONEY

[1972] A theorem on bounded analytic functions.
Pacific J.Math.43(1972) 457-463.

Michael MÜRMANN

[1967] Zur Theorie der abstrakten H^p-Räume.
Diplomarbeit,Universität des Saarlandes,Saarbrücken 1967.

Paul S.MUHLY

[1972] Maximal weak*Dirichlet algebras.
Proc.Amer.Math.Soc.36(1972) 515-518.

Takahiko NAKAZI

[1975] Invariant subspaces of weak*Dirichlet algebras.
Preprint.

[1976] Nonmaximal weak*Dirichlet algebras.
Hokkaido Math.J.5(1976) 88-96.

Rolf NEVANLINNA

[1953] Eindeutige analytische Funktionen.
Berlin 1953.

B.V.O'NEILL

[1968] Parts and one-dimensional analytic spaces.
Amer.J.Math.90(1968) 84-97.

W.ORLICZ (see MAZUR)

Alexander PELCZYNSKI

[1974] Sur certaines propriétés isomorphiques nouvelles des espaces de Banach de fonctions holomorphes A et H^{∞}.
C.R.Acad.Sci.Paris Sér.A.279(1974) 9-12.

S.K.PICHORIDES

[1972] On the best values of the constants in the theorems of M.Riesz, Zygmund and Kolmogorov.
Studia Math.44(1972) 165-179.

George PIRANIAN,Allen L.SHIELDS and James H.WELLS

[1967] Bounded analytic functions and absolutely continuous measures.
Proc.Amer.Math.Soc.18(1967) 818-826.

I.I.PRIWALOW

[1956] Randeigenschaften analytischer Funktionen.
Berlin 1956.

Vlastimil PTAK

[1956] On a theorem of Mazur and Orlicz.
Studia Math.15(1956) 365-366.

Heydar RADJAVI and Peter ROSENTHAL

[1973] Invariant subspaces.
Berlin 1973.

John RAINWATER

[1969] A note on the preceding paper.
Duke Math.J.36(1969) 799-800.

Richard ROCHBERG (see HIRSCHMAN)

Hugo ROSSI (see HOFFMAN)

Donald SARASON (see AHERN)

Luis SALINAS

[1976] Maximalitätssätze in der abstrakten Hardy-Räume-Theorie.
Dissertation,Universität des Saarlandes,Saarbrücken 1976.

G.L.SEEVER (see also KÖNIG)

[1973] Algebras of continuous functions on hyperstonian spaces.
Arch.Math.24(1973) 648-660.

Allen L.SHIELDS (see PIRANIAN)

T.P.SRINIVASAN and Ju-Kwei WANG

[1966] Weak*Dirichlet algebras.
Function algebras. Proceedings of an international symposium on function algebras held at Tulane University,1965,edited by Frank T.Birtel,Glenview 1966,216-249.

Edgar Lee STOUT

[1971] The theory of uniform algebras.
Tarrytown-on-Hudson 1971.

Ion SUCIU

[1973] Function algebras.
Bucharest 1973.

M.TSUJI

[1959] Potential theory in modern function theory.
Tokyo 1959.

A.G.VITUSHKIN

[1975] Uniform approximations by holomorphic functions.
J.Functional Analysis 20(1975) 149-157.

J.L.WALSH

[1928] Über die Entwicklung einer harmonischen Funktion nach harmonischen Polynomen.
J.Reine Angew.Math.(Crelle) 159(1928) 197-209.

[1929] The approximation of harmonic functions by harmonic polynomials and by harmonic rational functions.
Bull.Amer.Math.Soc.35(1929) 499-544.

Ju-Kwei WANG (see SRINIVASAN)

Max L.WEISS (see BEAR)

James H.WELLS (see PIRANIAN)

John WERMER (see also GLICKSBERG)

[1960] Dirichlet algebras.
Duke Math.J.27(1960) 373-381.

[1961] Banach algebras and analytic functions.
Advances in Math.1(1965) 51-102.

[1964] Seminar über Funktionen-Algebren.
Lectures Notes Vol.1,Berlin 1964.

David Vernon WIDDER

[1946] The Laplace transform.
Princeton 1946.

Donald R.WILKEN

[1967] Lebesgue measure of parts for R(X).
Proc.Amer.Math.Soc.18(1967) 508-512.

[1968a] Representing measures for harmonic functions.
Duke Math.J.35(1968) 383-389.

[1968b] The support of representing measures for R(X).
Pacific J.Math.26(1968) 621-626.

Kôzô YABUTA

[1973a] Funktionen mit nichtnegativem Realteil in abstrakten Hardy-algebren.
Arch.Math.24(1973) 164-168.

[1973b] On the distribution of values of functions in some function classes in the abstract Hardy space theory.
Tohoku Math.J.25(1973) 89-102.

[1974a] On bounded functions in the abstract Hardy space theory.
Tohoku Math.J.26(1974) 77-84.

[1974b] On bounded functions in the abstract Hardy space theory II.
Tohoku Math.J.26(1974) 513-533.

[1975] On bounded functions in the abstract Hardy space theory III.
Tohoku Math.J.27(1975) 111-128.

[1976] On the distributions of bounded functions in the abstract Hardy space theory and some of their applications.
Preprint.

[1977] M.Riesz's theorem in the abstract Hardy space theory.
Arch.Math.28(1977).

Kosaku YOSIDA (see HEWITT)

Lawrence ZALCMAN

[1968] Analytic capacity and rational approximation.
Lecture Notes Vol.50, Berlin 1968.

A.ZYGMUND

[1968] Trigonometric series.
Cambridge 1968.

Notation Index

$\mathbb{N}$ $\mathbb{Z}$ $\mathbb{Q}$ $\mathbb{R}$ $\mathbb{C}$	the usual number systems
$B^{\perp}$	the annihilator of the set B in a dual system
$B^{\times}$	the set of invertible elements of the algebra B

Subject Index